Biopolymers in Sustainable Corrosion Inhibition

Biopolymers in Sustainable Corrosion Inhibition covers the fundamentals, properties, and applications of biopolymers and considers their superiorities over traditional alternatives. It explores the synthesis, characterization, inhibition mechanism, and applications of biopolymeric anticorrosive materials.

Focusing on environmentally friendly corrosion prevention methods, this book demonstrates how biopolymers slow the corrosion rate and avoid economic losses owing to the metallic corrosion on industrial liners, tools, or surfaces. This book covers the sustainable corrosion inhibition potential of biopolymers and their derivatives, including chitosan, cellulose, chitin, starch, and natural gums.

This book will be a valuable reference for undergraduate and graduate students and academic researchers in the fields of biopolymers, corrosion science and engineering, environmental science, chemical engineering, green chemistry, and mechanical/industrial engineering.

Biopolymers in Sustainable Corrosion Inhibition

Edited by
Saman Zehra, Mohammad Mobin,
and Chandrabhan Verma

CRC Press
Taylor & Francis Group
Boca Raton London New York

CRC Press is an imprint of the
Taylor & Francis Group, an **informa** business

MATLAB® is a trademark of The MathWorks, Inc. and is used with permission. The MathWorks does not warrant the accuracy of the text or exercises in this book. This book's use or discussion of MATLAB® software or related products does not constitute endorsement or sponsorship by The MathWorks of a particular pedagogical approach or particular use of the MATLAB® software.

Designed cover image: Shutterstock

First edition published 2024
by CRC Press
2385 NW Executive Center Drive, Suite 320, Boca Raton FL 33431

and by CRC Press
4 Park Square, Milton Park, Abingdon, Oxon, OX14 4RN

CRC Press is an imprint of Taylor & Francis Group, LLC

ISBN: 978-1-032-50860-3 (hbk)
ISBN: 978-1-032-50872-6 (pbk)
ISBN: 978-1-003-40005-9 (ebk)

DOI: 10.1201/9781003400059

Typeset in Times
by codeMantra

Contents

Preface

Metallic materials are extensively employed in many applications, including building, particularly in petroleum, oil, and gas industries. But as most metals in their pure state are thermodynamically unstable, they easily succumb to corrosive disintegration due to interactions with other elements in the environment. Both safety and economic loss are significantly harmed by corrosion. According to a recent estimate by the NACE (National Association of Corrosion Engineers), the cost of corrosion globally is estimated to be over US$2.5 trillion, or roughly 3.5% of global GDP. Biopolymers are polymers made from natural resources, either wholly or partially by living organisms, or chemically from biological material. For many years, the utilization of biopolymers from various sources has been researched for pharmacological and biological purposes. Biopolymers and their derivatives are widely used in sustainable corrosion protection. At the moment, organic compounds are recognized as one of the most successful and profitable corrosion inhibition techniques. However, their anticorrosive application is restricted due to the rising popularity of green chemistry and sustainable advancements. Biopolymers are considered environmentally beneficial substitutes for conventional hazardous corrosion inhibitors because of their natural (animal and plant) origin. The biopolymers are non-bioaccumulative, eco-friendly, and biodegradable.

The present book describes the collections of sustainable corrosion inhibition using biopolymers. This book is divided into three parts. Part 1 contains four chapters describing fundamentals, properties, and applications of biopolymers and their superiorities over traditional toxic alternatives. Part 2 contains ten chapters describing the corrosion inhibition potential of different biopolymers and their derivatives, including starch, cellulose, chitosan, natural gums, collagen, hyaluronic acid, alginates, lignin, pectin, and their derivatives. Lastly, Part 3 describes the sustainable corrosion inhibition potential of biopolymer composites such as cellulose and chitosan composites as corrosion inhibitors.

This book is primarily intended for academic and professional researchers, practicing corrosion engineers, materials science students, and applied and engineering chemists. The editor and contributors are distinguished academic and industrial researchers, scientists, and true professionals. We want to express our sincere gratitude to all the chapter authors on behalf of the CRC for their tremendous and enthusiastic work on this book. Thanks to Dr. Allison Shatkin (Senior Publisher) and Kyra Lindholm (Technical Editorial at CRC) for their tireless assistance and support during this endeavor. Finally, a big thank you to the CRC for releasing this book.

Thanks and Regards
Drs. Saman Zehra, Mohammad Mobin, and Chandrabhan Verma
(editors)

About the Editors

Dr. Saman Zehra

Corrosion Research Laboratory, Department of Applied Chemistry, Faculty of Engineering and Technology, Aligarh Muslim University, Aligarh, 202 002 (India).
drsamanzehra90@gmail.com

Dr. Saman Zehra, Ph.D., is currently working at the Women Scientist Program (WOS-A) of DST, New Delhi, in the Department of Applied Chemistry of Aligarh Muslim University, Aligarh, India. Her area of research broadly covers stimuli-responsive smart functional anticorrosion coatings and corrosion inhibitors. She received her Ph.D. and M.Sc. from the same university. She has published 43 articles in international peer-reviewed journals, including reviews and book chapters. She has also presented her research work at national and international conferences.

Mohammad Mobin

Corrosion Research Laboratory, Department of Applied Chemistry, Faculty of Engineering and Technology, Aligarh Muslim University, Aligarh, 202 002 (India).
drmmobin@hotmail.com

Professor Mohammad Mobin, Ph.D., has 34 years of an extraordinarily active and productive career, working at Aligarh Muslim University and the multi-national Seawater Desalination Research Institute, Saline Water Conversion Corporation (SWCC), Kingdom of Saudi Arabia. At Aligarh Muslim University, he is working as a Professor in the Department of Applied Chemistry and is involved in teaching and researching materials and corrosion. Dr. Mobin has completed 28 root cause failure investigations and eight major and three minor research projects on materials and corrosion, including two projects under international collaboration. Dr. Mobin has a Ph.D. in corrosion, supervised 14 students for the award of the Ph.D. degree, and authored 180 research papers. He has also attended 35 national/international conferences/workshops.

Chandrabhan Verma

Interdisciplinary Research Center for Advanced Materials (ARC-AMs), King Fahd University of Petroleum and Minerals, Dhahran 31261, Saudi Arabia.
Chandraverma.rs.apc@itbhu.ac.in

Chandrabhan Verma, Ph.D., works at the Centre of Research Excellence in Corrosion, Research Institute, King Fahd University of Petroleum and Minerals, Dhahran 31261, Saudi Arabia.

He is a member of the American Chemical Society (ACS). His research is mainly focused on the synthesis and designing of environmentally friendly corrosion inhibitors useful for several industrial applications. Dr Verma is the author of several research and review articles in peer-reviewed international journals of ACS, Elsevier, RSC, Wiley and Springer, etc. He has a total of more than 8250 citations with an H-index of 52 and an i-10 index of 132. Dr Verma received several national and international awards for his academic achievements.

Contributors

Khamdam Akbarov
Faculty of Chemistry
National University of Uzbekistan
Tashkent, Uzbekistan

K. Alaoui
Advanced Materials and Process
  Engineering, Faculty of Sciences
Ibn Tofaïl University
Kenitra, Morocco

Humira Assad
Department of Chemistry, School of
  Chemical Engineering and Physical
  Sciences
Lovely Professional University
Phagwara, India

Lazizbek Azimov
Faculty of Chemistry
National University of Uzbekistan
Tashkent, Uzbekistan

Elyor Berdimurodov
Faculty of Chemistry
National University of Uzbekistan
Tashkent, Uzbekistan

Khasan Berdimuradov
Faculty of Industrial Viticulture and
  Food Production Technology
Shahrisabz Branch of Tashkent Institute
  of Chemical Technology
Shahrisabz,
  Uzbekistan

Kaartik Chandan
Department of Chemistry
K. E. T.'S. V. G. Vaze College
  (Autonomous)
Mumbai, India

Fatima Choudhary
Department of Chemistry
K. E. T.'S. V. G. Vaze College
  (Autonomous)
Mumbai, India

Omar Dagdag
Department of Mechanical Engineering
Gachon University
Seongnam, Republic of Korea

K. Dahmani
Laboratory of Organic Chemistry,
  Catalysis and Environment, Faculty
  of Sciences
Ibn Tofaïl University
Kenitra, Morocco

Kesavan Devarayan
College of Fisheries Engineering
Tamil Nadu Dr. J. Jayalalitha Fisheries
  University
Nagapattinam, India

Eno E. Ebenso
Centre for Material Science, College
  of Science, Engineering and
  Technology
University of South Africa
Johannesburg, South Africa

Ilyos Eliboev
Faculty of Chemistry
National University of Uzbekistan
Tashkent, Uzbekistan

M. Galai
Advanced Materials and Process
  Engineering, Faculty of Sciences
Ibn Tofaïl University
Kenitra, Morocco

Richika Ganjoo
Department of Chemistry, School of
   Chemical Engineering and Physical
   Sciences
Lovely Professional University
Phagwara, India

Rajesh Haldhar
School of Chemical Engineering
Yeungnam University
Gyeongsan, Republic of Korea

Y. El Kacimi
Advanced Materials and Process
   Engineering, Faculty of Sciences
Ibn Tofaïl University
Kenitra, Morocco

Babar Khan
Corrosion Research Laboratory,
   Department of Applied Chemistry,
   Faculty of Engineering and
   Technology
Aligarh Muslim University
Aligarh, India

Abduvali Kholikov
Faculty of Chemistry
National University of Uzbekistan
Tashkent, Uzbekistan

Hansang Kim
Department of Mechanical Engineering
Gachon University
Seongnam, Republic of
   Korea

Seong-Cheol Kim
School of Chemical Engineering
Yeungnam University
Gyeongsan, Republic of Korea

Ashish Kumar
Department of Science and Technology
Government of Bihar
Patna, India

Rahayu Kusumastuti
Research Center for Metallurgy
National Research and Innovation
   Agency
Jakarta, Republic of Indonesia

B. Lakhrissi
Laboratory of Organic Chemistry,
   Catalysis and Environment, Faculty
   of Sciences
Ibn Tofaïl University
Kenitra, Morocco

Jaykhun Mamatov
Faculty of Chemistry
National University of Uzbekistan
Tashkent, Uzbekistan

Sheerin Masroor
Department of Chemistry, A.N. College
Patliputra University
Patna, India

Oybek Mikhliev
Karshi Engineering Economics Institute
Karshi, Uzbekistan

Mohammad Mobin
Corrosion Research Laboratory,
   Department of Applied Chemistry,
   Faculty of Engineering and
   Technology
Aligarh Muslim University
Aligarh, India

Paresh More
Department of Chemistry
K. E. T.'S. V. G. Vaze College
   (Autonomous)
Mumbai, India

Siti Musabikha
Research Center for Metallurgy
National Research and Innovation
   Agency
Jakarta, Republic of Indonesia

Anju Nair
Department of Chemistry
K. E. T'.S. V. G. Vaze College
  (Autonomous)
Mumbai, India

Shahab A. A. Nami
Department of Industrial Chemistry,
  Faculty of Science
Aligarh Muslim University
Aligarh, India

Ravisankar Natarajamani
Department of Civil Engineering,
  Faculty of Engineering and
  Technology
Annamalai University
Chidambaram, India

Arini Nikitasari
Research Center for Metallurgy
National Research and Innovation
  Agency
Jakarta, Republic of Indonesia

S. Poongothai
Department of Civil Engineering,
  Faculty of Engineering and
  Technology
Annamalai University
Chidambaram, India

Siska Prifiharni
Research Center for Metallurgy
National Research and Innovation
  Agency
Jakarta, Republic of Indonesia

Satarupa Priyadarshani
Department of Chemistry
Dibrugarh University
Dibrugarh, India

Gadang Priyotomo
Research Center for Metallurgy
National Research and Innovation
  Agency
Jakarta, Republic of Indonesia

Taiwo W. Quadri
Centre for Material Science, College
  of Science, Engineering and
  Technology
University of South Africa
Johannesburg, South Africa

Yusufboy Rajabov
Faculty of Chemistry
National University of Uzbekistan
Tashkent, Uzbekistan

Sonish Rashid
Department of Industrial Chemistry,
  Faculty of Science
Aligarh Muslim University
Aligarh, India

M. Rbaa
Laboratory of Organic Chemistry,
  Catalysis and Environment, Faculty
  of Sciences
Ibn Tofaïl University
Kenitra, Morocco

A. Rouifi
Advanced Materials and Process
  Engineering, Faculty of Sciences
Ibn Tofaïl University
Kenitra, Morocco

Z. Rouifi
Laboratory of Organic Chemistry,
  Catalysis and Environment, Faculty
  of Sciences
Ibn Tofaïl University
Kenitra, Morocco

Sanghamitra Sarmah
Department of Chemistry
Dibrugarh University
Dibrugarh, India

Gitalee Sharma
Department of Chemistry
Dibrugarh University Institute of
    Engineering and Technology
Dibrugarh, India

Shveta Sharma
Department of Chemistry, Government
    College Una
Himachal Pradesh University
Una, India

Amita Somya
Department of Applied Chemistry
Amity School of Engineering and
    Technology
Amity University
Bengaluru, India

Monikandon Sukumaran
Department of Civil Engineering,
    Faculty of Engineering and
    Technology
Annamalai University
Chidambaram, India
College of Fisheries Engineering
Tamil Nadu Dr. J. Jayalalitha Fisheries
    University
Nagapattinam, India

Abhinay Thakur
Department of Chemistry, School of
    Chemical Engineering and Physical
    Sciences
Lovely Professional University
Phagwara, India

Jenis Tripathi
Department of Chemistry
K. E. T.'S. V. G. Vaze College
    (Autonomous)
Mumbai, India

Saman Zehra
Corrosion Research Laboratory,
    Department of Applied Chemistry,
    Faculty of Engineering and
    Technology
Aligarh Muslim University
Aligarh, India

1 Overview of Biopolymers
Properties and Applications

Sheerin Masroor
Patliputra University

1.1 INTRODUCTION

Monomers are simple building blocks. A substance that is comprised of many high-molecular-weight molecules is referred to as a polymer. The repeating unit of a monomer chain, known as a polymer, can be produced either naturally (natural polymers) or intentionally (manmade polymers or synthetically derived polymers). Biopolymers are organic polymers that can be found in living things. Being a biodegradable molecule, a biopolymer is a long-chain molecule comprised of monomers that are covalently bonded to one another. The major sources of biopolymers are plants, trees, microorganisms, and other naturally occurring entities. Compared with biopolymers, which are intricate molecules with clearly defined three-dimensional structures, synthetic polymers are less complex and more arbitrary [1]. Numerous biopolymers with different physical and chemical properties are synthesized using renewable resources. Lignin, starch, cellulose, hemicelluloses, and numerous other biopolymers can be found in nature [2]. Manufacturers of new materials are facing tremendous demands for biopolymers. However, the cost-effectiveness of biopolymers should increase as their usage is primarily focused on sustainable development. There is disagreement regarding the precise definition of generic terms degradable, biodegradable, bio-based, compostable, and biopolymer in the literature and patents, which seem to have several overlapping meanings.

A generic term used to describe polymers or plastics that break down through a variety of processes, including physical breakdown, chemical breakdown, and biodegradation through biological mechanisms, is "degradable." According to this definition, a polymer needs to be degradable, but not necessarily biodegradable.

A polymer's ability to disintegrate when subjected to the actions of microorganisms such as mould, fungi, and bacteria over the course of a specified amount of time and in a particular environment is referred to as "biodegradability," a term that focuses on the functionality of a polymer. The word "biodegradable" is meaningless and ambiguous on its own. The withdrawn ASTM D5488-94del standard defines biodegradable polymers as those that can "undergo decomposition into carbon dioxide, methane, water, inorganic compounds, or biomass, wherein the predominant mechanism is the enzymatic action of microorganisms that can be measured by standard tests, over a specific period of time, reflecting available disposal conditions."

The characteristic of a material that may be microbiologically broken down into ultimate products of carbon dioxide and water, which are then recycled in nature, is referred to as being biodegradable, according to the Japan Bioplastics Association (JBPA). Disintegration, which refers to the substance being split into smaller, independent fragments, is different from biodegradation. Plastics' biodegradability is assessed using predetermined criteria and is determined using International Organisation for Standardisation (ISO) methodologies. Only biodegradable materials that meet strict requirements, such as those related to heavy metal content and safe intermediate reaction products, can be branded as Green-Pla® [3].

The following legally binding international standards are used to certify biodegradable polymers [4]:

ISO 17088:2012
EN 13432:2000, EN 14995:2006
ASTM D6400-12

The term "bio-based" refers to the use of raw ingredients and denotes polymers made from renewable resources. If natural processes replenish raw materials at rates that are equal to or greater than their rate of consumption, then the term "renewable" is used [5].

The U.S. Secretary of Agriculture has determined that bio-based products are "commercial or industrial goods—other than food or feed—composed in whole or in a significant part of biological products, forestry materials, or renewable domestic agricultural materials, including plant, animal, or marine materials," as defined by the Farm Security and Rural Investment Act of 2002 (FSRIA) [6,7].

The biodegradability of biopolymers is associated with the presence of specific microbes and enzymes with particular degradable properties [50]. As biopolymers have oxygen and nitrogen atoms in their skeletal backbone, they are easily biodegradable. During biodegradation, the biopolymer is transformed into CO_2, water, biomass, humid water, and other elements of nature. Applications of biodegradable polymers in bioengineering are numerous and include tissue engineering, drug delivery systems, and wound dressing, among others [8]. Because of their rigidity, charge-free chains, great resilience to salt and cold, and characteristic helical structure, biopolymers thicken and show better stability under challenging pool conditions [9].

1.2 CLASSIFICATION OF BIOPOLYMERS

Three types of biopolymers are available:

a. biodegradable biopolymers synthesized from renewable raw materials (bio-based);
b. non-biodegradable biopolymers synthesized from renewable raw materials (bio-based); and
c. biodegradable biopolymers synthesized from fossil fuels.

Biodegradable bio-based biopolymers can be synthesized chemically from biological building blocks such as corn, sugar, starch, and other biological starting materials, or they can be synthesized by biological systems such as bacteria, plants, and animals. Some examples of this type of biopolymers include (1) synthetic polymers manufactured from renewable resources, such as poly(lactic acid); (2) biopolymers produced by microorganisms, such as PHAs; and (3) naturally occurring biopolymers, such as starch and proteins. By definition, natural polymers are produced through different biosynthetic pathways in the biosphere. Starch and PHAs are the two most popular biodegradable polymers.

Non-biodegradable bio-based biopolymers can be synthesized from biomass or renewable resources. Some examples of this type of biopolymers include (1) synthetic polymers manufactured from renewable resources, such as specific polyamides from castor oil (polyamide 11), specific polyesters based on biopropanediol, biopolyethylene (bio-LDPE, bio-HDPE), biopolypropylene (bio-PP), or biopoly(vinyl chloride) (bio-PVC) based on bioethanol (e.g. from sugar cane), etc.

Biodegradable biopolymers synthesized from fossil fuels, such as synthetic aliphatic polyesters derived from crude oil or natural gas, are approved for use in composting and biodegradation. Despite being partially derived from fossil fuels, polymers such as poly(caprolactone), poly(butylene succinate), and some "aliphatic–aromatic" copolyesters are susceptible to microbial degradation.

1.3   MANUFACTURE OF SYNTHETIC BIOPOLYMERS

Synthetic biopolymers are biopolymers that have undergone biotic chemical modification artificially. They can be synthesized using numerous methods. Below is a list of common methods for synthesizing biopolymers with well-generated structural and mechanical properties:

1. Esterification.
2. Dehydration.
3. Polycondensation.
4. Hydrolysis.
5. Granulation.
    i. **Esterification**: The chemical process of esterification takes place during the synthesis of esters. In this process, alcohol (ROH) is mixed with an organic acid (RCOOH), thus yielding an ester (RCOOR) and water. This chemical reaction results in the esterification of a carboxylic acid with an alcohol to yield at least one ester product. Some of the esterification reactions for starch, cellulose, and hemicelluloses are as follows:
        a. **Esterification of starch**: Acetic acid and acetic anhydride are used in the esterification process to create starch acetate. Acetyl groups are incorporated into starch molecules depending on a number of factors, such as pH, duration, and the presence of a catalytic agent. The degree of substitution varies from 0.01 to 0.2 for low (0.01–0.2), medium (0.2–1.5), and high (1.5–3) acetyl starch. This alkenyl succinic anhydride is called DDSA. A water-resistant product that is widely used in the paper and film industries is created by esterifying DDSA starch in an aqueous solution [10].

b. **Esterification of cellulose**: To synthesize cellulose ester, cellulose is esterified either at the outside thread of the cellulose molecule or throughout the polymeric chain. The surface is changed when large quantities of cellulose are produced by a homogeneous or heterogeneous process. Cellulose nitrate is the inorganic cellulose ester with the highest production in the industry. It can be found frequently in a variety of products, including propellants, explosives, coatings, and polymers. The esterification of cellulose nitrate is carried out in heterogeneous equilibrium using cellulose and a solution of nitric acid and sulphuric acid with a degree of substitution ranging from 1.8 to 2.8. The heterogeneous direct reaction between cellulose and sulphuric acid is termed "Cellulose Sulphate (CS) esterification." An ether or ester group in cellulose is removed to create synthetic CS.

c. **Esterification of hemicellulose**: Plant cells and cell walls contain a polymer called hemicellulose. Hemicelluloses, a biopolymer used to create films and hydrogels, have been used in biomedical applications such as drug delivery and tissue engineering. Furfural, xylitol, ethanol, and lactic acid are produced through the chemical synthesis of hemicellulose. Biomaterials such as cationic hemicelluloses, carboxymethyl hemicelluloses, lauroylated hemicelluloses, and acylated hemicelluloses are homogeneous upgrades of hemicellulose and have properties such as eco-friendliness, reduced hydrophobicity, higher electrochemical and thermal balance, and faster reaction rates.

ii. **Dehydration**: The process of mixing several monomers to produce polymers is known as polycondensation.

Water, alcohol, and salt are a few low-molecular-weight subsidiary products that are typically released along the process. Polycondensation offers a versatile method for biopolymer synthesis. Phosphorus-containing polymers can be created by polycondensing H-phosphonic acid diols and diesters, phosphoric dihalides, or phosphoric acid. Dihydroxy(oligolactide) and ethyl dichlorophosphate react in solution during polycondensation to produce bulk polilactofate [11].

Water-soluble melamine and formaldehyde polycondense in acidic environments, forming a gel with a sizable spherical hollow. To synthesize more flexible foam and fibre resins, melamine polycondensation and melamine substitution with formaldehyde can also be used. In addition, polycondensation can be used to synthesize monodisperse melamine/formaldehyde microspheres [12].

iii. **Hydrolysis**: Synthetic biopolymers do not break down into hydrocarbons or carbon dioxides but instead into water. This property gives them non-toxic characteristics. Carbon dioxide, which is harmful to both humans and the environment, is the end product. Through hydrolytic modification, Azotobacter plays a crucial role in the production of the synthetic biopolymer poly-3-hydroxybutyrate (POB). At 70°C, POB hydrolyses into a film in the presence of a phosphate buffer. This process is dependent on various factors such as molecular weight, temperature, incubation medium, chemical make-up of the biopolymer, and hydrolytic breakdown of POB. During hydrolysis, PBA is hydrolysed at the surface [13]. In contrast to hemicellulose, which is random and amorphous with minimal cohesion, cellulose has a strong intermolecular arrangement in its structural backbone that gives it great strength and resilience against hydrolysis. Thermal cleavage of the cellulose biopolymeric chain at one end is induced by dehydration, followed by a hydrolysis reaction at a critical water condition. The bond breaks when cellulose is hydrolysed in a nitrogenic atmosphere at temperatures between 200°C and 350°C, with or without a catalyst. Biopolymers such as lignin and hemicellulose can be created by the process of hydrolysis, in which bonds surrounding critical water are broken down.

iv. **Granulation**: How a biopolymer is made into granules is speculative and unsure. The biogranulation of biopolymers predominates the projected applicability, which comprises, among other things, enhanced biomass withholding, higher toxicity tolerance, treatment of pollutants with high-loaded carbon/nutrient loads, and high-grade settleability [14]. Proteins, lipids, humic acids, polysaccharides, and nucleic acids are components of extracellular polymeric substances (EPS) produced during granulation, which hold microorganisms together. A lengthy period of time—roughly 308 days—is needed for the biopolymer to fully granulate, and this time period is accelerated through stages 1 and 2. Phase 1 takes day 0~182–220 for low granular mechanism, whereas phase 2 takes pale after 182–220 days for steady granular mechanism. The stability of Alginate-like extracellular polymers (ALE) is increased through granulation under the influence of the local microbial composition, which is helpful for later qualitative study for industrial usage [15].

1.4   BIOPOLYMER COMPOSITES

Conglomerated materials called biocomposites are made of a biopolymer outer layer in the form of resin reinforced with natural fibres as fillers [48]. Different techniques can be used to synthesize biopolymer composites, including grafting [49], various types of moulding [16], extrusion [17], intercalation [18], phase separation [19], filament winding [20], melt blending [21], electrospinning [22], laser printing [23], and film stacking [24]. When appropriate plasticizers and solvents are used, biopolymer nanocomposites with better thermal stability, toughness, rigidity, and elastic modulus of the matrix can be obtained [25]. Using spectroscopic investigations, the physical, chemical, thermal, and structural changes in the synthetic natural graft copolymers can be determined. Compared with natural composites made from natural polymers, those made from raw polymers have better tensile and mechanical properties [26].

1.5   TYPES OF BIOPOLYMER COMPOSITES

Biopolymer composites are made of biodegradable polymers with natural fibres enclosed as fillers in the biopolymer matrix. They can be categorized into two types: matrix-reinforced fibres and natural fibres. Hemp, jute, sisal, etc. are examples of natural fibres, whereas cellulose, PHA, PVA, and thermoplastic starch are examples of matrix-reinforced fibres [27]. In addition, several naturally occurring biocomposites have been classified as having wood fibres or non-wood fibres based on the percentage of wood in them. These naturally occurring composites have significant amounts of cellulose and lignin, which gives them high tensile strength and crystallinity. Biocomposites are also classified based on their use in different industries: green, hybrid, and textile composites. Biocomposites can be easily synthesized using electrospinning because of the ready availability of cellulose fibres. Due to their biodegradable and inert characteristics, various nanocomposites have been developed, which are used in a wide range of applications in many fields [28].

1.6   PREFERENCE OF BIOCOMPOSITES OVER BIOPOLYMERS

Due to their higher dimensional stability, which is lacking in natural fibres, biocomposites are preferred over biopolymers [29]. This might be due to the presence of hydroxyl groups in biocomposites. Because they are sustainable and environmentally benign, polymers made from natural resources, for example, have attracted increased attention.

Surface modification, on the other hand, is a crucial component in the synthesis of biopolymer composites since it plays a significant role in giving distinctive features to these composites, thus making them useful for applications in industries [30].

1.7   CHARACTERIZATION TECHNIQUES USED FOR BIOPOLYMERS AND BIOPOLYMER COMPOSITES

Biopolymers are synthesized using a variety of techniques and resources. After characterization, the biodegradability and environmental friendliness of biopolymers are assessed in accordance with ASTM standards to ensure that they meet the required quality criteria as outlined in Figure 1.1 [31].

1.8   GENERAL PROPERTIES OF BIOPOLYMERS AND BIOPOLYMER COMPOSITES

The properties of biopolymers can be categorized into relative, synthesis, and component properties. The term "relative properties" describes the underlying and innate characteristics of a polymer that are independent of its structure and chemical

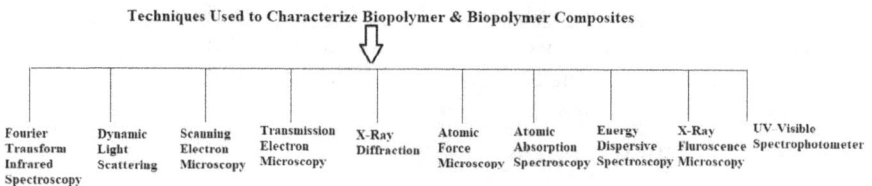

Techniques Used to Characterize Biopolymer & Biopolymer Composites

| Fourier Transform Infrared Spectroscopy | Dynamic Light Scattering | Scanning Electron Microscopy | Transmission Electron Microscopy | X-Ray Diffraction | Atomic Force Microscopy | Atomic Absorption Spectroscopy | Energy Dispersive Spectroscopy | X-Ray Fluroscence Microscopy | UV Visible Spectrophotometer |

FIGURE 1.1   General characterization techniques for biopolymers and biopolymer composites.

make-up. Relative properties denote the polymer's density, solubility, transparency, and permeability. Characteristics such as viscosity, optical purity, mechanical qualities, stability, and molecular weight are considered synthesis properties. Component properties, which are related to the functioning and performance of the biopolymer, are a combination of the first two properties [32]. The effectiveness of fibres in controlled drug release, tissue engineering, and other ensuing medical applications depends on various stereo-complexes of biopolymers. Different biopolymers, both biodegradable and non-biodegradable, have been characterized and visualized to demonstrate the complexity of interactions brought on by stereo-selective van der Waals forces, which change the physical properties of biopolymers [33]. In the following, we will discuss the physical, thermal, mechanical, and optical features of biopolymers that are produced spontaneously.

1.8.1 PHYSICAL PROPERTIES

Melting and boiling points, together with form, density, and viscosity, are the primary physical characteristics of biopolymers and biopolymer composites. It has been determined that biopolymers' internal structures are altered by the interaction of water molecules with them, making them more susceptible to moisture. In hydrophilic biopolymers cellulose and hemicellulose, the area of adsorption of water molecules increases with swelling. This decrease in the strength of hydrogen bonds weakens the material's physical and mechanical properties. In addition, the rubber-like behaviour brought on by high moisture is reduced, and the mechanical response of the biopolymer is controlled by interfering with the hydrogen bonding mechanism [34, 35].

Through the technique of melt blending, two biocomposite systems based on poly(L-lactic acid) with lignin as a filler and Poly(ethylene glycol) (PEG) as a plasticizer have been examined. This process has shown that the poly(L-lactic acid) biocomposite has higher stiffness and flexibility [36].

Natamycin and rosemary have been used to create antifungal biopolymers utilizing wheat gluten and methylcellulose. Using scanning electron microscopy, an Rosemary extract (RE) has been found to interact with bioactive antifungal substances on the biopolymer. The incorporation of these substances does not significantly alter the characteristics of films; however, the crystallinity of natamycin is increased. Biodegradable sugar palm starch (SPS) has been synthesized using glycerol as a plasticizer, and a novel biopolymer from Arenga pinnata (sugar palm) trees has been added. The plasticized SPS has then been supplemented with increasing weight quantities of glycerol. The eco-friendly samples thus resulted have shown optimal density, moisture content, water absorption, and thickness swelling, demonstrating an inverse relationship between SPS and glycerol concentration [37,38].

Chitosan is a potent green rival for its synthetic commercial counterparts in the film industry because of its excellent antibacterial, non-toxic, and biodegradable properties. As a result of its poor thermal and mechanical qualities, its use and commercialization are constrained; instead, it is combined with cellulose nanofibers to reach the highest level of compatibility in the synthesis of nanofibers and nanowhiskers. These cellulose nanofibers with chitosan reinforcement show improved physical properties, which has made them useful in medicinal applications [39].

1.8.2 THERMAL PROPERTIES

Thermal stability and thermal conductivity are the two primary factors that determine a biopolymer's thermal characteristics. When analysing biomaterials using thermal analysis, it is important to include variables such as the rate of thermal degradation, stability, glass transition temperature, melting temperature, crystallization temperature, degree of crystallinity, and heat of fusion. Analytical methods such as differential scanning calorimetry and thermal gravimetric analysis can be used to calculate these values [40]. As lignin has intra- and intermolecular hydrogen bonds, which allow for its applications as a thermosetting material, it behaves like a thermoplastic. Due to its tendency towards thermal instability, lignin has been used to create mixes, composites, and copolymers with other polymers, thus altering their thermal properties. Meanwhile, the thermal characteristics of biopolymers and biocomposites are significantly influenced by the chemical structure and physical characteristics of lignin [41]. Using potassium persulphate to mercerize ethyl acrylate (EA) on cellulose polymers, cellulose graft copolymers are synthesized. The thermal stability of the natural polymer ethyl acrylate is higher than that of raw mercerized cellulosic polymers when potassium persulphate is utilized as a promoter to decrease the water absorptivity. Thermogravimetric analysis and differential thermal analysis are used to characterize graft polymers, which show improved moisture, chemical, and thermal resistance [42].

1.8.3 MECHANICAL PROPERTIES

The use of biopolymer composites depends on temperature, applied force, deformation, heating, and cooling rates. These factors account for the mechanical properties of biopolymers and biopolymer composites. In industrial sectors, techniques like Young's modulus, tensile strength, Poisson's ratio, rheology, and elongation at break are required to design biopolymer materials with significant mechanical properties [43]. The most important properties for characterizing fibre-reinforced composites are tensile strength and impact strength.

Glycerol and sorbitol are used as plasticizers in varying amounts to synthesize kefiran, a novel edible biopolymer. It has been thoroughly researched in terms of its physical, mechanical, and water vapour permeability characteristics. The water vapour permeability of glycerol shows that it is an appropriate plasticizer by enhancing flexibility. Glycerol plays a significant role as a plasticizer in ensuring the qualities of a film, making it suitable for food technology-based applications, as it detects even a slight change in the puncture deformation and tensile strength [44].

1.8.4 OPTICAL PROPERTIES

For the industrial application of biopolymers as coatings, glazing agents, plastics, and transparent materials, their optical characteristics should be determined. The most crucial optical properties that have been the subject of numerous research studies include colour, transparency, polarizability, light absorption/transmission, absorption coefficient, dielectric constant, and refractive index. While some polymers are colourless and transparent, others are coloured and opaque. Each of these polymers is intended for use in the food and pharmaceutical industries. Furthermore, the refractive index demonstrates how well a polymer can bend and reflect light. The molecular weight and identity of the biopolymer are represented by polarizability [45].

Finally, the mechanical, thermal, elasticity, flexibility, stability, and adsorption properties of these biopolymers as well as their composites are examined in order to improve their applications. Applications of water and moisture content to biopolymers and biocomposites have shown enormous responses from both.

1.9   BIOPOLYMER APPLICATIONS

Due to their various uses in the biomedical and industrial sectors, the applications of biopolymers can be categorized into two main categories.

1.9.1   BIOMEDICAL APPLICATIONS

Due to their biocompatibility, biopolymers are extensively used in the tissue engineering, medical device, and pharmaceutical industries. This is because one of the key goals of biomedical engineering is to replicate body parts to preserve normal bodily processes [46]. Due to their mechanical properties, several biopolymers can be used in tissue engineering, medication delivery, regenerative medicine, and other medical applications. They offer qualities such as the ability to heal wounds, catalysing bioactivity, and being non-toxic [47]. Many biopolymers are typically better at integrating into the body than synthetic polymers because the former have more complicated structural similarities to the human body, which helps reduce immunogenic rejection and toxicity following decomposition. More precisely, because they are accessible and affordable, polypeptides such as collagen and silk are used in cutting-edge research as biocompatible materials. Gelatin polymer is frequently used as an adhesive while treating wounds. Gelatin-based scaffolds and films enable scaffolds to contain medications and other nutrients that can be supplied to a wound to aid in healing.

Collagen and chitosan are the most popular biopolymers used in biomedical science as follows [48]:

- Drug delivery systems based on collagen and chitosan.
- Sponges made from collagen.
- Use of collagen in haemostatic balance.
- Antimicrobial agents based on chitosan.
- Tissue engineering based on chitosan composites.

1.9.2   INDUSTRIAL APPLICATIONS [49]

- Biopolymers in food.
- Biopolymers in packaging.
- Biopolymers in water purifications.
- Biopolymers as plastic materials.

1.10 CONCLUSION

The synthesis of eco-friendly materials has been the focus of recent advances in sustainable energy and environmental practices. One such resource is the discovery and synthesis of biopolymers and their composites, which are on the verge of fully replacing synthetic polymers due to their excellent properties. In order to compete in the expanding market as eco-friendly alternatives, the biocompatibility and non-toxic biodegradability characteristics of biopolymers need to be enhanced. After synthesis, the structural details of biopolymers and biopolymer composites are determined using characterization methods based on microscopic, spectroscopic, and diffraction examinations. Both biopolymers and their composites demonstrate reliability and effectiveness as drug delivery systems; organ and tissue scaffolds in tissue engineering; gels, ointments, and wound dressing materials in wound healing; and implantable devices in the medical and pharmaceutical industries. Their use as stabilizers and gelling and emulsifying agents in food processing, as well as edible films and packaging materials in the food preservation sectors, has also been established. They have excellent physical properties such as flexibility, viscosity, density, and high melting and boiling points; thermal properties such as thermal stability, conductivity, and degree of crystallinity; mechanical properties such as shear strength and elasticity; and necessary optical properties for their applications in food and pharmaceutical products. Now that a larger proportion of society is concerned about sustainable growth, the market for biopolymer materials is rapidly expanding. As a result of the high cost of synthesizing biopolymers, researchers are now focusing on the synthesis of biomass substrates that can be used to manufacture large quantities of biopolymers at low cost. The improvement needed in the properties of biopolymers that are comparable to those of synthetic polymers, particularly in terms of their quality and performance, is another major challenge. A bright future for these environmentally beneficial chemicals and their potential commercialization is feasible by having a vision to adopt biopolymers in industries by introducing acceptable modifications or combinations of biopolymers with other components. In conclusion, these biopolymers show potential for useful applications, calling for more attention to the complete eradication of synthetic polymers [50].

REFERENCES

[1] Mohan S, Oluwafemi OS, Kalarikkal N, Thomas S, Songca SP (2016) Biopolymers-application in nanoscience and nanotechnology. *Recent Adv Biopolym* 1(1):47–66.
[2] Hernández N, Williams RC, Cochran EW (2014) The battle for the "green" polymer. Different approaches for biopolymer synthesis: Advantaged vs. replacement. *Org Biomol Chem* 12(18):2834–2849.
[3] Japan Bioplastics Association (JBPA). *GreenPla(r)*. https://www.jbpaweb.net/english/english.htm.
[4] Niaounakis M (2013) Economic evaluation and environmental impacts; section 10.3.2: Source of feedstocks. In: Michael Niaounakis (ed.), *Biopolymers Reuse, Recycling, and Disposal*. Plastics Design Library (PDL) (1st ed., Amsterdam, Elsevier,), p. 432.
[5] Ravenstijn J (2010) Bioplastics in consumer electronics. *Ind Biotechnol* 6:252–63.
[6] Public law 107-171-May 13 (2002) *Farm Security and Rural Investment Act of 2002.* https://www.congress.gov/107/plaws/publ171/PLAW-107publ171.pdf.

[7] United Sates Department of Agriculture (USDA). *Biopreferred Program. USDA Biopreferred Brand Guide Addendum-Definitions and Frequently Asked Questions.* https://www.biopreferred.gov/FILES/ Defintions_and_FAQs.pdf.

[8] Yadav P, Yadav H, Shah VG, Shah G, Dhaka G (2015) Biomedical biopolymers, their origin and evolution in biomedical sciences: A systematic review. *J Clin Diagn Res JCDR* 9(9):21.

[9] Pu W, Shen C, Wei B, Yang Y, Li Y (2018) A comprehensive review of polysaccharide biopolymers for enhanced oil recovery (EOR) from flask to field. *J Ind Eng Chem* 61:1–11.

[10] Sun Y, Gu J, Tan H, Zhang Y, Huo P (2018) Physicochemical properties of starch adhesives enhanced by esterification modification with dodecyl succinic anhydride. *Int J Biol Macromol* 112:1257–1263.

[11] Koseva N, Mitova V, Todorova Z, Tsacheva I (2019) Nanomaterials derived from phosphorus-containing polymers: Diversity of structures and applications. In: Cornelia Vasile (ed.), *Polymeric Nanomaterials in Nanotherapeutics* (Amsterdam, Elsevier), pp. 183–233.

[12] Fink JK (2013) *Reactive Polymers Fundamentals and Applications: A Concise Guide to Industrial Polymers* (William Andrew, Norwich).

[13] Boskhomdzhiev AP, Bonartsev AP, Ivanov EA, Makhina TK, Myshkina VL, Bagrov DV, Iordanskii AL et al (2010) Hydrolytic degradation of biopolymer systems based on poly-3-hydroxybutyrate: Kinetic and structural aspects. *Int Polym Sci Technol* 37(11):25–30.

[14] Feng C, Lotti T, Canziani R, Lin Y, Tagliabue C, Malpei F (2021) Extracellular biopolymers recovered as raw biomaterials from waste granular sludge and potential applications: A critical review. *Sci Total Environ* 753:142051.

[15] Schambeck CM, Magnus BS, de Souza LCR, Leite WRM, Derlon N, Guimaraes LB, da Costa RHR (2020) Biopolymers recovery: Dynamics and characterization of alginate-like exopolymers in an aerobic granular sludge system treating municipal wastewater without sludge inoculum. *J Environ Manag* 263:110394.

[16] Yano H, Nakahara S (2004) Bio-composites produced from plant microfiber bundles with a nanometer unit web-like network. *J Mater Sci* 39:1635–1638, doi:10.1023/B:JMSC.0000016162.43897.0a.

[17] Chaitanya S, Singh I (2017) Processing of PLA/sisal fiber biocomposites using direct and extrusion-injection molding. *Mater Manuf Process* 32:468–474.

[18] Sharma B, Malik P, Jain P (2018) Biopolymer reinforced nanocomposites: A comprehensive review, *Mater Today Commun* 16:353–363, doi:10.1016/j.mtcomm.2018.07.004.

[19] Cardon LK, Ragaert KJ, De Santis R, Gloria A (2017) *Design and Fabrication Methods for Biocomposites* (2nd ed., Elsevier, Amsterdam), doi:10.1016/b978-0-08-100752-5.00002-0.

[20] Misri S, Ishak MR, Sapuan SM, Leman Z (2015) Filament winding process for Kenaf fibre reinforced polymer composites. In: Mohd Sapuan Salit, Mohammad Jawaid, Nukman Bin Yusoff, M. Enamul Hoque (eds.), *Manufacturing of Natural Fiber-Reinforced Polymer Composites* (Springer, New York), pp. 369–383.

[21] Chieng BW, Ibrahim NA, Then YY, Loo YY (2014) Epoxidized vegetable oils plasticized poly(lactic acid) biocomposites: Mechanical, thermal and morphology properties. *Molecules* 19:16024–16038, doi:10.3390/molecules191016024.

[22] Deng L, Kang X, Liu Y, Feng F, Zhang H (2018) *Characterization of Gelatin/Zein Films Fabricated by Electrospinning vs Solvent Casting* (Elsevier Ltd, Amsterdam), doi:10.1016/j.foodhyd.2017.08.023.

[23] Savelyev MS, Gerasimenko AY, Vasilevsky PN, Fedorova YO, Groth T, Ten GN, Telyshev DV (2020) Spectral analysis combined with nonlinear optical measurement of laser printed biopolymer composites comprising chitosan/SWCNT. *Anal Biochem* 598:113710.

[24] Bourmaud A, Baley C (2010) Effects of thermo mechanical processing on the mechanical properties of biocomposite flax fibers evaluated by nanoindentation. *Polym Degrad Stab* 95:1488–1494, doi:10.1016/j.polymdegradstab.2010.06.022.

[25] Wang SF, Shen L, Tong YJ, Chen L, Phang IY, Lim PQ, Liu TX (2005) Biopolymer chitosan/montmorillonite nanocomposites: Preparation and characterization. *Polym Degrad Stab* 90:123–131, doi:10.1016/j.polymdegradstab.2005.03.001.

[26] Thakur VK, Thakur MK, Gupta RK (2014) Graft copolymers of natural fibers for green composites. *Carbohydr Polym* 104:87–93, doi:10.1016/j.carbpol.2014.01.016.

[27] Payal R (2019) Reliable natural-fibre augmented biodegraded polymer composites. In: Inamuddin, Sabu Thomas, Raghvendra Kumar Mishra, Abdullah M. Asiri (eds.), Sustainable Polymer Composite and Nanocomposites (Springer, New York), pp. 961–975.

[28] John M, Thomas S (2008) Biofibres and biocomposites. *Carbohydr Polym* 71:343–364, doi:10.1016/j.carbpol.2007.05.040.

[29] Fahim M, Chand N (2008) *Tribology of Natural Fiber Polymer Composites* (Elsevier, Amsterdam).

[30] Gurunathan T, Mohanty S, Nayak SK (2015) A review of the recent developments in biocomposites based on natural fibres and their application perspectives. *Compos A Appl Sci Manuf* 77:1–25, doi:10.1016/j. compositesa.2015.06.007.

[31] Okoro AM, Oladele IO, Khoathane MC (2016) Synthesis and characterization of the mechanical properties of high-density polyethylene-based composites reinforced with animal fibers. *Leonardo J Sci* 29:99–112.

[32] George A, Sanjay MR, Srisuk R, Parameswaranpillai J, Siengchin S (2020) A comprehensive review on chemical properties and applications of biopolymers and their composites. *Int J Biol Macromol* 154:329–338, doi:10.1016/j.ijbiomac.2020.03.120.

[33] Slager J, Domb AJ (2003) Biopolymer stereocomplexes. *Adv Drug Deliv Rev* 55:549–583, doi:10.1016/S0169-409X(03)00042-5.

[34] Hubbell DS, Cooper SL (1977) The physical properties and morphology of poly-ε-caprolactone polymer blends. *J Appl Polym Sci* 21:3035–3061.

[35] Kulasinski K, Guyer R, Keten S, Derome D, Carmeliet J (2015) Impact of moisture adsorption on structure and physical properties of amorphous biopolymers. *Macromolecules* 48:2793–2800.

[36] Rahman MA, De Santis D, Spagnoli G, Ramorino G, Penco M, Phuong VT, Lazzeri A (2013) Biocomposites based on lignin and plasticized poly(L-lactic acid). *J Appl Polym Sci* 129:202–214, doi:10.1002/app.38705.

[37] Pucciariello R, Villani V, Bonini C, D'Auria M., Vetere T (2004) Physical properties of straw lignin-based polymer blends. *Polymer* 45:4159–4169.

[38] Sahari J, Sapuan SM, Zainudin ES, Maleque MA (2012) A new approach to use Arenga pinnata as sustainable biopolymer: Effects of plasticizers on physical properties. *Proc Chem* 4:254–259.

[39] HPS AK, Saurabh CK, Adnan AS, Fazita MRN, Syakir MI, Davoudpour Y, Rafatullah M, Abdullah CK, Haafiz MKM, Dungani R (2016) A review on chitosan cellulose blends and nanocellulose reinforced chitosan biocomposites: Properties and their applications. *Carbohydr Polym* 150:216–226.

[40] Gan PG, Sam ST, bin Abdullah MF, Omar MF (2020) Thermal properties of nanocellulose-reinforced composites: A review. *J Appl Polym Sci* 137:48544.

[41] Sen S, Patil S, Argyropoulos DS (2015) Thermal properties of lignin in copolymers, blends, and composites: A review. *Green Chem* 17:4862–4887.

[42] Thakur VK, Singha AS, Thakur MK (2012) Surface modification of natural polymers to impart low water absorbency. *Int J Polym Anal Charact* 17:133–143, doi:10.1080/1023666X.2012.640455.

[43] Sadasivuni KK, Saha P, Adhikari J, Deshmukh K, Ahamed MB, Cabibihan J (2020) Recent advances in mechanical properties of biopolymer composites: A review. *Polym Compos* 41:32–59.

[44] Ghasemlou M, Khodaiyan F, Oromiehie A (2011) Physical, mechanical, barrier, and thermal properties of the polyol-plasticized biodegradable edible film made from kefiran. *Carbohydr Polym* 84:477–483.

[45] Aziz SB, Brza MA, Nofal MM, Abdulwahid RT, Hussen SA, Hussein AM, Karim WO (2020) A comprehensive review on optical properties of polymer electrolytes and composites. *Materials* 13:3675.

[46] Yadav P, Yadav H, Shah VG, Shah G, Dhaka G (2015) Biomedical biopolymers, their origin and evolution in biomedical sciences: A systematic review. *J Clin Diag Res* 9(9):ZE21–ZE25. doi:10.7860/JCDR/2015/13907.6565.

[47] Rebelo R, Fernandes M, Fangueiro R (2017) Biopolymers in medical implants: A brief review. In: *Procedia Engineering. 3rd International Conference on Natural Fibers: Advanced Materials for a Greener World, ICNF 2017*, 21–23 June 2017, Braga, Portugal, vol. 200, pp. 236–243. doi:10.1016/j.proeng.2017.07.034.

[48] Yadav P, Yadav H, Shah VG, Shah G, Dhaka G (2015) Biomedical biopolymers, their origin and evolution in biomedical sciences: A systematic review. *J Clin Diag Res* 9(9):ZE21–ZE25. doi:10.7860/JCDR/2015/13907.6565.

[49] Desbrières J, Guibal E (2018) Chitosan for wastewater treatment. *Polym Int* 67(1):7–14. doi:10.1002/pi.5464.

[50] Rao MG, Akila RM (2014) A comprehensive review on biopolymers. *Sci Revs Chem Commun* 4(2):61–68.

2 Challenges and Advantages

*Siti Musabikha, Rahayu Kusumastuti, Arini
Nikitasari, Siska Prifiharni, and Gadang Priyotomo*
National Research and Innovation Agency

2.1 INTRODUCTION

Corrosion is a natural process that occurs when metals and alloys react with the environment [1–3]. It can cause stresses and damage to metal structures and equipment [4–7], leading to expensive repairs and mitigations [8]. An inhibitor can protect metals from corroding, especially in acidic environments [9,10]. Inorganic corrosion inhibitors, such as chromates, cobalt, cadmium, tetraborates, and other heavy metals, and organic inhibitors, such as imidazole, thiazole, pyrazole, pyrimidine, and hydrazine, have been effective in preventing corrosion [1,9,11]. However, these inhibitors are toxic and have harmful effects on the environment and living organisms [9,11,12]. Therefore, there has been a growing interest in using biopolymers as alternative corrosion inhibitors [13].

Biopolymers are polymers derived from natural sources, namely plants, animals, and microbes such as guar gum, starch, chitin, lignin, cellulose, agar, carrageenan, alginates, polyhydroxyalkanoates, polyhydroxybutyrate, and proteins [10,14–28]. Biopolymers have a wide range of applications in various industries, including medicine, food, and agriculture [15,16,21,23–25,29]. The use of biopolymers as corrosion inhibitors has been extensively studied in recent years due to their unique properties and sustainability [1].

Biopolymers have molecules/heteroatoms with π-electrons that function as adsorption centers [30,31]. Biopolymer inhibitors interact with metal surfaces via chemical and physical adsorption [14,32]. In physical adsorption, electrostatic forces draw charged inhibitor molecules to charged metal surfaces. In chemical adsorption, the inhibitor molecule's electron pair shares the vacant d-orbital of the metal. Physisorption is the predominant adsorption mechanism if the inhibitor species protonates. Under corrosive conditions, neutral inhibitor compounds prefer the chemisorption process. However, both processes can take place at the same time on a metal surface [32].

Biopolymers are classified into three major groups (Figure 2.1) based on monomeric structures and composition, namely polynucleotides (DNA, RNA, and linear plasmid DNA), polysaccharides (starch, pectin, lignin, cellulose, chitin, chitosan, alginate, agarose, and dextran), and polypeptides (proteins, gelatin, collagen, and wheat gluten) [9,15,16,19,23,24,28,29,33,34]. The most widely used biopolymer corrosion inhibitors are chitosan, cellulose, and their derivatives [24,35].

DOI: 10.1201/9781003400059-2

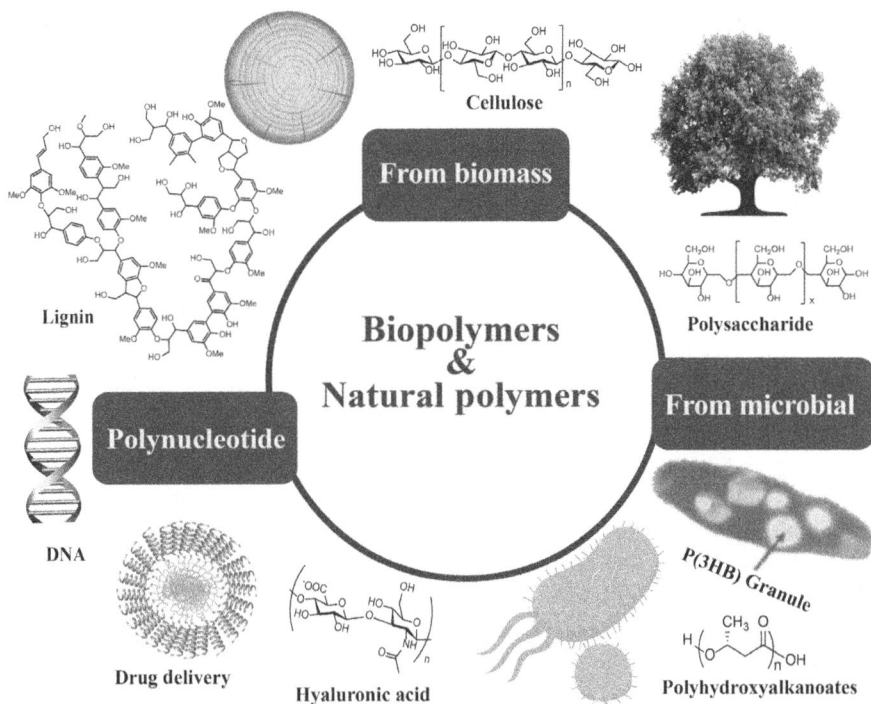

FIGURE 2.1 Sources of biopolymers: cellulose, polysaccharide, hyaluronic acid, and lignin from biomass; polynucleotide (DNA); and polyhydroxyalkanoate from microbial sources [29].

Chitosan is a cationic polysaccharide that can be produced from N-deacetylated chitin by combining D-glucosamine and *N*-acetyl-D-glucosamine that are randomly spread and joined by a β-1,4-glycosidic bond, and it is commonly found in the shells of shrimp, clams, and crabs [16,24,35]. Chitosan has polar substituents such as hydroxyl group (−OH), ether group (−O −), acetyl group (−NHCOCH$_3$), hydroxymethyl group (−CH$_2$OH), and amine group (−NH$_2$), which can bond electrostatically or covalently with metals [35]. Cellulose is one of the most abundant polymers found on the Earth [24,35]. It is a linear chain polysaccharide with a β→1−4-glycosidic linkage that connects with D-glucose units [33,35]. Celluloses have the general formula (C$_5$H$_{10}$O$_5$)$_n$ and can be found in plant walls, algae, and oomycetes [33,35].

The use of biopolymers as corrosion inhibitors has been extensively studied in recent years [20,36]. Unmodified natural chitosan has been tested for its ability to prevent corrosion of the mild steel in 0.1 M HCl solution [37]. Even at low concentrations, the chitosan biopolymer inhibits corrosion. The inhibition efficiency increases to 96% at 60°C, but it decreases to 93% at 70°C at 4.0 μM chitosan concentration. Chitosan acts as a mixed biopolymer inhibitor and follows the Langmuir adsorption isotherm. Jmiai et al. [38] showed that chitosan's inhibition efficiency is 87% at 10^{-1} mg/L, which is higher than that of sodium alginate (81%) when both compounds are evaluated on copper in a 1 M HCl solution using electrochemical measurements. This difference between chitosan and sodium alginate may be due to their distinct charge transfer and binding bond energies characteristics.

The inhibition efficiency of a biopolymer blend of starch and pectin (1:3) has also been evaluated using electrochemical techniques [31]. The inhibition efficiency for mild steel corrosion control in HCl at 303 K is found to be 74%. Later, the corrosion control is demonstrated in 0.025 M HCl media by inulin, a polysaccharide from herbaceous plant extract, on 6061 aluminum-15%(v) SiC composite material with a maximum efficiency of 93% for 1 g/L at 303 K [27]. Inulin acts as a mixed-type biopolymer inhibitor, is adsorbed through physisorption, and follows the Langmuir adsorption isotherm. However, its inhibitor efficiency is decreased with an increase in temperature.

Inhibition efficiency can be improved by modifying or combining the biopolymer inhibitor with other materials. The modified alginate (8-hydroxyquinoline-*grafted*-alginate) has been examined as a biopolymer inhibitor of mild steel in 1 M HCl aqueous media via electrochemical measurement and simulation analysis [10]. This analysis showed that the inhibition efficiency is enhanced by modifying the alginate biopolymer up to 92.6% at 298 K with optimal concentration, and it functions as a mixed-type inhibitor. The experimental data support the molecular dynamics simulation and quantum chemical parameters, which provide additional information on the inhibition process. In a 15% sulfuric acid solution, Solomon et al. investigated the effectiveness and stability of carboxymethyl cellulose (CMC) integrated with silver nanoparticle (AgNP) composite on low-carbon steel St37 [39]. Compared with CMC, CMC/AgNPs performed better. At 25°C, 1,000 ppm CMC/AgNPs with a longer immersion duration of 15 hours resulted in an excellent inhibition efficiency of 93.94%. CMC/AgNPs inhibited both the cathodic and anodic reactions, and their adsorption followed the Langmuir adsorption isotherm.

In another research, Biswas et al. [9] investigated corrosion inhibition of natural biopolymer dextrin and dextrin-*graft*-polyvinyl acetate as its derivatives in a 15% HCl solution. The results showed that the biopolymer dextrin-based graft copolymer provides effective protection against corrosion with an inhibitor efficiency of 98.39%, compared with native dextrin's inhibitor efficiency of 84.56%. This is attributable due to its big molecules, which have more reactive centers and promote stronger adsorption on the metallic surface. Therefore, it is important to discuss biopolymer corrosion inhibitors in detail. Furthermore, the advantages and challenges that must be fixed in developing biopolymer corrosion inhibitors also need to be investigated.

2.2   ADVANTAGES OF USING BIOPOLYMER CORROSION INHIBITORS

Biopolymers have been extensively studied as an alternative to traditional corrosion inhibitors due to their numerous advantages:

1. **Non-toxicity**: Biopolymers are non-toxic [10,11,15,16,27,31] and safe to handle, and they do not pose risks to human health and the environment. This makes them attractive alternatives to inorganic inhibitors and an ideal choice for use in industries where worker safety is a primary concern [40].
2. **Biodegradability**: Biopolymers are polymers derived from nature that are biodegradable, meaning that they can decompose naturally without causing

harm to the environment [10,11,15–17,20,21,28,31]. This property makes them an environmentally friendly alternative to traditional inhibitors that can leave harmful residues [23].

3. **Cost-effectiveness**: Biopolymers can be extracted from natural sources, such as plant wastes. Thus, expenses can be lower, and it may reduce economic costs. The lower cost of biopolymers compared with traditional inorganic inhibitors is another advantage, making them an attractive option for industries looking to reduce their costs [22,23,25,41]. Additionally, using biopolymers as corrosion inhibitors can reduce maintenance costs and extend product lifespans, as they provide long-term protection against corrosion. This reduces the need for frequent repairs and replacements, resulting in cost savings for industries that use them [6,15].

4. **Versatility**: Biopolymers can be modified to enhance their corrosion inhibition properties, making them suitable for a wide range of applications [16,27,29]. They can be modified ability to serve specific purposes, which makes them a desirable choice for industries seeking corrosion inhibitors that meet their specific needs [42]. Also, biopolymers have been successfully used as corrosion inhibitors in various environments. They are effective in acidic and alkaline environments, as well as in seawater and other harsh environments [21,31]. This makes biopolymers a versatile and reliable option for corrosion inhibition.

5. **Long-term protection**: Biopolymers can provide long-term protection against corrosion due to the dense, stable, and protective film they form on the metal surface, which protects over an extended period and prevents further damage due to abrasion or erosion [24,43].

6. **Renewable, sustainable, and environmentally friendly characteristics**: Natural sources used to produce biopolymers are renewable and sustainable sources, making them a more environmentally friendly alternative to traditional synthetic inhibitors [11,15,20,25,28,30], which are often derived from non-renewable sources [2]. Some biopolymers can be produced using waste materials, such as agricultural and food wastes, which are a sustainable and environmentally friendly choice for companies trying to reduce waste and reuse materials [44]. Also, biopolymers have a lower carbon footprint than synthetic inhibitors, making them a more sustainable option for businesses aiming to minimize their environmental impact [26].

7. **Biocompatibility**: Biopolymers are biocompatible [11,15,16,21,24], making them suitable for use in medical devices, drug delivery systems, and implant biomaterials. The use of biopolymers as corrosion inhibitors in medical devices and implants can provide long-term protection against corrosion without posing risks to patient health [16,22,43].

8. **Self-assembling properties**: Biopolymers can form a self-assembled monolayer on the metal surface, providing excellent protection against corrosion. The film can also be self-healing, ensuring continued protection [36]. This property ensures that the inhibitor adheres well to the metal surface and is not easily washed off by the environment [36].

2.3 CHALLENGES OF USING BIOPOLYMER CORROSION INHIBITORS

Though biopolymer corrosion inhibitors have many advantages, several challenges limit their use. Here are some challenges facing biopolymers as corrosion inhibitors:

1. **Poor solubility**: Some biopolymer corrosion inhibitors have poor solubility in polar solvents [35,39,45–56]. This is due to their hydrophobic character, aromatic rings, and non-polar hydrocarbon chains in biopolymers [35,50]. The poor solubility can also result from depolymerization reactions that take place during the extraction stage [55]. In addition, their high molecular weights, which enable considerable surface coverage, also result in poor solubility in corrosive solutions [46,47,57]. Solubility is a crucial prerequisite for efficient corrosion inhibitors. A chemical that dissolves in its application environment acts as a better inhibitor than one that only partially or not at all dissolves [52].

2. **Limited stability and mechanical properties**: Biopolymer corrosion inhibitors have low mechanical properties and lower stability over days [19,25,29,58]. It means that after a certain amount of time, they may stop functioning effectively in an extremely hostile solution [51]. It can easily age and be biologically damaged [50]. Biopolymers can be unstable in certain environments due to the combined effects of atmospheric oxygen, humidity, oxidative stress, mechanical stress, heat, and UV light [18]. Also, they may not provide sufficient protection against corrosion in applications with mechanical stress, such as in moving parts and high-speed environments [33]. The properties of biopolymers may also degrade and decay over time, reducing their lifespan and effectiveness as inhibitors [18,34].

3. **Limited thermal resistances**: The corrosive environment's temperature significantly affects the metal–inhibitor interaction and its inhibition efficiency [59]. Thus, inhibitors must have good thermal stability, especially for sections exposed to high temperatures [60]. A biopolymer corrosion inhibitor may perform well at room temperature [50]. However, its efficiency will decrease with an increase in temperature [15,18,25,29,35,39,50,51,58,61–63]. This is due to the inhibitor molecule's acid- or base-catalyzed decomposition and/or regrouping [35]. Besides, with an increase in temperature, the ionic mobility or the kinetic energy to the metal surface increases [60], and the attractive force between the inhibitor and the metal substrate surface naturally decreases [35].

   The decrease in corrosion inhibition effectiveness caused by increasing the temperature is aligned with the physical adsorption and its electrostatic bonding. The chemisorption mechanism increases the inhibition efficiency under high-temperature environments. Moreover, most organic inhibitors adsorb using mixed adsorption or physio-chemisorption [35]. For instance, chitosan inhibition efficiency was found to be increased to 96% at 333 K in an acidic solution [37]. However, as the temperature increased further, a decline in the inhibition efficiency was observed. It might be caused by the degradation of the inhibitor's adsorption–desorption balance [60].

4. **Long extraction process**: Due to the complexity of biopolymers, the extraction method requires several processes and can be adjusted depending on the kind of biopolymer [54,55]. Additionally, the extraction process is very laborious [54], expensive, time-consuming [64], and too complicated to be used on a large scale in industries [65]. Furthermore, the extraction method used can affect their corrosion inhibition quantities and qualities [55,58,65]. It is also reported that extending the extraction duration from 2 to 6 days can increase the concentrations, and the solvent system may affect the compositions [55]. Moreover, producing biopolymer extracts frequently involves using hazardous solvents that can have a negative impact when released into the environment and are harmful to other living things [54,64].

5. **Limited in industrial applications**: The use of natural biopolymer corrosion inhibitors in large-scale plants is limited [31,66]. There is a gap regarding the application of green corrosion inhibitors in the industry. Today, almost no industry uses green corrosion inhibitors. The extract from green corrosion inhibitors used in research so far is in the form of a mixture, not one active substance. Sometimes, a combination of one substance with another substance in the crude extracts can show synergy with each other, but there is a possibility that it will show the opposite antagonistic effect. Moreover, the processes of extraction, isolation, and synthesis into its derivatives of active ingredients in charge of corrosion inhibition are quite long and tough tasks [31].

   On the other hand, little research is being performed on materials that are readily accessible, such as biowaste, whereas more research is being conducted on corrosion inhibitors extracted from rare and hard-to-find sources. Thus, it is getting more difficult to utilize green materials as inhibitors on an industrial scale [66]. Therefore, research on green corrosion inhibitors should pay more attention to identifying and describing active substituents in green materials, which are abundant and suitable for implementation in industrial applications [67].

6. **Required in high concentrations**: To increase the corrosion inhibition rate, high concentrations of biopolymer corrosion inhibitors are needed [13,49,50,56,60,61]. The concentration can range between 500 and 1,000 ppm, which may cause the solutions to become turbid or even endanger the effectiveness of machinery in practical and industrial uses [51].

To improve the stability and inhibition efficiency of biopolymer corrosion inhibitors, numerous attempts have been made, such as copolymerization, mixing, cross-linking, injection of a small amount of metallic material into the matrix of macromolecules, and combination with compounds (i.e., halide ions and surfactants) that have a synergistic impact with several synthesis techniques [13,39,47,48,50,53,62,68,69]. Synthesis of new biopolymer derivatives can also enhance their anticorrosion properties and solubility in corrosive solutions. Biopolymer chitin has also been chemically altered to create a novel derivative called phosphorylated chitin, which has a maximum efficacy of 92% with 200 ppm concentrations [47].

Halide ions have been utilized to increase the effectiveness of organic compounds' ability to suppress corrosion. Moreover, the iodide ion (I^-) has a higher synergistic effect (compared with Cl^- and Br^-) due to its big size (216 pm) and straightforward polarizability [68,69]. Increased surface coverage by ion–pair interactions between the organic cation and the anion leads to synergistic corrosion inhibition [70].

The ionic radii and electronegativity of these ions significantly contribute to corrosion inhibition since molecule adsorption plays a major role in the corrosion inhibition mechanism. Owing to their ionic radii (pm), halide ions have a synergistic impact in the following order: I^- (206) > Br^- (182) > Cl^- (167). The initial halide ion adsorption on the metal surface helps the adsorption of organic inhibitor molecules via coulombic attraction, forming a steadier protective layer at the metal–solution interface [53]. The inhibition efficiency increases from 74% of chitosan alone to 90% of chitosan (200 ppm) combined with 5 ppm of KI for medium steel in acidic media [57].

Halide ions may generate synergism and antagonism mechanisms on the biopolymer's ability to inhibit. The chloride ion works against the inhibition process, while the bromide and iodide ions support the biopolymer's ability to prevent corrosion [52]. This characteristic of the chloride ion might be explained by the development of soluble adsorption intermediates on the substrate surface, which show rapid metal dissolution rates, particularly at higher temperatures [52].

In addition, surfactant interactions with biopolymers have demonstrated a synergistic effect by altering the characteristics of surfaces and interfaces [69,71]. Surfactant is known as a surface-active substance with both hydrophilic and hydrophobic groups in each molecule [57]. Biopolymers are bound to ionic surfactants, which have charged groups that might resist each other and result in the growth of the biopolymer's backbone. It provides a better defense against an aggressive medium due to the expansion of the polymer molecules, which allows it to occupy an even bigger surface area upon adsorption to the metal surface [13]. For instance, the use of surfactants, namely sodium dodecyl sulfate and cetyl trimethyl ammonium bromide, as a synergistic additive considerably improved the inhibitor efficiency of starch at 200 ppm from 66.21% to 68.75% (5 ppm sodium dodecyl sulfate) and 66.62% (1 ppm cetyl trimethyl ammonium bromide) in 30°C 0.1 M H_2SO_4 acidic solution [72].

Chemical modification of the molecular structure of biopolymer corrosion inhibitors may also be used strategy [46,47,51]. For example, modified hydrophobic carboxymethyl chitosan derivatives have been synthesized and their ability to prevent corrosion of carbon steel in an acidic media tested. These derivatives have a higher inhibition efficiency of 93.23% compared with that of pure chitosan of 67.56% (at 250 ppm and 25°C) due to their higher surface activity and more active sites [73].

2.4 FUTURE PREDICTIONS AND SUGGESTIONS

Previous studies have demonstrated that the modification or synthesis of biopolymers results in increased efficiency of inhibitors. Further studies can attempt more incorporations through modifications, such as combining nanoparticles with biopolymer molecular structures and exploring synthesis techniques that are more straightforward, eco-friendly, and cost-effective [39].

Additionally, computational studies can provide further significant support through quantum chemical modeling to examine the molecular structure of biopolymer inhibitor reaction pathways and their adsorption mechanisms [51]. Surface adhesion problems can be understood by using mechanistic predictions of pertinent parameters in theoretically based modeling studies to clarify potential phases, bonding locations, and the energy of inhibitory reactions in complex systems [51,53]. Besides, the composition of biopolymers varies, making it more challenging to analyze how they inhibit corrosion, but this offers opportunities for further research.

Molecular dynamics simulation could be a useful tool to investigate the adsorption of key chemical components on metal surfaces, to better understand the complex inhibitor–metal interface interactions, and to provide several perspectives on potential dynamic molecular reactions at the microscopic level [52,53]. Also, density functional theory can be used to study the energy gap and the energies linked to the border orbitals of molecules by computing the energies of the highest occupied molecular orbital and lowest unoccupied molecular orbital [53]. Moreover, material computing has a significant role in biopolymer inhibitor research as it can be applied in simulating the interaction between biopolymer inhibitors and metal surfaces; investigating the effectiveness of biopolymer inhibitors under different environmental conditions; developing novel biopolymers; optimizing their structure and functional properties; and predicting the effect of environmental factors on biopolymer inhibition process. These advances will likely lead to the development of more applicable, sustainable, and efficient corrosion control mechanisms for various industries.

REFERENCES

[1] R. Kusumastuti, R. I. Pramana, and J. W. Soedarsono, "The use of morinda citrifolia as a green corrosion inhibitor for low carbon steel in 3.5% NaCl solution," *AIP Conf. Proc.*, vol. 1823, pp. 020012-1–020012-9, 2017. doi:10.1063/1.4978085.

[2] R. I. Pramana, R. Kusumastuti, J. W. Soedarsono, and A. Rustandi, "Corrosion inhibition of low carbon steel by Pluchea indica less in 3.5% NaCL solution," *Adv. Mater. Res.*, vol. 785–786, pp. 20–24, 2013, doi:10.4028/www.scientific.net/AMR.785-786.20.

[3] G. R. Sunaryo, R. Kusumastuti, and Sriyono, "Corrosion surveillance program for tank, fuel cladding and supporting structure of 30 MW Indonesian RSG GAS research reactor," *Kerntechnik*, vol. 86, no. 3, pp. 236–243, 2021, doi:10.1515/kern-2020-0083.

[4] S. Musabikha, I. K. A. P. Utama, and Mukhtasor, "Characterisation and performance evaluation of marine coatings used for submerged ocean energy application," *J. Eng. Sci. Technol.*, vol. 16, no. 5, pp. 4309–4324, 2021.

[5] S. Musabikha, I. K. A. P. Utama, and Mukhtasor, "State of the art in protection of erosion-corrosion on vertical axis tidal current turbine," *AIP Conf. Proc.*, vol. 1964, pp. 020047-1–020047-8, 2018. doi:10.1063/1.5038329.

[6] R. Aslam, M. Mobin, M. Shoeb, and J. Aslam, "Novel ZrO2-glycine nanocomposite as eco-friendly high temperature corrosion inhibitor for mild steel in hydrochloric acid solution," *Sci. Rep.*, vol. 12, no. 1, pp. 1–19, 2022, doi:10.1038/s41598-022-13359-y.

[7] A. Royani, S. Prifiharni, G. Priyotomo, A. Nikitasari, and Sundjono, "The corrosion performance of low carbon steel in feedwater for cooling water systems," *AIP Conf. Proc.* vol. 2382, pp. 060003-1–060003-10, 2021. doi:10.1063/5.0060105.

[8] S. Musabikha, I. K. A. P. Utama, Mukhtasor, J. A. Wharton, and R. J. K. Wood, "Effects of nickel-aluminum bronze pre-oxidized films on the cathodic kinetics of oxygen reduction," *Anal. Lett.*, vol. 53, no. 8, pp. 1218–1232, 2020, doi:10.1080/00032719.2 019.1700515.

[9] A. Biswas, D. Das, H. Lgaz, S. Pal, and U. G. Nair, "Biopolymer dextrin and poly (vinyl acetate) based graft copolymer as an efficient corrosion inhibitor for mild steel in hydrochloric acid: Electrochemical, surface morphological and theoretical studies," *J. Mol. Liq.*, vol. 275, pp. 867–878, 2019, doi:10.1016/j.molliq.2018.11.095.

[10] M. Fardioui et al., "Bio-active corrosion inhibitor based on 8-hydroxyquinoline-grafted-Alginate: Experimental and computational approaches," *J. Mol. Liq.*, vol. 323, no. 114615, pp. 1–13, 2021, doi:10.1016/j.molliq.2020.114615.

[11] M. Mobin and M. Rizvi, "Polysaccharide from Plantago as a green corrosion inhibitor for carbon steel in 1 M HCl solution," *Carbohydr. Polym.*, vol. 160, pp. 172–183, 2017, doi:10.1016/j.carbpol.2016.12.056.

[12] G. Priyotomo et al., "Corrosion inhibition of carbon steel in sulfuric acid using cymbopogon citratus as a green corrosion inhibitor," *Corros. Sci. Technol.*, vol. 21, no. 6, pp. 423–433, 2022, doi:10.14773/CST.2022.21.6.423.

[13] M. Mobin, M. Rizvi, L. O. Olasunkanmi, and E. E. Ebenso, "Biopolymer from tragacanth gum as a green corrosion inhibitor for carbon steel in 1 M HCl solution," *ACS Omega*, vol. 2, no. 7, pp. 3997–4008, 2017, doi:10.1021/acsomega.7b00436.

[14] C. Shen, V. Alvarez, J. D. B. Koenig, and J. L. Luo, "Gum Arabic as corrosion inhibitor in the oil industry: Experimental and theoretical studies," *Corros. Eng. Sci. Technol.*, vol. 54, no. 5, pp. 444–454, 2019, doi:10.1080/1478422X.2019.1613780.

[15] G. P. Udayakumar et al., "Biopolymers and composites: Properties, characterization and their applications in food, medical and pharmaceutical industries," *J. Environ. Chem. Eng.*, vol. 9, no. 4, pp. 1–22, 2021, doi:10.1016/j.jece.2021.105322.

[16] R. Rebelo, M. Fernandes, and R. Fangueiro, "Biopolymers in medical implants: A brief review," *Proc. Eng.*, vol. 200, pp. 236–243, 2017. doi:10.1016/j.proeng.2017.07.034.

[17] E. Kabir, R. Kaur, J. Lee, K. H. Kim, and E. E. Kwon, "Prospects of biopolymer technology as an alternative option for non-degradable plastics and sustainable management of plastic wastes," *J. Clean. Prod.*, vol. 258, pp. 1–15, 2020, doi:10.1016/j.jclepro.2020.120536.

[18] N. T. Dintcheva, G. Infurna, M. Baiamonte, and F. D'Anna, "Natural compounds as sustainable additives for biopolymers," *Polymers (Basel)*, vol. 12, no. 4, pp. 1–24, 2020, doi:10.3390/polym12040732.

[19] M. G. A. Vieira, M. A. Da Silva, L. O. Dos Santos, and M. M. Beppu, "Natural-based plasticizers and biopolymer films: A review," *Eur. Polym. J.*, vol. 47, no. 3, pp. 254–263, 2011, doi:10.1016/j.eurpolymj.2010.12.011.

[20] M. Keramatinia, B. Ramezanzadeh, and M. Mahdavian, "Green production of bioactive components from herbal origins through one-pot oxidation/polymerization reactions and application as a corrosion inhibitor for mild steel in HCl solution," *J. Taiwan Inst. Chem. Eng.*, vol. 105, pp. 134–149, 2019, doi:10.1016/j.jtice.2019.10.005.

[21] S. Ranganathan, S. Dutta, J. A. Moses, and C. Anandharamakrishnan, "Utilization of food waste streams for the production of biopolymers," *Heliyon*, vol. 6, no. 9, pp. 1–13, 2020, doi:10.1016/j.heliyon.2020.e04891.

[22] B. Imre and B. Pukánszky, "Compatibilization in bio-based and biodegradable polymer blends," *Eur. Polym. J.*, vol. 49, no. 6, pp. 1215–1233, 2013, doi:10.1016/j.eurpolymj.2013.01.019.

[23] R. K. Gupta, P. Guha, and P. P. Srivastav, "Natural polymers in bio-degradable/edible film: A review on environmental concerns, cold plasma technology and nanotechnology application on food packaging: A recent trends," *Food Chem. Adv.*, vol. 1, no. 100135, pp. 1–15, 2022, doi:10.1016/j.focha.2022.100135.

[24] M. C. Biswas et al., "Recent advancement of biopolymers and their potential bio-medical applications," *J. Polym. Environ.*, vol. 30, no. 1, pp. 51–74, 2022, doi:10.1007/s10924-021-02199-y.

[25] J. Baranwal, B. Barse, A. Fais, G. L. Delogu, and A. Kumar, "Biopolymer: A sustainable material for food and medical applications," *Polymers (Basel)*, vol. 14, no. 5, pp. 1–22, 2022, doi:10.3390/polym14050983.

[26] K. Amulya, R. Katakojwala, S. Ramakrishna, and S. Venkata Mohan, "Low carbon biodegradable polymer matrices for sustainable future," *Compos. C Open Access*, vol. 4, no. 100111, pp. 1–13, 2021, doi:10.1016/j.jcomc.2021.100111.

[27] B. P. Charitha and P. Rao, "Carbohydrate biopolymer for corrosion control of 6061 Al-alloy and 6061Aluminum-15%(v) SiC(P) composite-Green approach," *Carbohydr. Polym.*, vol. 168, pp. 337–345, 2017, doi:10.1016/j.carbpol.2017.03.098.

[28] F. Samalens et al., "Progresses and future prospects in biodegradation of marine biopolymers and emerging biopolymer-based materials for sustainable marine ecosystems," *Green Chem.*, vol. 24, no. 5, pp. 1762–1779, 2022, doi:10.1039/d1gc04327g.

[29] S. K. Kumaran, M. Chopra, E. Oh, and H. J. Choi, "Biopolymers and natural polymers," In: R. Narain (Ed.), *Polymer Science and Nanotechnology: Fundamentals and Applications*, 2020, pp. 245–256. doi:10.1016/B978-0-12-816806-6.00011-X.

[30] M. Pais and P. Rao, "Electrochemical approaches for material conservation: Experimental and theoretical insights using a biopolymer," *J. Bio- Tribo-Corrosion*, vol. 6, no. 4, pp. 1–13, 2020, doi:10.1007/s40735-020-00412-4.

[31] S. Y. and P. Rao, "Material conservation and surface coating enhancement with starch-pectin biopolymer blend: A way towards green," *Surf. Interfaces*, vol. 16, pp. 67–75, 2019, doi:10.1016/j.surfin.2019.04.011.

[32] S. A. Umoren and M. M. Solomon, "Biopolymer composites and nanocomposites for corrosion protection of industrial metal substrates," In: H. M. A. Amin and A. Galal (Eds.), *Corrosion Protection of Metals and Alloys Using Graphene and Biopolymer Based Nanocomposites*, 1st edn., CRC Press, Taylor & Francis Group, Boca Raton, FL, 2021, pp. 16–31. doi:10.1201/9781315171364

[33] M. S. B. Reddy, D. Ponnamma, R. Choudhary, and K. K. Sadasivuni, "A comparative review of natural and synthetic biopolymer composite scaffolds," *Polymers (Basel)*, vol. 13, no. 7, pp. 1–51, 2021, doi:10.3390/polym13071105.

[34] C. M. Runnels et al., "Folding, assembly, and persistence: The essential nature and origins of biopolymers," *J. Mol. Evol.* vol. 86, no. 9, pp. 598–610, 2018, doi:10.1007/s00239-018-9876-2.

[35] C. Verma, E. E. Ebenso, M. A. Quraishi, and C. M. Hussain, "Recent developments in sustainable corrosion inhibitors: Design, performance and industrial scale applications," *Mater. Adv.*, vol. 2, no. 12, pp. 3806–3850, 2021, doi:10.1039/d0ma00681e.

[36] M. L. Zheludkevich et al., "Self-healing protective coatings with 'green' chitosan based pre-layer reservoir of corrosion inhibitor," *J. Mater. Chem.*, vol. 21, no. 13, pp. 4805–4812, 2011, doi:10.1039/C1JM10304K.

[37] S. A. Umoren, M. J. Banera, T. Alonso-Garcia, C. A. Gervasi, and M. V Mirífico, "Inhibition of mild steel corrosion in HCl solution using chitosan," *Cellulose*, vol. 20, no. 5, pp. 2529–2545, 2013, doi:10.1007/s10570-013-0021-5.

[38] A. Jmiai et al., "The effect of the two biopolymers 'sodium alginate and chitosan' on the inhibition of copper corrosion in 1 M hydrochloric acid," *Mater. Today Proc.*, vol. 22, pp. 12–15, 2020, doi:10.1016/j.matpr.2019.08.057.

[39] M. M. Solomon, H. Gerengi, and S. A. Umoren, "Carboxymethyl cellulose/silver nanoparticles composite: Synthesis, characterization and application as a benign corrosion inhibitor for St37 steel in 15% H2SO4 medium," *ACS Appl. Mater. Interfaces,* vol. 9, no. 7, pp. 6376–6389, 2017, doi:10.1021/acsami.6b14153.

[40] W. Zhang, Y. Ma, L. Chen, L.-J. Wang, Y.-C. Wu, and H.-J. Li, "Aloe polysaccharide as an eco-friendly corrosion inhibitor for mild steel in simulated acidic oilfield water: Experimental and theoretical approaches," *J. Mol. Liq.*, vol. 307, p. 112950, 2020, doi:10.1016/j.molliq.2020.112950.

[41] H. A. Fetouh, A. Hefnawy, A. M. Attia, and E. Ali, "Facile and low-cost green synthesis of eco-friendly chitosan-silver nanocomposite as novel and promising corrosion inhibitor for mild steel in chilled water circuits," *J. Mol. Liq.*, vol. 319, p. 114355, 2020, doi:10.1016/j.molliq.2020.114355.

[42] F. Sardabi, M. H. Azizi, H. A. Gavlighi, and A. Rashidinejad, "Potential benefits of Moringa peregrina defatted seed: Effect of processing on nutritional and anti-nutritional properties, antioxidant capacity, in vitro digestibility of protein and starch, and inhibition of α-glucosidase and α-amylase enzymes," *Food Chem. Adv.*, vol. 1, no. 100034, pp. 1–9, 2022, doi:10.1016/j.focha.2022.100034.

[43] Q. Yang et al., "Biopolymer coating for particle surface engineering and their biomedical applications," *Mater. Today Bio.*, vol. 16, no. 100407, pp. 1–17, 2022, doi:10.1016/j.mtbio.2022.100407.

[44] A. Ronzano, R. Stefanini, G. Borghesi, and G. Vignali, "Agricultural waste as a source of innovative and compostable composite biopolymers for food packaging: A scientific review," In: *Proceedings of the 7th International Food Operations and Processing Simulation Workshop (FoodOPS 2021)*, 2021, pp. 29–36. doi:10.46354/i3m.2021.foodops.005.

[45] J. Haque, V. Srivastava, D. S. Chauhan, H. Lgaz, and M. A. Quraishi, "Microwave-induced synthesis of chitosan schiff bases and their application as novel and green corrosion inhibitors: Experimental and theoretical approach," *ACS Omega*, vol. 3, no. 5, pp. 5654–5668, 2018, doi:10.1021/acsomega.8b00455.

[46] D. S. Chauhan, M. A. Quraishi, H. Al-Qahtani, and M. A. Jafar Mazumder, "Green polymeric corrosion inhibitors: Design, synthesis, and characterization," In: M. J. Mazumder, M. Quraishi, and A. Al-Ahmed (Eds.), *Polymeric Corrosion Inhibitors for Greening the Chemical and Petrochemical Industry,* John Wiley & Sons, Ltd, New York, 2022, pp. 1–22. doi:10.1002/9783527835621.ch1.

[47] B. El Ibrahimi, "Sustainable corrosion inhibitors for copper and its alloys," *Mat. Res. Found.*, 2021, pp. 175–203. doi:10.21741/9781644901496-8.

[48] Y. Ren, Y. Luo, K. Zhang, G. Zhu, and X. Tan, "Lignin terpolymer for corrosion inhibition of mild steel in 10% hydrochloric acid medium," *Corros. Sci.*, vol. 50, no. 11, pp. 3147–3153, 2008, doi:10.1016/j.corsci.2008.08.019.

[49] N. A. Aljeaban et al., "Polymers decorated with functional motifs for mitigation of steel corrosion: An overview," *Int. J. Polym. Sci.*, vol. 2020, 2020, doi:10.1155/2020/9512680.

[50] Z. Yihang, "Application of water-soluble polymer inhibitor in metal corrosion protection: Progress and challenges," *Front. Energy Res.*, vol. 10, pp. 1–13, 2022, doi:10.3389/fenrg.2022.997107.

[51] M. Rizvi, "Natural polymers as corrosion inhibitors," In: *Organic Corrosion Inhibitors*, John Wiley & Sons, Ltd, New York, 2021, pp. 411–434. doi:10.1002/9781119794516.ch18

[52] S. A. Umoren and M. M. Solomon, "Effect of halide ions on the corrosion inhibition efficiency of different organic species: A review," *J. Ind. Eng. Chem.*, vol. 21, pp. 81–100, 2015, doi:10.1016/j.jiec.2014.09.033.

[53] S. A. Umoren and U. M. Eduok, "Application of carbohydrate polymers as corrosion inhibitors for metal substrates in different media: A review," *Carbohydr. Polym.*, vol. 140, pp. 314–341, 2016, doi:10.1016/j.carbpol.2015.12.038.

[54] S. H. Alrefaee, K. Y. Rhee, C. Verma, M. A. Quraishi, and E. E. Ebenso, "Challenges and advantages of using plant extract as inhibitors in modern corrosion inhibition systems: Recent advancements," *J. Mol. Liq.*, vol. 321, 2021, doi:10.1016/j.molliq.2020.114666.

[55] M. H. Shahini, B. Ramezanzadeh, and H. E. Mohammadloo, "Recent advances in bio-polymers/carbohydrate polymers as effective corrosion inhibitive macro-molecules: A review study from experimental and theoretical views," *J. Mol. Liq.*, vol. 325, 2021, doi:10.1016/j.molliq.2020.115110.

[56] B. El Ibrahimi, "Corrosion inhibitors for oil and gas systems," In: L. Guo, C. Verma and D. Zhang (Eds.), *Eco-Friendly Corrosion Inhibitors: Principles, Designing and Applications,* Elsevier, Amsterdam, 2022, pp. 111–126, doi:10.1016/B9784.00025-8.

[57] B. El Ibrahimi, L. Guo, J. V. Nardeli, and R. Oukhrib, "The application of chito-san-based compounds against metallic corrosion," In: M. Berrada (Ed.), *Chitin and Chitosan*, IntechOpen, Rijeka, 2021, doi:10.5772/intechopen.96046.

[58] M. M. Abe et al., "Advantages and disadvantages of bioplastics production from starch and lignocellulosic components," *Polymers (Basel)*, vol. 13, no. 15, pp. 1–25, 2021, doi:10.3390/polym13152484.

[59] P. R. Prabhu, D. Prabhu, A. Chaturvedi, and P. Kishore Dodhia, "Corrosion inhibition of ferrite bainite AISI1040 steel in H2SO4 using biopolymer," *Cogent Eng.*, vol. 8, no. 1, 2021, doi:10.1080/23311916.2021.1950304.

[60] E. S. Güler, "Corrosion protection using organic and natural polymeric inhibitors," In: H. M. A. Amin and A. Galal (Eds.), *Corrosion Protection of Metals and Alloys Using Graphene and Biopolymer Based Nanocomposites*, CRC Press, Taylor & Francis Group, Boca Raton, FL, 2021, pp. 64–77. doi:10.1201/9781315171364.

[61] U. Z. Husna, K. A. Elraies, J. A. B. M. Shuhili, and A. A. Elryes, "A review: The utilization potency of biopolymer as an eco-friendly scale inhibitors," *J. Pet. Explor. Prod. Technol.*, vol. 12, no. 4, pp. 1075–1094, 2022, doi:10.1007/s13202-021-01370-4.

[62] A. A. Fathima Sabirneeza, R. Geethanjali, and S. Subhashini, "Polymeric corrosion inhibitors for iron and its alloys: A review," *Chem. Eng. Commun.*, vol. 202, no. 2, pp. 232–244, 2015, doi:10.1080/00986445.2014.934448.

[63] K. El Mouaden et al., "Chitosan polymer as a green corrosion inhibitor for copper in sulfide-containing synthetic seawater," *Int. J. Biol. Macromol.*, vol. 119, pp. 1311–1323, 2018, doi:10.1016/j.ijbiomac.2018.07.182.

[64] O. S. Shehata, L. A. Korshed, and A. Attia, "Green corrosion inhibitors, past, present, and future," In: M. Aliofkhazraei (Ed.), *Corrosion Inhibitors, Principles and Recent Applications,* IntechOpen, Rijeka, 2017. doi:10.5772/intechopen.72753.

[65] P. Kumari and M. Lavanya, "Plant extracts as corrosion inhibitors for aluminum alloy in NaCl environment: Recent review," *J. Chil. Chem. Soc.*, vol. 67, no. 2, pp. 5490–5495, 2022, doi:10.4067/S0717-97072022000205490.

[66] S. Marzorati, L. Verotta, and S. P. Trasatti, "Green corrosion inhibitors from natural sources and biomass wastes," *Molecules*, vol. 24, no. 1, pp. 1–24, 2019, doi:10.3390/molecules24010048.

[67] V. Argyropoulos, S. C. Boyatzis, M. Giannoulaki, E. Guilminot, and A. Zacharopoulou, "Organic green corrosion inhibitors derived from natural and/or biological sources for conservation of metals cultural heritage BT: Microorganisms in the deterioration and preservation of cultural heritage," In: E. Joseph (Ed.), *Microorganisms in the Deterioration and Preservation of Cultural Heritage,* Springer International Publishing, Cham, 2021, pp. 341–367, doi:10.1007/9781_15.

[68] N. K. Gupta, P. G. Joshi, V. Srivastava, and M. A. Quraishi, "Chitosan: A macromolecule as green corrosion inhibitor for mild steel in sulfamic acid useful for sugar industry," *Int. J. Biol. Macromol.*, vol. 106, pp. 704–711, 2018, doi:10.1016/j.ijbiomac.2017.08.064.

[69] M. Mobin and M. Rizvi, "Inhibitory effect of xanthan gum and synergistic surfactant additives for mild steel corrosion in 1 M HCl," *Carbohydr. Polym.*, vol. 136, pp. 384–393, 2016, doi:10.1016/j.carbpol.2015.09.027.

[70] S. A. Umoren, O. Ogbobe, I. O. Igwe, and E. E. Ebenso, "Inhibition of mild steel corrosion in acidic medium using synthetic and naturally occurring polymers and synergistic halide additives," *Corros. Sci.*, vol. 50, no. 7, pp. 1998–2006, 2008, doi:10.1016/j.corsci.2008.04.015.

[71] V. Pokhmurs'kyi et al., "Inhibition of the corrosion of carbon steel by xanthan biopolymer," *Mater. Sci.*, vol. 55, no. 4, pp. 522–528, 2020, doi:10.1007/s11003-020-00334-z.

[72] M. Mobin, M. A. Khan, and M. Parveen, "Inhibition of mild steel corrosion in acidic medium using starch and surfactants additives," *J. Appl. Polym. Sci.*, vol. 121, no. 3, pp. 1558–1565, 2011, doi:https://doi.org/10.1002/app.33714.

[73] A. M. Alsabagh, M. Z. Elsabee, Y. M. Moustafa, A. Elfky, and R. E. Morsi, "Corrosion inhibition efficiency of some hydrophobically modified chitosan surfactants in relation to their surface active properties," *Egypt. J. Pet.*, vol. 23, no. 4, pp. 349–359, 2014, doi:10.1016/j.ejpe.2014.09.001.

3 Properties and Applications of Biopolymers

Gitalee Sharma
Dibrugarh University Institute
of Engineering and Technology

Satarupa Priyadarshani and Sanghamitra Sarmah
Dibrugarh University

3.1 INTRODUCTION

The term "biopolymers" refers to a type of polymer synthesized by living things throughout their normal development cycles. Consequently, they are also known as "natural polymers". Complex metabolic processes take place within cells to produce them.[1] Sugars, amino acids, and nucleotides are the monomeric building blocks of biopolymers.[2] Covalent bonds are used to link the monomer units of these polymers to create large structures.[3] Biopolymers include substances such as cellulose, starch, chitin, proteins, peptides, DNA, and RNA. The diversity, abundance, and significance of biopolymers and their derivatives are vital to life.[2] Biopolymers, such as DNA and RNA, can pass genetic information from generation to generation. The primary components that plants use as building blocks are cellulose and starch. In contrast to artificial polymers, these biopolymers have been present on the Earth for billions of years. Biopolymers are structured in a particular fashion as opposed to the similar and random configurations of synthetic polymers. Therefore, biopolymers involve the disintegration of long chains into shorter chains caused by biological processes. They often have low C–C bond energies, making them vulnerable to degradation by enzymes, moisture, heat, and other causes. The nature of many biopolymers is hydrophilic; they are readily degraded after absorbing water. For the characterization of deterioration, no test technique is available. They can also be created by the hydrolysis and condensation of substances like carbohydrates, proteins, etc.[3] They include a wide variety of materials, such as chitosan derived from crustaceans or plant derivatives (xanthol, gums, starch, etc.). Proteins (casein and gelatin) are also a part of them. Due to the overuse of fossil fuels and the current oil crisis, a switch to the use of biopolymers is required. The development of biopolymers as an alternative source of synthetic raw materials has begun in a number of polymer sectors. "BIOPOL" is the name of a program started by Imperial Chemical Industries to produce thermoplastic biopolymers. In the past, the synthesis of biopolymers

DOI: 10.1201/9781003400059-3

has consumed more energy than the manufacture of synthetic polymers. However, modern science is moving in the direction of producing biopolymers that are both energy-efficient and marketable.[4]

3.2 CLASSIFICATION

Natural biopolymers: Based on the types of monomer units that are repeatedly arranged, biopolymers can be classified into three broad classes: polysaccharides, polynucleotides, and polypeptides. These are natural biopolymers.

Polysaccharides: Glycosidic linkage connects monosaccharide units to form polysaccharides, a type of biopolymer. Common examples of polysaccharides include starch, cellulose, glycogen, chitin/chitosan, pectin, and alginate[5] (Figure 3.1a). Polysaccharide biopolymers are further classified into four categories: those based on sugar, starch, cellulose, and lignin. Polyhydroxybutyrate is produced using starch or sucrose as a feedstock. Lactose is processed from potatoes, wheat, and sugar beet to produce lactic acid polymers (polyactides). Techniques like vacuum forming, blowing, and injection molding were employed to synthesize these sugar-based biopolymers.[5] Wheat, tapioca, and potatoes all include starch, a biopolymer that functions as a natural polymer. Plants can only store the substance in their tissues as one-way carbs. Melting starch yields this sugar, which is mostly glucose. Furthermore, dextran is a class of low-molecular-weight carbohydrates derived from starch hydrolysis.[5] Cellulose, the main component of plant cell walls, is made of glucose. Cellulose-based biopolymers are usually used in the packaging of cigarettes, compact discs, and candies.[6] Another significant biopolymer, lignin is not derived from polysaccharides, but rather from cellulose.[6]

Polynucleotides: Thirteen or more nucleotide units form the biopolymer backbone in the form of a chain of polynucleotides like DNA and RNA. A polynucleotide chain is devoid of branches. Alternate sugar and phosphate groups constitute each strand of a polynucleotide.[7] N-glycosidic linkages connect the purine or pyrimidine bases to the polynucleotide backbone's repeating sugar–phosphate group. The polynucleotide DNA consists of two polynucleotide chains entwined into a helix and has a diameter of ~2 nm (Figure 3.1b). The sequence of nucleotides determines the instruction for a specific cell. Gene therapy and DNA sequencing are the two important applications of polynucleotides.[8]

Polypeptides: Peptide or amide linkages connect amino acid monomers to form polypeptide biopolymers (Figure 3.1c). In both polymer science and protein research, polypeptides are among the most crucial polymers. Chemical composition determines the polypeptide three-dimensional framework. Protected amino acids make polypeptides.[12] Peptic sequences significantly cluster due to inter- and intramolecular hydrogen bonding, resulting in incomplete acylation/deprotection processes. This in turn results in the production of slag in polymerization reactions. Polypeptides

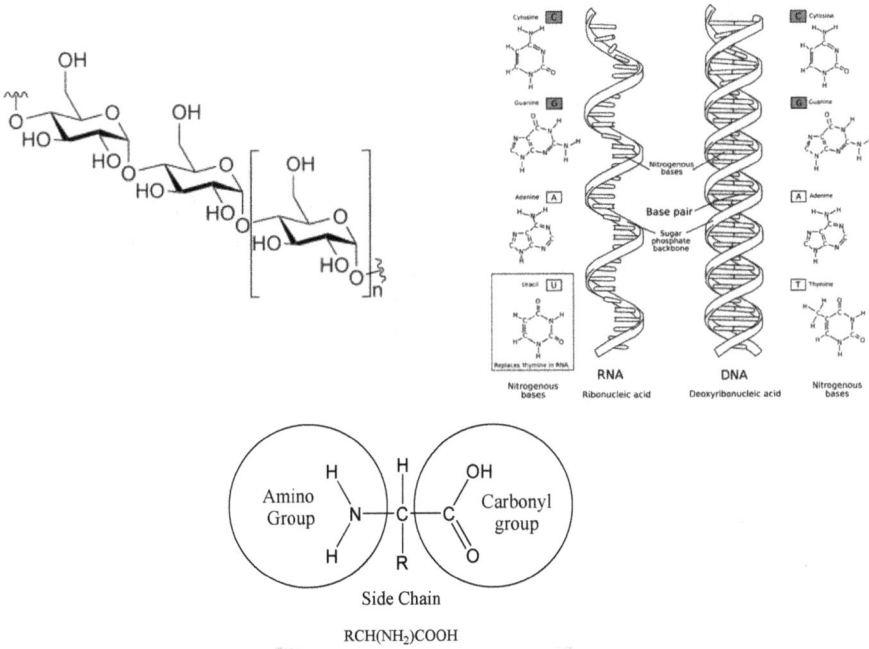

FIGURE 3.1 Examples of (a) polysaccharides,[9] (b) polynucleotides,[10] and (c) polypeptides.[11]

are useful in applications in medication, gene transfer, and tissue scaffolds exhibit biocompatibility and biological activity.[12] Biopolymers of this class of polypeptides include silk, collagen, and keratin. Silk peptide is a medicinal substance made from cocoon silk protein hydrolysate. It shows anti-inflammatory, immunoregulatory, antitumor antiviral, and antibacterial activities. Silk peptide has been shown to stimulate macrophages and pro-inflammatory cytokines although its effect on natural killer cells is unknown.[13] Collagen peptides are involved in health nourishment. They nourish bones, joints, and skin. Collagen has three polypeptide chains. Triple-helical macromolecules are formed by wrapping α-chains around each other. Every third collagenous residue is glycine (Gly). The three chains form a triple-helical structure. Hence, all collagens share a sequence (Gly-X-Y)*n, where Pro and Hyp are proline and hydroxyproline, respectively. This sequence allows collagen to form fibrils and fibers, giving conjunctive tissue extracellular matrix unequaled structural integrity.[14] Keratins are stable, insoluble structural proteins in vertebrate epidermis and appendages like feathers and hair. Keratin is mechanically stable and resistant to proteolytic enzymes like pepsin, trypsin, and papain because its α-helix or β-sheet chains are tightly packed into a supercoiled polypeptide chain heavily crosslinked with disulfide bridges, hydrogen bonds, and hydrophobic contacts.[15]

Synthetic biodegradable biopolymers: Biopolymers derived from synthetic chemicals are employed in the production of biodegradable polymers and are very often referred to as bioplastics.[5] Despite being made of synthetic materials, they are fully compostable and biodegradable. Bio-based biopolymers that are not resistant to degradation are vastly outnumbered by their biodegradable counterparts. Synthetic polymers made from renewable resources, including poly(lactic acid), are one type of biodegradable bio-based biopolymer.[2] Starch and polyhydroxyalkanoates (PHAs) are two common types of bio-based biodegradable polymers. A few common examples of biodegradable synthetic biopolymers are polyhydroxybutyrate (PHB), polyhydroxyvalerate, polyglycolide acid (PGA), polylactide acid, poly(lactide-co-glycolide), polydiaxanone, polyurethane, polyanhydride, etc.[2] The structural representation of these biodegradable biopolymers is depicted in Figure 3.2. Biodegradable biopolymers are also derived from fossil fuels. These biopolymers are certified biodegradable and compostable and are made from fossil fuels like synthetic aliphatic polyesters generated from crude oil or natural gas. Microbes are capable of biodegrading biopolymers such as poly(butylene succinate) and certain "aliphatic–aromatic" co-polyesters.[5]

Biodegradable biopolymers can also be divided into four types based on their place of origin and method of synthesis[6] (Figure 3.3).

Those are – (1) natural biopolymers (e.g., cellulose, hemicellulose, lignin), (2) biopolymers obtained through the modification of naturally occurring polymers (e.g., nanocellulose and carboxymethylcellulose), (3) biopolymers synthesized by microorganisms (e.g., PHAs), and (4) biopolymers obtained through the polymerization of biomonomers.[6]

Nonbiodegradable biopolymers: These biopolymers are made from renewable resources like biomass but are not biodegradable. Biopolymers made from renewable resources that are not biodegradable include synthetic polymers like bio-based polyethylene (bio-low density polyethylene (LDPE), bio-high density polyethylene (HDPE)), bio-based polypropylene, and bio-based polyvinyl chloride made from ethanol (from sugarcane) and biopolymers found in nature, including amber or natural rubber.[7]

3.3 PRODUCTION AND SYNTHESIS

Bioplastics: Bioplastics are made from proteins, lipids, and polysaccharides. They resolve environmental pollution and wildlife suffering caused by conventional plastics' nonrecyclable, nonrenewable, and nonbiodegradable characteristics.[16] Starch- and cellulose-based bioplastics cannot be made without synthetic polymers. Initially, cellulose is chemically processed and then reinforced with polymer composites, while starch is fluidized only in the presence of plasticizers like water, glycerol, and sorbitol at high temperatures. However, polysaccharide-based biodegradable polymers from agro-resources cannot replace plastic in packaging. Bioplastics from microorganisms could be a potential solution to petrochemical resource shortages. *Alcaligenes eutrophus* bacteria can polymerize volatile fatty acids (acetic, propionic, and butyric acids) from starchy wastewater to produce polyesters (PHAs) with mechanical properties similar to polypropylene and biodegradable characteristic.[17] *Aeromonas caviae* produce PHA from soybean oil. Poly-3-hydroxybutyrate (3-PHB), a promising microbial polyester, is equivalent to petrochemical-derived thermoplastics. Alkanes, alkenes, alkanols, and carboxylic acids have been added to PHA to synthesize a wide spectrum of homo- and co-polyesters with improved physico-mechanical properties.

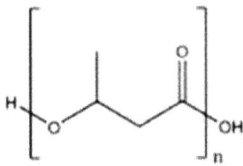

Polyhydroxybutyrate (PHB)

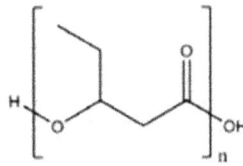

Polyhydroxyvalerate (PHV)

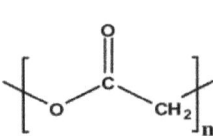

Polyglycolide acid

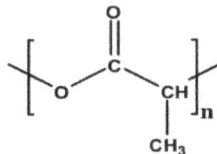

Polylactide acid

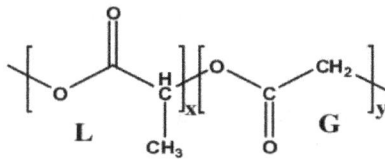

Poly (lactide-co-glycolide)

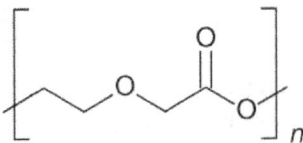

Polydioxane

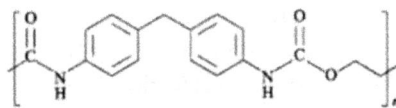

Polyurethane

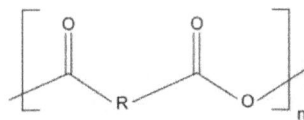

Polyanhydride

FIGURE 3.2 Examples of biodegradable biopolymers.[5]

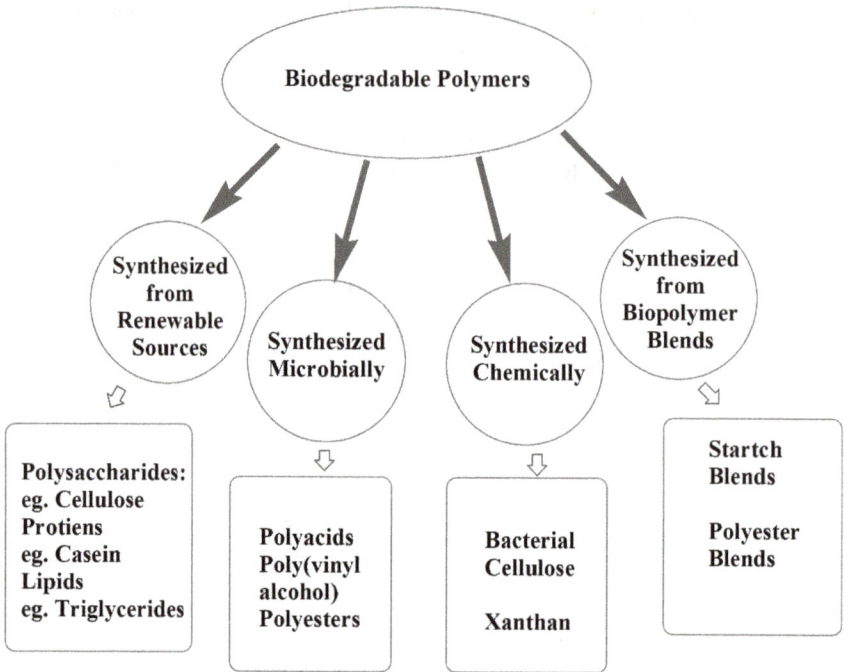

FIGURE 3.3 Classification of biodegradable biopolymers.[6]

Poly(3-hydroxybutyrate-co-3-valerate) is made from biodegradable bacterial polyester and low-cost agricultural byproducts like maize straw, soy stalk, or wheat straw. Natural monomer bioplastics such as polylactic acid, an aromatic polyester made from monomers of lactic acid, are sustainable alternatives to petrochemical-derived products and are widely marketed.[18] Even though the majority of bio-polysaccharides are obtained from plants (e.g., cellulose), they may also be obtained from animals (e.g., chitin/chitosan) and microbes (e.g., bacterial cellulose).[19,20]

Microbial polysaccharide: In recent years, microbial polysaccharides have been gaining attention. Bacteria, yeast, fungi, and microalgae synthesize polysaccharides. The synthesis of microbial polysaccharides takes less time compared with that of plant ones. These organisms may also convert industrial wastes into useful compounds, making them suitable for wastewater nutrient removal. Due to their role in the survival and pathogenicity of bacteria, scientists have spent decades studying the production processes of bacterial polymers. These polymeric materials can serve as storage molecules, as protective capsular layers around cells, and as significant matrix elements of biofilms, which are responsible for 60%–80% of all bacterial infections in humans.[19] Microbes generate many polysaccharides with amazing physicochemical characteristics. Microbial polysaccharides are used as thickeners, stabilizers, and gelling agents. In this context, exopolysaccharides (EPSs) and capsular polysaccharides (CPSs) are of interest.[21] EPSs and CPSs can be homopolymers or heteropolymers. Bacteria like *Leuconostoc* sp. manufacture dextran, a glucose-based homopolysaccharide. Levan is another extracellular EPS made of d-fructofuranosyl units. *Bacillus subtilis* and *Zymomonas mobilis* generate it. Xanthan is another widely studied heteropolysaccharide.[21] CPSs form an extracellular envelope by covalently linking to cell surface lipids or phospholipids, unlike EPSs. The capsule of some harmful bacteria determines their immunological characteristics. Biofilms are mostly CPSs and EPSs. Multicellular aggregates interact with their surroundings through biofilms. Polysaccharides help microorganisms stick to surfaces and each other. *Escherichia coli* and *Klebsiella pneumoniae* synthesize many CPSs and are pathogenic.[22]

In vitro synthesis using isolated enzymes: Another way to synthesize biopolymers is by in vitro synthesis using isolated enzymes in cell-free environments. The use of heat-stable DNA polymerases in the polymerase chain reaction to create monodisperse specified DNA molecules is one example. Dextran is another example, which may be made on a technological scale using isolated dextran sucrase. Biopolymers produced by fermentation are employed in industry, such as polysaccharides. Biopolymer formation in biotechnology can take place either intracellularly or extracellularly. Currently, the main constraints is on the upstream and downstream processes to produce the biopolymers in the pure condition, which has several serious consequences.[1]

3.3.1 MECHANISM OF BACTERIAL POLYMER BIOSYNTHESIS

Almost all microorganisms can produce biopolymers as intracellular and extracellular inclusions. There are numerous routes for producing biopolymers. Figure 3.4 depicts in detail the mechanism of bacterial biosynthesis for biopolymers in particular bioplastics and polysaccharides.

Synthesis of bacterial bioplastics: Three natural biosynthetic processes for PHAs are known. Most bacteria follow route I. It generates 3-hydroxybutyryl (3HB) monomers from sugar-derived acetyl CoA using a set of enzymes, including PhaA (3-ketothiolase, which converts acetyl CoA to acetoacetyl-CoA), PhaB (NADPH-dependent acetoacetyl-CoA reductase, which produces 3HB-CoA), and PhaC (PHA synthase, which polymerizes 3HB-CoA to the final monomers). Pseudomonas typically follow routes II and III. They convert sugars or fatty acids to acetyl CoA or acyl CoA, producing mcl-(R)-hydroxyacyl (3HA) monomers. PhaB's role in route I to get 3HA-CoA is comparable to the role of PhaJ and PhaG.[18,23]

Synthesis of bacterial polysaccharide: Bacteria produce EPSs (e.g., xanthan, dextran, alginate, cellulose, hyaluronic acid, and colonic acid), CPSs (e.g., the K30 antigen), and Intracellular polysaccharide (IPS) (glycogen). Biosynthesis gene clusters are organized differently because polysaccharides with different structures and contents require distinct enzymes and proteins. Two-component signal transduction pathways, quorum sensing, alternative RNA polymerase factors and antifactors, integration host factor-dependent, and cyclic di-GMP-dependent processes regulate EPS and CPS biosynthesis gene clusters. ADP-glucose, GDP-mannuronic acid, and UDP-N-acetylglucosamine are direct precursors for bacterial polysaccharide biosynthesis. EPS and CPS polymerization and secretion can limit carbon flux to high-molecular-mass exopolymers. Most EPSs are polymerized and released by

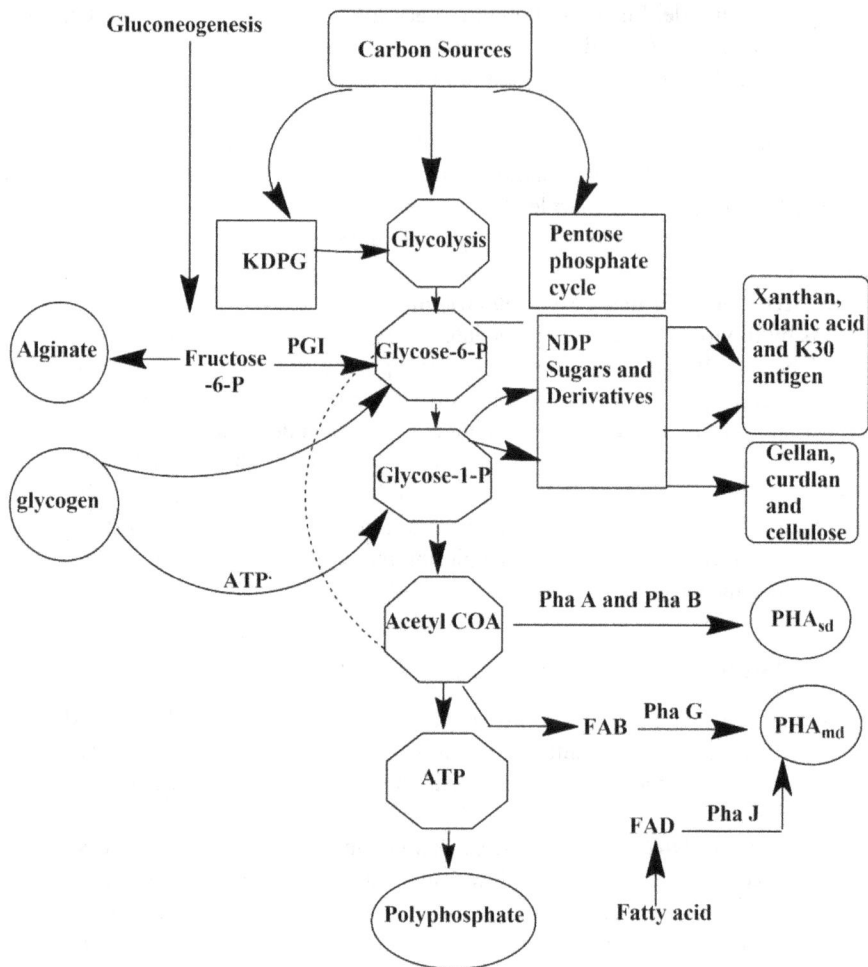

FIGURE 3.4   Mechanism of bacterial polymer biosynthesis.[18]

membrane-spanning multiprotein complexes, while hyaluronic acid is produced by a single protein, Hyaluronic Acid (HA) synthase (HasA). Complex-mediated biosynthesis has two main routes. Wzy-dependent polymerization and secretion in xanthan production require a lipid carrier to carry the repeat unit oligosaccharide across the cytoplasmic membrane. The second process produces nonrepeat EPS like lignite and cellulose without Wzy or a lipid carrier.[18,24,25]

3.4   CURRENT APPLICATIONS

Medical applications: The early uses of biopolymers were in the medical area, where their high starting costs were justified by their excellent value. To ensure the desired functionalities in this sector, their mechanical strength, biocompatibility, and bioresorbability features are crucial. These days, medical disciplines utilize a variety of biopolymers, including PHA, PGA, polylactides, cellulose, etc. Agar, glucose, and fructose are utilized as diluents in the production of tablets and syrups, while glucose is employed as a physiological saline solution. Furthermore, the coating on tablets occasionally consists of gelled starch. In addition to shielding the active component from stomach assault and protecting pills against atmospheric and mechanical agents, this process also covers up offensive tastes and odors. The manufacturing of medical supplies and pharmaceutical products can benefit from the qualities of amylose and its derivatives, which include nontoxic filaments and fibers for medical sutures. Amylose sponges can also be employed during surgical procedures due to their propensity for absorption.[26]

Agricultural applications: According to recent studies, biopolymers are used with their many benefits for soil improvement and stabilization. Due to its capacity to absorb water molecules via hydrogen bonding, xanthan gum is frequently used as a viscosity thicker. By blocking soil pores, xanthan gum is used in geotechnical engineering to make sandy soils less permeable and to make soils more resistant to soil erosion. Agar gels offer hard textures and have been used as a stabilizer. Sand and silt become less permeable when guar gum has been included, while sand's cohesion stress remains doubled.[27]

Application in the food industry: The pharmaceutical, food, biomedical, chemical, and waste treatment areas have all exploited biopolymers as matrices for the microencapsulation, immobilization, or controlled release of a variety of active substances. To enhance the emulsifying and filmogenic capabilities during microencapsulation, the combination of two types of biopolymers is crucial. Consequently, the scientific community and industry have shown considerable interest in microencapsulation utilizing biopolymers, particularly for food applications.[28] In the food sector, natural polymers can be used in a variety of ways. Most extensive studies have been conducted on polysaccharides, which are employed as a material for food packaging.[29]

Application as corrosion inhibitor: In recent years, investigations on corrosion inhibition have begun to pay much more attention to biopolymers. This is because they are easily accessible, biocompatible, nontoxic, affordable, and completely risk free to use. The ability of naturally occurring polymers to form metal–polymer complexes that ultimately shield metal surfaces from contact with corrosive media is like that of other organic corrosion inhibitors.[29]

Application in surfactant removal: One of the intriguing research areas in the field of biophysical chemistry is the study of the interaction between biopolymers and surfactants. The interactions between several kinds of surfactants and biopolymers have been examined in a number of studies. However, the use of biopolymers for surfactant removal (pollutant abatement) is a relatively recent concept that has gained considerable interest during the past decade.[30]

Application in packaging: In a variety of industries and types of goods, biopolymers have been used in things like textiles, packaging, fast food containers and packaging, lawn and garden waste bags, paper coating, agriculture mulch films, toys, tubes, medical products, disposable wipes, biologically based resins, car parts, glass fiber agents, adhesives, and coatings.[31]

Application in the textile industry: The fundamental objective of the sustainable textile business is to find effective uses for eco-friendly textiles that have low operating costs and energy consumption, as well as the essential reducing, reusing, and recycling features. There are several ways to process biopolymers, including polymerization, crystallization, and manufacture, depending on the nature of the polymer and its intended function. The adoption of biopolymer materials as standard fare in the textile industry needs time, as is the case with all different biopolymer manufacturing methods. Biopolymers are more environmentally sensitive than most contemporary technologies since they have the potential to drastically lower costs, energy use, and material used for coming generations.[32]

3.5 CONCLUSIONS

Polymers are stable, but biotic degradation cycles are infinite. As synthetic polymers pollute the environment, researchers around the globe are seeking biodegradable natural polymers to eliminate this risk. Biopolymers are polymers generated naturally during the growth cycle of all species. Medicine, food, and petroleum industries use biopolymers extensively. Petrochemically synthesized polymers, including thermoplastics and thermoset plastics, have dominated the commodity materials to date. However, in the recent era, renewable biopolymers are expected to replace a large portion of synthetic polymers due to environmental consciousness and fossil resource scarcity. Industry will increasingly value the ecologically friendly and highly processable synthesis of biodegradable and biocompatible polymers. It is noted that due to genetic systems and metabolic pathway engineering, bacteria are considered good polymer-manufacturing organisms. Though microorganisms excrete polysaccharides, their production is expensive. Their manufacturing is hindered by inefficient extraction and purification methods. However, research into bacterial polysaccharide manufacturing is developing and focuses on low-cost substrates, downstream processing, and metabolic engineering to produce polymers with fine-tuned characteristics. In addition, biopolymer biosynthesis enzyme inhibitors can be designed via molecular processes when the biopolymer is a virulence factor. Bacterial polysaccharides' ability to be customized chemically and structurally makes them useful in pharmaceuticals, medical equipment, and maquillages and thus shows high potential in their pertinence in a varied spectrum of applications.

REFERENCES

1. Rao, M. G., Bharathi, P., and Akila, R. M. A comprehensive review on biopolymers. *Sci. Rev. Chem. Commun.* 4, 61–68 (2014).
2. Hassan, M. E., Bai, J., and Dou, D. Q. Biopolymers: Definition, classification and applications. *Egypt. J. Chem.* 62, 1725–1737 (2019).
3. Singh, R., Gautam, S., Sharma, B., Jain, P., and Chauhan, K. D. Biopolymers and their classifications. In: S. Thomas, S. Gopi, and A. Amalraj (Eds.), *Biopolymers and their Industrial Applications*, Elsevier, Amsterdam, 2021, pp. 21–44.
4. Leja, K., and Lewandowicz, G. Polymer biodegradation and biodegradable polymers-a review. *Pol. J. Environ. Stud.* 19, 255–266 (2010).

5. Savitha, K. S., et al. "Polybutylene succinate, a potential bio-degradable polymer: synthesis, copolymerization and bio-degradation." *Polymer Chemistry* 13, 3562-3612 (2022).

6. Avinash Ingle, P., Kumar Chandel, A., and Silvério da Silva, S. *Lignocellulosic Biorefining Technologies*, Wiley-Blackwell, 1st ed., 2020.

7. Javaherdashti, R., and Alasvand, K. *Biological Treatment of Microbial Corrosion: Opportunities and Challenges*, Elsiver, Amsterdam, 2019, pp. 101–144.

8. Abbasi, Y. F., Panda, P., Arora, S., Layek, B., and Bera, H. *Tailor-Made and Functionalized Biopolymer Systems*, 1st edn., Elsiver, Amsterdam, 2021, pp. 1–31.

9. https://upload.wikimedia.org/wikipedia/commons/thumb/d/d4/Amylose_3Dprojection. svg/1200px-Amylose_3Dprojection.svg.png

10. https://ars.els-cdn.com/content/image/3-s2.0-B9780128167519000039-f03-15-9780128167519.jpg

11. https://cdn-acgla.nitrocdn.com/bvIhcJyiWKFqlMsfAAXRLitDZjWdRlLX/assets/ static/ optimized/rev-5131b73/wp-content/uploads/2021/01/aminoacidstructure.jpg

12. Ivanova, E. P., Bazaka, K., and Crawford, R. J. Advanced synthetic polymer biomaterials derived from organic sources. In: E. P. Ivanova, K. Bazaka, and R. J. Crawford (Eds.), *New Functional Biomaterials for Medicine and Healthcare*, 2014, pp. 71–99.

13. Jang, S.-H., Oh, M.-H., Baek, H.-I., Ha, K.-C., Lee, J.-Y., and Jang, Y.-S. Oral administration of silk peptide enhances the maturation and cytolytic activity of natural killer cells. *Immune Netw.* 18(5), e37 (2018). doi: 10.4110/in.2018.18.e37

14. https://www.gelita.com/en/knowledge/collagen-peptides/what-are-collagen-peptides.

15. Daroit, D. J., Correa, A. P. F., and Brandelli, A. Keratinolytic potential of a novel Bacillus sp. P45 isolated from the Amazon basin fish *Piaractus mesopotamicus*. *Int. Biodeter. Biodegrad.* 63, 358–363 (2009).

16. Koller, M., Salerno, A., de Sousa, M., Dias, M., Reiterer, A., and Braunegg, G. Modern biotechnological polymer synthesis: A review. *Food Technol. Biotechnol.* 48, 255–269 (2010).

17. Ahankari, S. S., Mohanty, A. K., and Misra, M. Mechanical behaviour of agro-residue reinforced poly(3-hydroxybutyrate-co-3-hydroxyvalerate), (PHBV) green composites: Comparison with traditional polypropylene composites. *Compos. Sci. Technol.* 71, 653–657 (2011).

18. Chaabouni, E., Gassara, F., and Brar, S. K. Biopolymers synthesis and application. In: S. Brar, G. Dhillon, and C. Soccol (Eds.), *Biotransformation of Waste Biomass into High Value* Biochemicals, Springer, New York, 2014.

19. Moradali, M. F., and Rehm, B. H. A. Bacterial biopolymers: From pathogenesis to advanced materials. *Nat. Rev. Microbiol.* 18, 195–210 (2020).

20. Balart, R., Garcia-Garcia, D., Fombuena, V., Quiles-Carrillo, L., and Arrieta, M. P. Biopolymers from natural resources. *Polymer* 13, 2532 (2021).

21. Donot, F., Fontana, A., and Baccou, J. C. Microbial exopolysaccharides: Main examples of synthesis, excretion, genetics and extraction. *Carbohydr. Polym.* 87, 951–962 (2012).

22. Cescutti, P. Bacterial capsular polysaccharides and exopolysaccharides. In: A. P. Moran, O. Holst, P. J. Brennan, M. von Itzstein (Eds.), *Microbial Glycobiology: Structures, Relevance and Applications*, Elsevier, Amsterdam, 2009, pp. 93–115.

23. Hofer, P., Choi, Y. J., Osborne, M. J., Miguez, C. B., Vermette, P., and Groleau, D. Production of functionalized polyhydroxyalkanoates by genetically modified methylobacterium extorquens strains. *Microb. Cell Fact.* 9, 70 (2010).

24. Rehm, B. H. A. Bacterial polymers: Biosynthesis, modifications and applications. *Nat. Rev. Microbiol.* 8, 578–592 (2010).

25. Brown, S. H., and Pummill, P. E. Recombinant production of hyaluronic acid. *Curr. Pharm. Biotechnol.* 9, 239–241 (2008).

26. Abdelhak, M. A. Review: Application of biopolymers in the pharmaceutical formulation. *J. Adv. Biopharm. Pharmacovigil.* 1, 15–25 (2019).

27. Jang, J., and Jia, P. A. Review of the application of biopolymers on geotechnical engineering and the strengthening mechanisms between typical biopolymers and soils. *Adv. Mater. Sci. Eng.* 2020, p. 1465709 (2020).

28. Estevinho, B. N., and Rocha, F. Application of biopolymers in microencapsulation processes. In: A. M. Grumezescu and A. M. Holban (Eds.), *Biopolymers for Food Design*, Elsevier, Amsterdam, 2018, pp. 191–222.

29. Grujic, R., Vukic, M., and Gojkovic, V. Application of biopolymers in the food industry. In: E. Pellicer, D. Nikolic, J. Sort, M. Baró, F. Zivic, N. Grujovic, R. Grujic, and S. Pelemis (Eds.), *Advances in Applications of Industrial Biomaterials*, Springer, New York, 2017, pp. 103–119.

30. Gowraraju, N. D., Jagadeesan, S., Ayyasamy, K., Olasunkanmi, L. O., Ebenso, E. E., and Subramanian, C. Adsorption characteristics of Iota-carrageenan and inulin biopolymers as potential corrosion inhibitors at mild steel/sulphuric acid interface. *J. Mol. Liq.* 232, 9–19 (2017).

31. Biswas, S., and Pal, A. Application of biopolymers as a new age sustainable material for surfactant adsorption: A brief review. *Carbohydr. Polym. Technol. Appl.* 2, 100145 (2021).

32. Younes, B. Classification, characterization, and the production processes of biopolymers used in the textiles industry. *J. Text. Inst.* 108, 674–682 (2017).

4 Green Corrosion Inhibitors and their Role in Corrosion Inhibition

Amita Somya
Amity University

4.1 INTRODUCTION

Corrosion is generally considered a forfeiture of a metal due to the impact of corrosive substances [1]. Nevertheless, corrosion is generally the devastating outcome of a chemical reaction between metals or metal alloys and their surroundings. The most frequent type of corrosion, known as general corrosion or uniform corrosion, results from electrochemical reactions in air or aqueous environments on a completely exposed metal surface and spreads uniformly to destroy the metal to the maximum extent. Although while discussing corrosion, only metals come to the fore, nonmetallic materials including plastic, ceramics, concrete, rubber, etc. are also susceptible to corrosion when they are exposed to various corrosive conditions. A primary factor driving the process of corrosion is the difference in the potential energies of the corrosion product and that of the corroding metal. To extract metals from naturally existing ores, a definite amount/extent of energy must be applied to them. Therefore, it makes sense that these metals have a tendency to return to the original condition from where they were derived when they were exposed to their environs. It is intriguing that every metal has a distinct energy requirement, energy storage capacity, and energy released during corrosion. The higher the energy required to extract a metal, the more thermodynamically unsteady it is, and hence, the shorter its interim existence in the metallic state. Corrosion is thus described as the opposite of the extraction of metals [1]. The most significant mechanism in corroding any metal is electrochemical disintegration, which forms the cornerstone of each and every category of localized and uniform corrosion. However, various types of corrosion, including oxidation, fret corrosion, melting salt corrosion, etc., can be explained without using electrochemistry. Regardless of the environment, electrochemistry always plays a role in determining how corrosion occurs in both air and aquatic environments. Through an electrolyte that is able to conduct ions, electrons move from specific regions of the metal surface to others, which is caused by the extraordinary propensity of metals to electrochemically react with water molecules, oxygen, and other chemicals in an aqueous environment. The cathode receives electrons produced by a corrosion reaction, while the anode is the portion of the metal surface that gets corroded owing to the loss of electrons in an electrochemical corrosion process.

We are all familiar with how corrosion affects metal objects like buildings, ships, metal pillars, home items, etc. However, corrosion also destroys the capabilities of the oil, water, and gas pipelines that run beneath our property or the water pipes within our homes where corrosion generally starts internally. In order to control significant collapses due to corrosion such as unplanned shutdowns, fatalities, individual injuries, and environmental contamination, successful organizations invest a lot of time and energy in corrosion control during the design and operational phases of their operations. Even the best strategy in design, however, is ineffective in accounting for all the circumstances that could lead to corrosion affecting the system's lifespan. Rebar, or steel-reinforced concrete, can corrode in concrete without being observed at all, resulting in the destruction of buildings, parking structures, bridges, transmission towers, highways, etc., thus endangering public safety and resulting in huge costs to restore them. Hence, it is important to regularly maintain the metallic components that are prone to corrosion.

4.2 IMPACT OF CORROSION

Although economic and ecological factors are the primary factors to consider when evaluating corrosion, damages from corrosion or the expenses associated with corrosion can be broken down into three basic categories, as shown in Figure 4.1.

4.2.1 MATERIALS AND ENERGY

In the industrial sector, the effects of corrosion on the machinery and its surroundings need a great deal of consideration. In the majority of industrialized countries, corrosion is one of the most difficult problems to solve. Corrosion of pipes, metal parts of machinery, bridges, marine ships, tanks, etc. can lead to significant material and financial losses. Furthermore, failure due to corrosion might endanger the functioning of

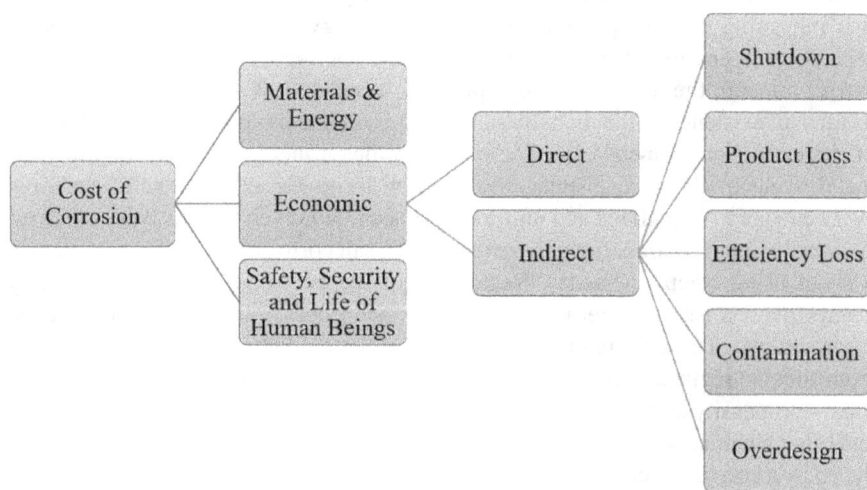

FIGURE 4.1 Impact of corrosion.

tools such as boilers, pressure containers, steel vessels, turbine blades, bridges, rotors, vehicle steering mechanisms, and airplane parts [2]. Furthermore, corrosion has a devastating effect on water, energy, and the process of making metal fixtures, in addition to the metals themselves. The energy needed to manufacture one tonne of steel is roughly equivalent to the energy used by a typical household over the course of three months; however, worldwide, one tonne of steel is said to rust every 90 seconds. Every tonne of steel produced across the world is replaced with new steel about 50% of the time [3].

4.2.2 Economic Losses

Economic losses due to corrosion can be categorized into direct and indirect losses. Direct loss includes the replacements of components of several types of machinery owing to rusting like pipelines, mufflers, metal roofing, and condenser tubes; refurbishing the surfaces of metal articles to avoid corrosion; maintenance of underground oil pipelines; replacements of domiciliary hot water tanks; usage of corrosion-resistant metals and alloys; and application of various processes to avoid or prevent corrosion such as galvanizing, Ni plating on steels, incorporation of corrosion inhibitors, and dehumidification process in metal-equipped warehousing rooms, which require additional labor and costs sometimes in millions. Although it might be challenging to estimate indirect losses, it is reported that they can increase direct losses by several billion dollars. A few examples of indirect losses are unexpected plant shutdowns; the loss of oil, water, and gas by corroded pipelines; the reduction in the efficiency of energy conversion systems; the pollution of water and foodstuffs in metal pipes and vessels; and overdesign, which call for equipment to be designed to be much heavier than necessary to withstand operating pressure or applied stress in order to increase the lifespan [4]. In 1949, Uhlig conducted the first comprehensive research on the effects of corrosion [5]. According to Uhlig's research, the yearly expenses due to corrosion in the United States were projected to be US$ 5.5 billion, or 2.1% of the gross national product (GNP), in 1949. The expenses associated with corrosion-induced sustentation and reinstatement incurred by holders and operators (direct) as well as those associated with customers (indirect) were added together in this study to determine the overall costs. Numerous major economies, including Finland, Australia, China, India, Kuwait, Germany, the United Kingdom, Japan, and the United States [6], have investigated corrosion expenses using a variety of methods, including the method developed by Uhlig in 1950 [5], the method developed by Hoar in 1976 [7], and the economic input/output model developed by the National Bureau of Standards (NBS) working with the Battelle Memorial Institute in 1978 [8]. These studies typically pointed out that the cost of corrosion varied between 1% and 5% of each country's GNP. The methodology followed in each research study and the unique characteristics of each country were the factors responsible for the variance in corrosion costs relative to GNP. According to an analysis conducted by the National Association of Corrosion Engineers as a part of their Worldwide Measures of Prevention, Application, and Economics of Corrosion Technologies Study (IMPACT), the cost of corrosion worldwide was estimated to be US$ 2.5 trillion in 2013, or 3.4% of the world's GDP [6]. In this study, the expenses estimated in corrosion studies were compared with the worldwide corrosion expenses using statistics from the World Bank's economic sector and GDP. According to the World Bank, the worldwide economy was split into economic zones with comparable economies in order to handle many economic sectors throughout the world. The World Bank described them as the following: the United States, European Region, India, Arab World, China, Japan, Four Asian Tigers plus Macau, and Rest of the World. However, the projected expenditures frequently do not take into account the effects on the environment or personal safety. It is important that more thorough and precise estimates of the worldwide expenses can be made after acquiring additional funding for corrosion research or the latest statistics on these studies.

4.2.3 SAFETY, SECURITY, AND LIFE OF HUMAN BEINGS

Corrosion may have an unimaginable impact on human life and safety. Because of liability issues or simply the nonexistence of proof after catastrophes, this destructive phenomenon has gone unnoticed as the primary cause of numerous deadly incidents. One of the most perilous and most talked-about corrosion catastrophes is the collapse of the Silver Bridge [9]. This bridge, which connected Point Pleasant, West Virginia, with Kanauga, Ohio, abruptly collapsed into the Ohio River on December 15, 1967, killing 46 people. This catastrophe was shown to be caused by corrosion fatigue and stress corrosion cracking (SCC). In terms of the number of people killed and injured, the Bhopal Tragedy, which occurred at night time on December 2–3, 1984, in Bhopal, India, is one of the deadliest industrial catastrophes. The exposure of methyl isocyanate and many other toxic reaction products into nearby regions resulted in the unfortunate seepage of water (500 L) from corroded valves, pipelines, and other safety tools at Union Carbide India Limited and caused the death of 3,000 people and injuries to an estimated number of 500,000 people [10]. Another noteworthy corrosion catastrophe that occurred in Uster, Switzerland, in 1985 was the swimming pool roof collapse. Twelve persons died when the stainless steel rods supporting the ceiling of this swimming pool broke due to stress corrosion cracking [11].

4.3 CORROSION INHIBITORS

A corrosion inhibitor is a chemical substance that, when applied to the metal environment in very little quantity, can reduce, prevent, and regulate corrosion. Corrosion inhibitors are thought of as the first line of defense against corrosion in the chemical and oil industries [12]. They are highly sought after to provide short-term protection for metals during the storage and transit and localized protection as well, to stop corrosion that may have been caused by the buildup of even the least quantity of aggressive media. When present in modest quantities, an efficient corrosion inhibitor should have the required effect, be economical, and be adaptable to the corrosive medium [12]. The three primary mechanisms by which corrosion inhibitors work are as follows:

   (i)   the formation of an adsorbed coating on the metal surface,
   (ii) the production of corrosion products, such as passivating iron sulfide, and
   (iii) the production of precipitates that can neutralize or inactivate an aggressive ingredient.

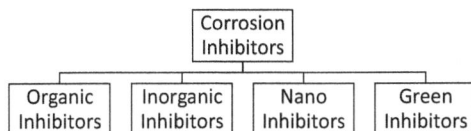

FIGURE 4.2 Different class of corrosion inhibitors.

These film-forming or interface inhibitors can be divided into anodic, cathodic, or mixed-type compounds depending on which electrochemical processes they inhibit [13]. Anodic inhibitors, often referred to as passivation inhibitors, slow down the pace of anodic reactions by forming a small amount of partially soluble layer like hydroxides, oxides, or salts in nearly neutral environments. In contrast, cathodic inhibitors lower the rate of cathodic or reduction processes by forming a barrier on cathodic sites that protects them from oxygen in alkaline settings and hydrogen in acidic situations. Mixed inhibitors form an adsorptive coating on the metal surface, which affects both cathodic and anodic reaction sites. The majority of organic inhibitors – about 80% – fall under this category. Inhibitors are categorized into organic, inorganic, nano, and green corrosion inhibitors (GCIs) based on their chemical composition and nature of the source, as shown in Figure 4.2. Organic and inorganic inhibitors can be further classified into neutralizing, scavenging, barrier- or film-producing, and other inhibitors based on their chemical compositions and mechanisms of action.

4.3.1 ORGANIC INHIBITORS

Organic inhibitors, which can be anodic, cathodic, or mixed inhibitors, form a coating on metal surfaces. Powerful interactions such as orbital adsorption, electrostatic adsorption, and chemisorption contribute to the formation creation of this protective coating and stopping of the corrosive species from corroding the metal surface [14]. In most cases, the above-mentioned adsorption is only one molecule layer thick, which usually does not reach the metal's core [15]. The following physicochemical parameters determine the adsorption process: aromaticity, functional groups, steric factors, electron density at donor atoms, orbital nature of donating electrons, and electronic structure of adsorbing molecules. The effectiveness of an organic inhibitor in preventing corrosion depends on both its capacity for adsorption and the structural, mechanical, and chemical properties of the adsorbed layers developed under specific conditions [14]. An effective organic inhibitor typically has a hydrophobic portion that repels the aqueous corroding species from the metallic surface and polar functional groups that consist of sulfur, oxygen, and/or nitrogen atoms in the molecule. However, it is suggested that the polar head is in charge of the formation of the adsorption layer. Chemical groups such as imidazolines, pyridines, 1,3-azoles, and fatty amides are a few examples of organic inhibitors [16].

4.3.2 INORGANIC INHIBITORS

Inhibitors using an inorganic component as the active ingredient are known as inorganic inhibitors. One of the easiest methods to increase a metal's passivity is to add electropositive salts of that metal to a corrosive solution. However, in order to protect metal ions to be discharged on a metal surface in necessity of protection, the inhibitor should have a redox potential that is higher than that of the metal that needs to be prevented and perhaps higher than the one needed for protons to be discharged. The coating of the preventive metal on the surface of the corrosion-prone metal causes cathodic depolarization via overvoltage reduction and the subsequent creation of an adhering deposit. The following metals can be used for this objective: mercury, palladium, rhodium, platinum, iridium, and rhenium. Numerous inorganic anions, including phosphates, chromates, silicates, nitrates, and molybdates, offer passivation preservation to metallic surfaces by their integration into the oxide layer [16]. While selecting an inhibitor, factors such as toxicity, cost, availability, and environmental friendliness need to be given due consideration. Traditional corrosion inhibitors released into the environment pose serious health risks due to their perniciousness, biodegradability, and bioaccumulation. The effects of commercial corrosion inhibitors on the environment may not yet be completely understood, but it is well established that their chemical constituents pose a risk [17]. Inorganic inhibitors such as arsenates, chromates, dichromates, and phosphates have demonstrated encouraging inhibitory effectiveness but have also been found to be intolerable because of the long-term risk they bring to human health and environment [18]. Similarly, bionomic and health dangers associated with organic inhibitors have compelled researchers to seek out or use nontoxic or (GCIs), which preserve metal surfaces to the highest extent while having the least negative effects on people and the environment.

4.3.3 NANOINHIBITORS

Nanotechnology is a tremendously intriguing, newly developing, and rapidly expanding branch of science and engineering that deals with the creation, processing, and use of nanoscale materials (1–100 nm). Nanomaterials are utilized extensively for a variety of industrial, biomedical, agricultural, food, and environmental applications due to their exceptional physical and chemical properties, including extremely high surface area, high chemical and optical reactivity, high stability, and high mechanical strength. They are also frequently employed as next-generation corrosion inhibitors for various metals and alloys used for different electrolytes [19,20]. It is important to take into account that the first phase of defense against metallic inhibition of corrosion is the application of organic and inorganic substances. However, these materials are costly and hazardous to both humans and the environment. Therefore, the typical inorganic and organic chemicals that have been used as corrosion inhibitors in recent times are inefficient. Therefore, there is growing interest in using nanomaterials and their derivatives as corrosion inhibitors. Nanomaterials and their derivatives offer huge surface coverage that manifests as excellent protection efficiency due to their extremely high surface-to-volume ratio as compared to standard macroscopic materials. The active sites that cause corrosive damage are blocked by nanomaterials once they have been adsorbed, resulting in the more effective prevention of corrosion. However, exposure of nanoinhibitors to the environment may lead to serious health issues. Exposures to nanomaterials can occur by ingesting, cutaneous contact, unintentional injection, and inhalation, and the risk increases with exposure time and nanoparticle concentrations in the sample or air. The biggest exposure risk comes from inhalation. However, inhalation exposure can happen when anything is soni-cated, shaken, stirred, poured, or sprayed. Therefore, there is a need to explore or design environmental-friendly inhibitors.

4.3.4 GREEN CORROSION INHIBITORS

Since protecting metals and equipment against corrosion is a common practice, corrosion inhibitors must be sustainable, nontoxic, and environmentally beneficial. Commercially available organic and inorganic inhibitors are expensive and possess negative side effects, which has caused a great deal of alertness in the area of cor-rosion alleviation. Therefore, corrosion researchers and technicians are very much motivated to consider the application of GCIs which are affordable, easily accessible, economical, ecologically acceptable, and sustainable. The phrase "green inhibitor" or "eco-friendly inhibitor" refers to compounds that are naturally biocompatible. By virtue of their biological source, the inhibitors, like plant extracts, are likely to have biocompatibility. Recent methods make use of chemical molecules that can be found in outdated pharmaceuticals, mushroom extracts, and even plant extracts [21]. There are several environmentally friendly organic substances that act as corrosion inhibitors and have excellent characteristics for shielding metal surfaces, such as chitosan derivatives [22], phenyl methanimine derivatives [23], quinolone derivatives [24], imidazoline derivatives [25], and ionic liquid derivatives [26]. As a result, these molecules take the place of the conventionally harmful corrosion inhibitors. These substances have led to the development of highly effective corrosion inhibitors, open-ing up new avenues for the recycling and reuse of pharmaceuticals as well as corro-sion inhibitors derived from environmentally friendly, sustainable, and sustainable sources, with plant extracts being a prominent subset [21]. These extracts are an addi-tional choice that is quite intriguing since they may provide a preliminary method for identifying the category of natural components that aid in inhibiting the corrosion process. The benefit is that producing an extract from any plant is thought of as an easy operation, allowing for higher efficiency in both the extraction and usage of these compounds for experiments. Different kinds of these inhibitors are presented in Table 4.1 with their efficacies in various aggressive media using different surfaces of metals such as mild steel (M.S.), aluminum (Al), carbon steel (C-steel), copper (Cu), zinc (Zn), iron (Fe), etc.

GCIs [27] protect the surface of metallic or alloys in corrosive environments when they are introduced in extremely low doses. When applied to many industrial systems, extracts of many plants are thought to be ample sources of naturally occur-ring chemicals that influence the rate of corrosion through adsorption of efficacious species on metal surfaces when applied in copious industrial processes by:

- Modifying cathodic and/or anodic reaction rates;
- Modifying the rate at which corrosive ions bind with metallic structures;
- Creating a layer (coat) on the metal surface to increase its electrical resistance.

In the corrosion process, an anodic process involves the movement of metal ions into the solution at active locations (the anode) and the transfer of electrons to an acceptor at less active areas (the cathode); the cathodic process necessitates the presence of an electron acceptor such as oxygen, oxidizing agents, and hydrogen (H^+) ions. The anodic or cathodic processes, or both, can be slowed down or stopped entirely to reduce corrosion. To lessen the oxidation and/or reduction of corrosive processes, inhibitors engage with anodic and/or cathodic reaction sites by being adsorbed on the metal surface and producing a protective barrier [27]. In corrosive media, there are a variety of typical cathodic reactions, including:

$$2\,H^+ + 2\,e^- \rightarrow H_2 \qquad\qquad (4.1)$$

$$O_2 + 4H^+ + 4e^- \rightarrow 2H_2O \qquad\qquad (4.2)$$

Hydrogen gas is evolved in the reduction reaction as given below:

$$2H^+ + 2e^- \rightarrow H_{ads} \rightarrow 2H_2 \qquad\qquad (4.3)$$

In Eq. (4.3), H^+- ions that have been adsorbed on the metal surface are catalyzed along with other H^+-ions to produce evolved hydrogen gas on the cathode surface; the amount of hydrogen bubble indicates the inhibitor's capacity to repress this reaction and protect the metal free from corrosion. Rather than H^+- ions being adsorbed off the metal surface, inhibitor molecules take the form of neutral molecules and adhere to an exposed metal surface, where they operate as inhibitors [28].

$$\text{Inhibitor} + n\text{H}_{ads} \rightarrow \text{Inhibitor}_{ads} + H_2 \qquad\qquad (4.4)$$

This process can also take place by the displacement of aqua molecules on the surface, as depicted in Eq. (4.2). Green inhibitors, sometimes referred to as site-blocking components, have adsorptive characteristics [29].

4.4 PLANT EXTRACTS AS CORROSION INHIBITORS

Phytochemicals are non-nutritive elements of plants that give them their flavor, aroma, and color. Plants are made up of many different types of chemicals. Terpenoids, polyphenols, and organic acids are only a few of the numerous classes of phytochemicals, and these classes can further be classified as shown in Table 4.1. Electronic structures of phytochemicals often correspond closely to those of traditional synthetic organic corrosion inhibitors. Thus, a variety of compounds can give steel corrosion resistance under corrosive conditions [75]. Most plants and their extracts are nontoxic, which makes them a good alternative to harmful organic inhibitors. They are particularly appealing since they are easily disposed of because of their biodegradable nature. The extraction procedure is also rather easy and affordable. It is possible to extract environmentally benign corrosion inhibitors from a variety of plant components, including fruits, leaves, bark, roots, seeds, or peels. The mixture of phytochemicals in a plant extract with different functional groups capable of adhering to a metal

TABLE 4.1
Summary of different kinds of green corrosion inhibitors

S.N.	Green Corrosion Inhibitors	Metals Investigated	Media	% Efficiency	References
1	*A. indica*	M.S.	Acid	88–96	[28]
2	*A. indica*	M.S.	Sodium chloride	86.1	[29]
3	*Acacia arabica*	M.S.	Acid	93–97	[30]
4	Acetyleugenol	Steel	Acid	91	[31]
5	*Alhagi maurorum* plant extract	Cu	Acid	33–83	[32]
6	*Allium cepa, Allium sativum, M. charantia*	M.S.	Acid	86–94	[33]
7	Aloe vera leaves	Steel, Al, Zn and, Cu	Acid	80-steel, 64-Al, 43-Zn, 43-Cu	[34]
8	*Artemisia pallens*	M.S.	Acid	93–98	[35]
9	*Auforpio turkiale* sap	M.S.	Acid	69	[36]
10	*Azadirachta indica*	M.S.	Acid	84	[36]
11	*Bambusa arundinacea*	Steel rebar	Chloride and nitrite	85	[37]
12	Berberine extracted from *Coptis chinensis*	M.S.	Acid	79.7	[38]
13	Black cumin	Steel	Acid	88.43	[39]
14	Black pepper	M.S. and Al	Neutral then acidic media	87.5-M.S., 65.8-Al	[40]
15	Caffeine and nicotine	C-steel	Chloride ions	80–90	[41]
16	Caffeine–Mn2+	C-steel	Chloride ions	50	[42]
17	*Calotropis procera*	M.S.	Acid	98	[36]
18	*Cassia occidentalis*	M.S.	Acid	94	[36]
19	Castor seeds	M.S. and Al	Neutral then acidic media	71.0-M.S., 45.9-Al	[40]
20	Catechu	M.S. and Al	Neutral then acidic media	50.0-M.S., 6.3 – Al	[40]
21	*Chamaerops humilis* leaves	Steel rebar	Alkaline	42.2	[43]
22	Chamomile	Steel	Acid	92.97	[39]
23	Chitosan	M.S. and Al	Acid	> 90	[44]
24	*Corchorus olitorius* stem	M.S.	Acid	94.34	[45]
25	*Datura stramonium* seeds	M.S.	Acid	93	[36]
26	Dodecanohydrazide, cis-9-octadecanohydrazide and 10-undecanohydrazide derived from fatty acids	M.S.	Acid	85	[46]
27	*Egyptian licorice* extract	Cu	Acid	89.55	[47]
28	Eugenol from cloves	Steel	Acid	80	[31]

(Continued)

TABLE 4.1 (*Continued*)

Summary of different kinds of green corrosion inhibitors

S.N.	Green Corrosion Inhibitors	Metals Investigated	Media	% Efficiency	References
29	Exudate gum from *Dacryodes edulis*	Al	Acid	42	[48]
30	Exudate gum from *Pachylobus edulis*	M.S.	Acid	56	[49]
31	Exudate gum from *Raphia hookeri*	Al	Acid + halide	56.3	[50]
32	Flour and yeast	Fe	Acid media	65–82	[51]
33	Fruit – peels of pomegranate	Al	Acid media	83	[52]
34	Guar gum	C-steel	Acid	75–93.88	[53]
35	Gum (acacia)	M.S. and Al	Neutral then acidic media	NA – M.S., 21.8-Al	[40]
36	Gum arabic	Al	Acid	80	[48]
37	Halfabar	Steel	Acid	90.5	[39]
38	Herbs (thyme, coriander, hibiscus, anis, black cumin, and garden cress)	Steel	Acid	37–92	[54]
39	*Hibiscus sabdariffa* extract	Al and Zn	Hydrochloric acid	85	[55]
40	Kidney bean	Steel	Acid	88.43	[39]
41	Mango peels	Steel, Al, Zn and Cu	Acid	82-steel, 82-Al, 80-Zn, 30-Cu	[34]
42	*Mangrove tannin*	Cu	Acid	82.4	[56]
43	*Mimosa tannin*	C-steel	Acid	66–87	[57]
44	*Momordica charantia*	M.S.	Sodium chloride	82.4	[29]
45	*Morinda lucida*	Steel rebar	Sodium chloride	92.8	[58]
46	*Myrtus communis*	Cu	Acid	>85	[59]
47	Natural honey	C-steel	Sodium chloride	82–91	[60]
48	*Nypa fructicans*	M.S.	Acid	75.11	[61]
49	Opium (0.1%)	M.S. and Al	Neutral then acidic media	NA – M.S., 42.7-Al	[40]
50	Opuntia extract	Steel, Al, Zn and Cu	Acid	75-steel, 70-Al, 55-Zn, 56-Cu	[34]
51	Opuntia extract	Aluminum	Acid	76–96	[62]
52	*Opuntia ficus indica*	C-steel	Acid	70–91	[63]

(*Continued*)

TABLE 4.1 (*Continued*)
Summary of different kinds of green corrosion inhibitors

S.N.	Green Corrosion Inhibitors	Metals Investigated	Media	% Efficiency	References
53	Orange peels	Steel, Al, Zn and Cu	Acid	80-steel, 59-Al, 59-Zn, NA – Cu	[34]
54	Papaia	M.S.	Acid	94	[36]
55	Parts of the kola tree (leaves, nuts, and bark) and tobacco	Steel rebar	Sodium chloride	70–91	[64]
56	*Poinciana pulcherrima*	M.S.	Acid	96	[36]
57	Pomegranate fruit shell	Steel, Al, Zn and Cu	Acid	65-steel, 81-Al, 71-Zn, 73-Cu	[34]
58	*Pongamia glabra, Annona squamosa*	M.S.	Acid	89–95	[65]
59	*Punica granatum*	M.S.	Sodium chloride	79.2	[29]
60	Reducing saccharides fructose and mannose	Al and Zn	Alkaline	92	[66]
61	*Ricinus communis* leaves	M.S.	Sodium chloride	43–84	[67]
62	*Rosmarinus officinalis*	Al–Mg alloy	Sodium chloride	75	[68]
63	*Sansevieria trifasciata*	Al	Acid and alkaline media + halides	94.3-HCl 95.3-KOH	[68]
64	Soya bean	M.S. and Al	Neutral then acidic media	65.5- steel, 27.1-Al	[40]
65	*Swertia angustifolia*	M.S.	Acid	75–96	[69]
66	*Tagetes erecta*	Cu	Acid	98.07	[69]
67	Tobacco leaves	M.S. and Al	Neutral then acidic media	87.5-steel, 77.5-Al	[40]
68	*Tridax procumbens Chromolaena odorata*	Stainless steel	Oilfield environment	82.03 95.6	[70]
69	*V. Amygdalina*	Steel rebar	Sodium chloride	90.8	[71]
70	Vanillin	C-steel	Acid	93–98	[72]
71	*Vernonia amygdalina*	Al alloys	Acid	49.5–72.5	[73]
72	*Zanthoxylum alatum*	M.S.	Acid	76–95	[74]

surface is often responsible for the extract's inhibitory effects. However, a complete knowledge of the corrosion inhibition mechanism of biomass extracts has not yet been established due to their complicated composition. Without separating the active components of any biomass extract, it is challenging to analyze each of the separate and/or mixed interactions of organic species on the metal surface.

4.5 GENERAL CHARACTERISTICS OF GCIS

GCIs have many of the same characteristics as "non-green" inhibitors. At room temperature, the majority of GCIs interact physically and chemically on the metal surface. At high temperatures, inhibition mostly happens as a result of chemisorption. The efficiency of GCIs in inhibiting corrosion changes depending on how long it is exposed to a corrosive environment. Information on the stability of the inhibitive behavior of GCIs on a time scale may be gleaned from the study of the impact of increased time on inhibition efficiency. Usually, the efficacy of an inhibitor declines with time, indicating that physical interactions are mostly responsible for the adhesion of inhibitor molecules to the metal surface [76]. Results of a few studies demonstrate an increase in the ability of GCIs to inhibit cells over time. For instance, studies on *Clematis gouriana* leaf extract for acid corrosion of mild steel showed that the effectiveness of the inhibitor increases with an increase in the time the steel specimen is immersed in the acid solution with 400 ppm concentration of the inhibitor [77]. This was explained by the durability and stability of the adsorbed inhibitor layer on the steel surface. Singh et al. [78] reported the trend of an increase in inhibition efficiency with time. Recently, *Citrus limetta* (Mousambi) peel extract [79] has been reported as a GCI with an inhibition efficiency of 93% on mild steel surfaces in aggressive acidic media. Researchers have also reported an increment in inhibition by increasing the concentration of inhibitor, whereas the inverse was observed by increasing temperatures. To achieve optimal efficiency, it is crucial to take the inhibitor's concentration into account. With regard to reaction factors, the inhibitory efficacies of the extract of the *Justicia gendarussa* plant [80] and Pennyroyal oil from *Mentha pulegium* [81] can be compared, despite the fact that the compositions and sources of the inhibitors differ. With 150 ppm of *J. gendarussa* plant extract, an inhibition efficiency of 93% was achieved, whereas Pennyroyal oil required 2.76 g/L (or around 2,760 ppm) to obtain 80% inhibition efficiency. In such a situation, taking synergism into account is crucial. This finding was supported by Oguzie's [82] research on the synergistic impact of halide derivatives on plant extracts' suppression of acid corrosion of mild steel. The synergistic effects of KCl, KBr, and KI increased the inhibition efficiency of 10 (v/v)% *Ocimum viridis* from 69% to 87%, 88%, and 95%, respectively. When compared to other inhibitors, those that are not green show greater inhibitory efficiency. For instance, triazole and benzimidazole derivatives are prominent corrosion inhibitors that have shown higher than 97% inhibition efficacy for acid corrosion of mild steel [83]. However, they are also renowned for their perniciousness [83–85, 87]. Regardless of their sources, one should be familiar with the EC50 (Effective Concentration 50) and LD50 (Lethal Dose 50) of inhibitors, even if it is assumed that biologically derived inhibitors are environmentally beneficial. These cytotoxicity assays can provide strong proof for the biocompatibility of inhibitors.

4.6 FACTORS INFLUENCING GCI EFFICACY

The adsorption properties of GCIs on a metal surface determine how well they prevent corrosion. The structure, concentration, temperature, and exposure period have a significant impact on the inhibition efficiency of GCIs. At a specific concentration value, an increment in the concentration of GCIs causes a decrease in the rate of corrosion and an increase in inhibition efficacy, which reaches a maximum level. This was due to the production of additional inhibitor molecules that were adsorbed on the metal surface, making it difficult for the electrolyte solution to corrode the metal. With an increased corrosion exposure time in the presence of GCIs, metals dissolve more readily. This is related to inhibitor molecules that had previously been adsorbed onto the metal surface as a result of partial desorption. An equilibrium between adsorption and GCIs' molecule desorption at the metal surface exists at a specific temperature since the rate of corrosion increases linearly with an increase in the temperature. The equilibrium shifts as a result of a higher desorption rate, which increases the temperature, until it is restored at different equilibrium constant values. As a result, the inhibition efficiency of GCIs decreases with increasing temperature. As established already, the structural behavior of GCIs greatly affects how effective they are under corrosive conditions. Through the establishment of an adsorptive bond by the Lewis acid–base process, in which GCIs and metal function as the electron donor and acceptor, respectively, the presence of a heteroatom in a GCI molecule increases their adsorption onto the metal surface. The electron density and polarizability of the reaction center determine how strong an adsorption bond is. Surface-active agents or, surfactants, when present in the media [84] even in very small concentrations enhance the inhibition efficiency of corrosion inhibitors [85]. In conclusion, research has demonstrated that surface-active GCI adsorption increases with an increasing molecular weight and dipole moment.

4.7   MEASUREMENT OF GCI EFFICACY

The preparation of a metal specimen to be checked for corrosion is necessary as the first step in determining the effectiveness of organic GCIs. The choice of metal coupons is crucial for determining the efficiency of an organic GCI since even little changes in the metal's composition or the availability of contaminants during manufacture might affect the results [86]. The metal composition needs to be as pertinent to metals involved in corrosion issues as is practicable. Weight loss measurements, electrochemical impedance spectroscopy, linear polarization resistance, and potentiodynamic polarization are the most popular approaches to evaluate the efficiency of organic GCIs that are currently available in the literature.

4.8   MECHANISM OF ACTION OF GCIS IN CORROSION CONTROL

The relative rate of corrosion [27] is correlated with the variation in the standard Gibbs free energy ($G°$), which is a spontaneous process. A higher spontaneous reaction, often known as a higher rate of corrosion, is correlated with a larger negative value of $G°$. When exposed to the environment, metals and alloys erode to produce stable corrosion products. To reduce corrosion rate, additional corrosion inhibitors must be used. Because they can build up on the surface and serve as physical protective barriers, corrosion products like rust and scale can also act as corrosion inhibitors. However, the rate ratio (RR) of corrosion of any specific metal is dependent on the Pilling–Bedworth ratio [87], which is applicable at high-temperature oxidation (or, corrosion). This ratio is depicted in Eq. (4.5).

$$M * d / n * m * D \qquad (4.5)$$

where m and d are the atomic weight and density of the metal, respectively; M and D are the molecular weight and density of the corroded product formed on the metal surface; and n represents the number of metal atoms in the corrosion product's molecular composition.

While determining whether a surface film will be protective or not, the value of the Pilling–Bedworth ratio might be utilized as an indicator. The corrosion product's surface layer has holes and cracks that are largely nonprotective when the volume of the corrosion product is less than that of the metal from which it is generated ($Md/nmD < 1$). In this case, the corrosion product's volume will be lower than the volume of the metal from which it is produced. When $Md/nmD > 1$ and the volume of the corrosion product is more than the volume of the metal, it is anticipated that the corrosion product's surface film is substantially more compressed and compact, which provides a more protective layer. On metallic surfaces at active locations, adsorption is the initial stage in the formation of a corrosion-protective layer or coat when there is an aggressive medium. Adsorption mechanism, chemical and electrical properties of the inhibitor, temperature, the kind of electrolyte used, steric effects, and the nature and surface charge of metals are some of the factors that affect the ability of the adsorption of inhibitor onto the metallic surface and isolate it. The interaction between the inhibitor and the metal surface can be best understood using the Langmuir adsorption isotherm. Adsorption on damaged surfaces approaches a steady state and can take the form of physical (physisorption), chemical (chemisorption), or combined adsorption mechanisms, which are thought to be the best for effectively inhibiting corrosion. The standard free energy of adsorption "Go ads" in an aqueous solution is related to physical adsorption. It is associated with an electrostatic contact between the charged centers of molecules and the charged metal surface, which results in a dipole interaction between the molecules and the metal surface, if its value is −20 kJ/mol or less negative. The development of a coordinating covalent bond, however, is the end consequence of the chemical adsorption process, which includes the transfer or sharing of electrons from the inhibitor to the metallic surface. In comparison with physical adsorption, when the value of "Go ads" is around "40 kJ/mol" or more negative, the bonding strength is far higher. GCI adsorption slows corrosion by removing the reactive area of metals from the surface, exposing inactive areas in corrosive environments. At low temperatures or room temperature, GCIs are generally more efficient, but in most situations, their efficiency decreases as temperature increases. Researchers have proposed a wide range of ideas to explain the mechanism of action of GCIs, which are dependent on the molecular composition of the active component. The active component produced by natural inhibitors differs from one plant species to another although their structures are quite similar to their organic counterparts. For instance, allyl propyl disulfide is present in garlic, pyrrolidine is present in carrot, and the alkaloid ricinine is present in castor seeds and mustard seeds. Berberine is an alkaloid with a long chain of aromatic rings and a N atom in the ring. Monomtrene-1,8-cineole is a component of eucalyptus oil. Lawsonia extract includes coumarin, gallic acid, sterols, 2-hydroxy-1,4-naphthoquinone resin, and tannin. Hexuronic acid, neutral sugar remnants, volatile monoterpenes, canaric and related triterpene acids, as well as reducing and nonreducing sugars, are present in gum exudate. Unsaturated fatty acids, bioflavonoids, and primary and secondary amines are present in garcinia kola seed. Ascorbic acid, amino acids, flavonoids, pigments, and carotene are present in calyx extract. These plant extracts' ability to suppress corrosion might be attributed to the heterocyclic components they contain, such as alkaloids, flavonoids, etc.

4.9   CONCLUSION

Corrosion is a damaging phenomenon that has frequently had a devastating impact on economies. It has both direct and indirect consequences that can no longer be disregarded on a global scale. Although it may be nearly difficult to completely eradicate corrosion, it is still avoidable; hence, it is not a required curse. GCIs, such as plant extracts, ionic liquids, drugs, amino acids, compounds of rare earth metals, surfactants, etc., not only are environmentally beneficial but also effectively prevent corrosion. Green inhibitors are intended to replace the hazardous organic and inorganic inhibitors now in use in a variety of industrial applications. Their use goal has been implemented, and this goal will continue to be accomplished. Most metals may be effectively inhibited by green inhibitors in a wide range of corrosive conditions. Two of the major benefits of these inhibitors are that they are nontoxic and biodegradable. But there are limitations to their performance as well. However, further study is required before green inhibitors can used extensively on an industrial scale, despite the fact that several studies have highlighted them as a possible option against corrosion in various conditions.

ACKNOWLEDGMENTS

The author is grateful and thanks to the Chancellor, Vice-Chancellor, Pro-Vice-Chancellor, Vice-Chancellor, and Dean, School of Engineering, Presidency University, Bengaluru, for the above project work. Thanks are also due to Mr. Amit Prakash Varshney, Miss. Anushi Varshney, and Mr. Anirudh Varshney for supporting the author to compile this work.

REFERENCES

[1] Goni L.K.M.O., Mazumder M.A.J. (2019). Green corrosion inhibitors. In: Singh A. (Ed.), *Corrosion Inhibitors*. Rijeka: IntechOpen. doi:10.5772/intechopen.81376.
[2] Revie R.W., Uhlig HH. (2008). *Corrosion and Corrosion Control: An Introduction to Corrosion Science and Engineering*, 4th edn., Hoboken, NJ: Wiley.
[3] Raja P.B., Ismail M., Ghoreishiamiri S., Mirza J., Ismail M.C., Kakooei S. et al. (2016). Reviews on corrosion inhibitors: A short view. *Chemical Engineering Communications*. 203:1145–1156.
[4] Javaherdashti R. (2000). How corrosion affects industry and life. *Anti-Corrosion Methods and Materials*. 47:30–34.

[5] Uhlig H.H. (1950). The cost of corrosion to the United States. *Corrosion*. 6:29–33.

[6] Koch G., Varney J., Thompson N., Moghissi O., Gould M., Payer J. (2016). *International Measures of Prevention, Application, and Economics of Corrosion Technologies Study*. Houston, TX: NACE International.

[7] Hoar T.P. (1976). Review lecture: Corrosion of metals: Its cost and control. *Proceedings of the Royal Society of London. Series A: Mathematical and Physical Sciences*. 348:1–18.

[8] Bennett L.H., Kruger J., Parker R.L., Passaglia E., Reimann C., Ruff A.W. et al. (1978). *Economic Effects of Metallic Corrosion in the United States*. A Report to the Congress by the National Bureau of Standards. National Bureau of Standards. Washington, DC: US Government Printing Office Publication.

[9] Lichtenstein A.G. (1994). The silver bridge collapse recounted. *Journal of Performance of Constructed Facilities*. 7:249–261.

[10] Bowonder B. (1987). The Bhopal accident. *Technological Forecasting and Social Change*. 32(2):169–182.

[11] Stress Corrosion Cracking Failure. (2018). https://corrosion-doctors.org/Forms-SCC/swimming.htm (accessed July 15, 2018).

[12] Taghavikish M., Dutta N.K., Choudhury N.R. (2017). Emerging corrosion inhibitors for interfacial coating. *Coatings*. 7(12):217–245.

[13] Dariva C.G., Galio A.F. (2014). Corrosion inhibitors-principles, mechanisms and applications. In: Aliofkhazraei M. (Ed.), *Developments in Corrosion Protection*, 1st edn., Rijeka: InTech, pp. 365–381. doi:10.5772/57255.

[14] Sowmyashree A.S., Somya A., Kumar S., Kumar P., Rao S. (2022). Discotic anthraquinones as novel corrosion inhibitor for mild steel surface, *Journal of Molecular Liquids*. 347:118194.

[15] Arthur D.E., Jonathan A., Ameh P.O., Anya C.A. (2013). Review on the assessment of polymeric materials used as corrosion inhibitor of metals and alloys. *International Journal of Industrial Chemistry*. 4(2):1–9.

[16] Palou R.M., Olivares-Xomelt O., Likhanova N.V. (2011). Environmentally friendly corrosion inhibitors. In: Sastri V.S. (Ed.), *Green Corrosion Inhibitors: Theory and Practice*, 1st edn., Hoboken, NJ: Wiley, pp. 257–303, doi:10.1002/9781118015438.ch7.

[17] Sastri V.S. (Ed.). (2011). *Green Corrosion Inhibitors: Theory and Practice*, 1st edn., Hoboken, NJ: Wiley, doi:10.1002/9781118015438.

[18] Obot I.B., Obi-Egbedi N.O., Umoren S.A., Ebenso EE. (2010). Synergistic and antagonistic effects of anions and ipomoea invulcrata as green corrosion inhibitor for aluminium dissolution in acidic medium. *International Journal of Electrochemical Science*. 5(7):994–1107.

[19] Sowmyashree A.S., Somya A., Kumar C.B.P., Rao S. (2021). Novel nano corrosion inhibitor, integrated zinc titanate nano particles: Synthesis, characterization, thermodynamic and electrochemical studies. *Surfaces and Interfaces*. 22:100812.

[20] Saji V.S., Cook R. (2012). *Corrosion Protection and Control Using Nanomaterials*. Amsterdam: Elsevier.

[21] Miralrio A., Espinoza Vázquez A. (2020). Plant extracts as green corrosion inhibitors for different metal surfaces and corrosive media: A review. *Processes*. 8(8):942.

[22] Zhao Q., Guo J., Cui G., Han T., Wu Y. (2020). Chitosan derivatives as green corrosion inhibitors for P110 steel in a carbon dioxide environment. *Colloids Surfaces B: Biointerfaces*. 194:111150.

[23] Machado Fernandes C., Pina V.G.S.S., Alvarez L.X., de Albuquerque A.C.F., dos Santos Júnior F.M., Barrios A.M., Velasco J.A.C., Ponzio E.A. (2020). Use of a theoretical prediction method and quantum chemical calculations for the design, synthesis and experimental evaluation of three green corrosion inhibitors for mild steel. *Colloids and Surfaces A: Physicochemical and Engineering Aspects*. 599:124857.

[24] Singh P., Srivastava V., Quraishi M.A. (2016). Novel quinoline derivatives as green corrosion inhibitors for mild steel in acidic medium: Electrochemical, SEM, AFM, and XPS studies. *Journal of Molecular Liquids*. 216(1):164–173.

[25] Munis A., Zhao T., Zheng M., Rehman A.U., Wang F.A. (2020). Newly synthesized green corrosion inhibitor imidazoline derivative for carbon steel in 7.5% NH4Cl solution. *Sustainable Chemistry and Pharmacy*. 16:100258.

[26] Verma C., Ebenso E.E., Quraishi M.A. (2017). Ionic liquids as green and sustainable corrosion inhibitors for metals and alloys: An overview. *Journal of Molecular Liquids*. 233:403–414.

[27] Shehata O.S., Korshed L.A., Attia A. (2018). Green corrosion inhibitors, past, present, and future corrosion inhibitors, principles and recent applications. In: Mahmood A. (Ed.), *Corrosion Inhibitors, Principles and Recent Applications*. Rijeka: InTech. doi:10.5772/intechopen.72753.

[28] Ekpe U.J., Ebenso E.E., Ibok U.J. (1994). Inhibitory action of Azadirachta indica leaves extract on the corrosion of mild steel in H_2SO_4. *Journal of West African Journal of Biological and Applied Chemistry*. 37:13–30.

[29] Quraishi M.A., Farooqi I.H., Saini P.A. (1999). Investigation of some green compounds as corrosion and scale inhibitors for cooling systems. *Corrosion*. 55(5):493–497.

[30] Verma S.A., Mehta G.N. (1999). Effect of acid extracts of Acacia arabica on acid corrosion of mild steel. *Bulletin of Electrochemistry*. 15(2):67–70.

[31] Chaieb E., Bouyanzer A., Hammouti B., Benkaddour M. (2005). Inhibition of the corrosion of steel in 1 M HCl by eugenol derivatives. *Applied Surface Science*. 246(1):199–206.

[32] Abd-El-Nabey B.A., El-Housseiny S., El-Naggar G.A., Matter E.A., Esmail G. (2015). Inhibitive action of *Alhagi maurorum* plant extract on the corrosion of copper in 0.5 M H2SO4 *Physical Chemistry*. 5(3):49–62.

[33] Parikh K.S., Joshi K.J. (2004). Natural compounds onion (*Allium cepa*), garlic (*Allium sativum*) and bitter gourd (*Momordica charantia*) as corrosion inhibitors for mild steel in hydrochloric acid. *Transactions-Society for the Advancement of Electrochemical Science and Technology*. 39(1/2):29–35.

[34] Saleh R.M., Ismail A.A., El Hosary A.A. (1982). Corrosion inhibition by naturally occurring substances: VII. The effect of aqueous extracts of some leaves and fruit-peels on the corrosion of steel, Al, Zn and Cu in acids. *British Corrosion Journal*. 17(3):131–135.

[35] Garai S., Garai S., Jaisankar P., Singh J.K., Elango A. (2012). A comprehensive study on crude methanolic extract of *Artemisia pallens* (Asteraceae) and its active component as effective corrosion inhibitors of mild steel in acid solution. *Corrosion Science*. 60:193–204.

[36] Zucchi F., Omar I.H. (1985). Plant extracts as corrosion inhibitors of mild steel in HCl solutions. *Surface Technology*. 24(4):391–399.

[37] Abdulrahman A.S., Ismail M. (2011). Green plant extract as a passivation-promoting inhibitor for reinforced concrete. *International Journal of Engineering Science and Technology*. 3:6484–6489.

[38] Li Y., Zhao P., Liang Q., Hou B. (2005). Berberine as a natural source inhibitor for mild steel in 1 M H_2SO_4. *Applied Surface Science*. 252(5):1245–1253.

[39] Abdel-Gaber A.M., Abd-El-Nabey B.A., Sidahmed I.M., El-Zayady A.M., Saadawy M. (2006). Inhibitive action of some plant extracts on the corrosion of steel in acidic media. *Corrosion Science*. 48(9):2765–2779.

[40] Srivastav K., Srivastava P. (1981). Studies on plant materials as corrosion inhibitors. *British Corrosion Journal*. 16(4):221–223.

[41] Subramanian A., Natesan M., Muralidharan V.S., Balakrishnan K., Vasudevan T. (2000). An overview: Vapor phase corrosion inhibitors. *Corrosion*. 56(2):144–155.

[42] Anthony N., Malarvizhi E., Maheshwari P., Rajendran S., Palaniswamy N. (2004). Corrosion inhibition of caffein-Mn2+ system. *Indian Journal Chemical Technology.* 11:346–350.

[43] Left D.B., Zertoubi M., Khoudali S., Benaissa M., Irhzo A., Azzi M. (2013). Effect of methanol extract of *Chamaerops humilis* L. leaves (MECHLL) on the protection performance of oxide film formed on reinforcement steel surface in concrete simulated pore solution. *International Journal of Electrochemical Science.* 8(10):11768–11781.

[44] Verma C., Kumar A.M., Mazumder M.A.J., Quraishi M.A. (2018). Chitosan-based green and sustainable corrosion inhibitors for carbon steel. In: Dongre R.S. (Ed.), *Chitin-Chitosan—Myriad Functionalities in Science and Technology.* Rijeka: InTech. doi:10.5772/intechopen.74989.

[45] Oyewole O., Oshin T.A., Atotuoma B.O. (2021). Corchorus olitorius stem as corrosion inhibitor on mild steel in sulphuric acid. *Heliyon.* 7(4):e06840.

[46] Quraishi M.A., Jamal D. (2002). Approach to corrosion inhibition by some green inhibitors based on oleochemicals. *Bulletin of Electrochemistry.* 18(7):289–294.

[47] Deyab M.A. (2015). Egyptian licorice extract as a green corrosion inhibitor for copper in hydrochloric acid solution. *Journal of Industrial and Engineering Chemistry.* 22:384–389.

[48] Umoren S.A., Obot I.B., Ebenso E.E., Obi-Egbedi N. (2008). Studies on the inhibitive effect of exudate gum from *Dacroydes edulis* on the acid corrosion of aluminium. *Portugaliae Electrochimica Acta.* 26(2):199–209

[49] Umoren S.A., Ekanem U.F. (2010). Inhibition of mild steel corrosion in H2SO4 using exudate gum from *Pachylobus edulis* and synergistic potassium halide additives. *Journal Chemical Engineering Communications.* 197(10):1339–1356.

[50] Umoren S.A., Ekanem U.F. (2010). Inhibition of mild steel corrosion in H2SO4 using exudate gum from *Pachylobus edulis* and synergistic potassium halide additives. *Journal Chemical Engineering Communications.* 197(10):1339–1356.

[51] Baldwin J. (1895). British Patent. 2327.

[52] Saleh R.M., El-Hosary A.A. (1972). Corrosion inhibition by naturally occurring substances. In: El-Hosary A.A. (Ed.), *Effect of Pomegranate Juice and the Aqueous Extracts of Pomegranate Fruit Shells, Tamarind Fruits and Tea Leaves on the Corrosion of Al: Electrochemistry,* 13th ed., Karaikudi, India: CECRI.

[53] Abdallah M. (2004). Guar gum as corrosion inhibitor for carbon steel in sulfuric acid solutions. *Portugaliae Electrochimica Acta.* 22(2):161–175.

[54] Khamis E., Alandis N. (2002). Herbs as new type of green inhibitors for acidic corrosion of steel. *Material Wissenschaft und Werkstofftechnik.* 33(9):550–554.

[55] El-Hosary A.A., Saleh R.M., El Din A.M.S. (1972). Corrosion inhibition by naturally occurring substances: The effect of *Hibiscus subdariffa* (karkade) extract on the dissolution of Al and Zn. *Corrosion Science.* 12(12):897–904.

[56] Shah A.M., Rahim A.A., Hamid S.A., Yahya S. (2013). Green inhibitors for copper corrosion by mangrove tannin. *International Journal of Electrochemical Science.* 8:2140–2153.

[57] Martinez S., Štern I. (2001). Inhibitory mechanism of low-carbon steel corrosion by mimosa tannin in sulphuric acid solutions. *Journal of Applied Electrochemistry.* 31(9):973–978.

[58] Okeniyi J.O., Loto C.A., Popoola A.P.I. (2014). Morinda lucida effects on steel-reinforced concrete in 3.5% NaCl: Implications for corrosion-protection of wind-energy structures in saline/marine environments. *Energy Procedia.* 50:421–428.

[59] Bozorg M., Shahrabi Farahani T., Neshati J., Chaghazardi Z., Mohammadi Ziarani G. (2014). Myrtus communis as green inhibitor of copper corrosion in sulfuric acid. *Industrial & Engineering Chemistry Research.* 53(11):4295–4303.

[60] El-Etre A.Y., Abdallah M. (2000). Natural honey as corrosion inhibitor for metals and alloys. II. C-steel in high saline water. *Corrosion Science.* 42:731–738.

[61] Orubite K.O., Oforka NC. (2004). Inhibition of the corrosion of mild steel in hydrochloric acid solutions by the extracts of leaves of Nypa fruticans Wurmb. *Materials Letters.* 58(11):1768–1772.

[62] El-Etre A.Y. (2003). Inhibition of aluminum corrosion using Opuntia extract. *Corrosion Science.* 45(11):2485–2495.

[63] Flores-De los Ríos J.P., Sánchez-Carrillo M., Nava-Dino C.G., Chacón-Nava J.G., GonzálezRodríguez, Huape-Padilla E., Neri-Flores M.A., Martínez-Villafañe A. (2015). Opuntia ficusindica extract as green corrosion inhibitor for carbon steel in 1 M HCl solution. *Journal of Spectroscopy.* 2015:14692–14701.

[64] Loto C.A., Loto R.T., Popoola A.P.I. (2011). Electrode potential monitoring of effect of plants extracts addition on the electrochemical corrosion behaviour of mild steel reinforcement in concrete. *International Journal of Electrochemical Science.* 6:3452–3465.

[65] Sakthivel P., Nirmala P.V., Umamaheswari S., Antony A.A.A., Paruthimal Kalaignan G., Gopalan A., Vasudevan T. (1999). Corrosion inhibition of mild steel by extracts of Pongamia glabra and Annona squamosa in acidic media. *Bulletin of Electrochemistry.* 15(2):83–86.

[66] Müller B. (2002). Corrosion inhibition of aluminium and zinc pigments by saccharides. *Corrosion Science.* 44(7):1583–1591.

[67] Sathiyanathan R., Maruthamuthu S., Selvanayagam M., Mohanan S., Palaniswamy N. (2005). Corrosion inhibition of mild steel by ethanolic extracts of Ricinus communis leaves. *Indian Journal Chemical Technology.* 12:356–360.

[68] Kliškic M., Radošević J., Gudic S., Katalinic V. (2000). Aqueous extract of Rosmarinus officinalis L. as inhibitor of Al-Mg alloy corrosion in chloride solution. *Journal of Applied Electrochemistry.* 30(7):823–830.

[69] Zakvi S.J., Mehta G.N. (1988). Acid corrosion of mild steel and its inhibition by Swertia angustifolia-study by electrochemical techniques. *Transactions of the SAEST.* 23(4):407–441.

[70] Aribo S., Olusegun S.J., Ibhadiyi L.J., Oyetunji A., Folorunso D.O. (2016). Green inhibitors for corrosion protection in acidizing oilfield environment. *Journal of the Association of Arab Universities for Basic and Applied Sciences.* 24(1):34–38.

[71] Eyu G. (2013). Effect of *Vernonia amygdalina* extract on corrosion inhibition of mild steel in simulated seawater. *Australian Journal of Basic and Applied Sciences.* 7(14):257–263.

[72] El-Etre A.Y. (2001). Inhibition of acid corrosion of aluminum using vaniline. *Corrosion Science.* 43(6):1031–1039.

[73] Avwiri G.O, Igho F.O. (2003). Inhibitive action of Vernonia amygdalina on the corrosion of aluminium alloys in acidic media. *Material Letters.* 57(22):3705–3711.

[74] Gunasekaran G., Chauhan L.R. (2004). Eco friendly inhibitor for corrosion inhibition of mild steel in phosphoric acid medium. *Electrochimica Acta.* 49(25):4387–4395.

[75] Fazal B.R., Becker T., Kinsella B. and Lepkova K. (2022). A review of plant extracts as green corrosion inhibitors for CO2 corrosion of carbon steel. *NPJ Mater Degrad.* 6:5.

[76] Devarayan K., Mayakrishnan G., Sulochana N. (2012). Green inhibitors for corrosion of metals: A review. *Chemical Science Review and Letters.* 1(1):1–8.

[77] Gopiraman M., Sakunthala P., Kanmani R., Alex Ramani V., Sulochana N. (2011). Inhibitive action of *Clematis gouriana* extract on the corrosion of mild steel in acidic medium. *Ionics.* doi:10.1007/s11581-011-5480584-9.

[78] Singh A., Ahamad I., Singh V.K., Quraishi M.A. (2010). Inhibition effect of environmentally benign karanj (*Pongamia pinnata*) seed extract on corrosion of mild steel in hydrochloric acid solution. *Journal of Solid State Electrochemistry.* 15:1087–1097.

[79] Sowmyashree A.S., Somya A., Rao S., Pradeep Kumar C.B., Al-Romaizan A.N., Hussein M.A., Khan A., Marwani H.M., Asiri A.M. (2023). Potential sustainable electrochemical corrosion inhibition study of *Citrus limetta* on mild steel surface in aggressive acidic media. *Journal of Materials Research and Technology.* 24:984–994.

[80] Satapathy A.K., Gunasekaran G., Sahoo S.C., Kumar A., Rodrigues P.V. (2009). Corrosion inhibition by *Justicia gendarussa* plant extract in hydrochloric acid solution. *Corrosion Science.* 51:2848–2856.

[81] Bouyanzer A., Hammouti B., Majidi L. (2006). Pennyroayl oil from Mentha pulegium as corrosion inhibitor for steel in 1 M HCl. *Material Letter.* 60:2840–2843.

[82] Oguzie E.E. (2008). Evaluation of the inhibitive effect of some plant extracts on the acid corrosion of mild steel. *Corrosion Science.* 50:2993–2998.

[83] Popova A., Sokolova E., Raicheva S., Christov M. (2003). AC and DC study of the temperature effect on mild steel corrosion in acid media in the presence of benzimidazole derivatives. *Corrosion Science.* 45:33–35.

[84] Somya A., (2019) Surfactant based hybrid ion exchangers, *Research Journal of Chemistry and Environment.* 23(3):96.

[85] Rafiquee M.Z.A., Khan S., Quraishi M. (2008). Influence of surfactants on the corrosion inhibition behaviour of 2-aminophenyl-5-mercapto-1-oxa-3,4-diazole (AMOD) on mild steel. *Materials Chemistry and Physics.* 107:528–533.

[86] Qian B., Wang J., Zheng M., Hou B. (2013). Synergistic effect of polyaspartic acid and iodide ion on corrosion inhibition of mild steel in H2 SO4. *Corrosion Science* 75:184–192.

[87] Bedworth R.E., Pilling N.B. (1923). The oxidation of metals at high temperatures. *Journal of the Institute of Metals.* 29(3):529–582.

5 Starch as Corrosion Inhibitors

*Fatima Choudhary, Anju Nair, Jenis Tripathi,
Kaartik Chandan, and Paresh More*
K. E. T'S, V. G. Vaze College (Autonomous)

5.1 INTRODUCTION

Corrosion of metals is an electrochemical process that greatly affects the final application of the metals and as a result the cost of maintenance and repairs for industrial applications. Coating plays a vital role in the modification of metal surfaces and protection from corrosion [1]. Understanding these aspects will allow for the use of appropriate control methods to reverse the kinetics of metal corrosion. The kinetics of metal corrosion is highly affected by the surface and substrate chemistries as well as some environmental impacts (e.g., solution concentration (pH), temperature, etc.). To prevent or at least reduce its spontaneous dynamics, researchers from all around the world are investigating solutions to deal with corrosion. As part of their typical mode of action, most inhibitor compounds normally function by producing a coating of passivation that prevents the entry of corrosive radicals to the metal surface. The environment to which the compound was applied, the fluid composition, the amount of water present, the type of metal used and the flow regime determine how effective the coating layer would be [2].

Carbohydrate polymers are frequently utilized as coatings and linings for metals. They represent a class of environmentally benign, biodegradable, chemically stable macromolecules with specialized inhibitory properties that protect metal surfaces and bulk in a mechanistic manner from corrosion [3]. According to the US Congress in 1993, biopolymers are a family of green corrosion inhibitors that are readily available and acceptable in the environment [4]. Biopolymers act as inhibitors and control corrosion formation by various mechanisms via absorption on the surface of metals as they form protective films on metal surfaces [5]. In the past few decades, organic inhibitors such as imidazole derivatives [6], thiazole derivatives [7], triazines, triazole derivatives [8], pyrazole derivatives [9], pyrimidine derivatives [10], hydrazine derivatives [11], Schiff bases [12], amino acids [13] and inorganic inhibitors, including cobalt complexes [14], have been investigated. Complexes of Mn (II), Co (II), Ni (II), Cu (II), Zn (II), [15,16] molybdate, cadmium complexes, chromate, dichromate, arsenates, tetraborates and phosphates [17] have been used as anticorrosion agents.

But these inhibitors are averted due to their toxicity and harmful effects on the environment and human beings [18]. Polymers exhibit better inhibitive capability and

DOI: 10.1201/9781003400059-5

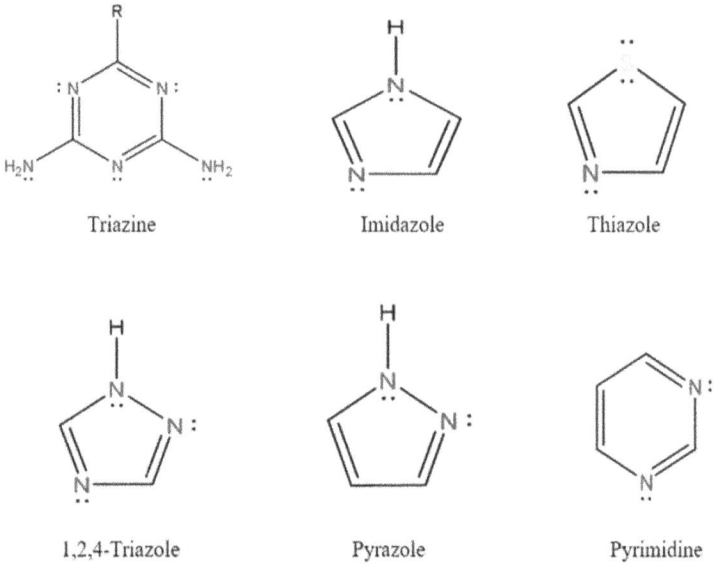

FIGURE 5.1 Organic inhibitors used as an anticorrosion agent.

are prone to adsorb on a greater metal surface area than the ones having low molar mass [19]. Other than adsorption, they have been used in the coating of materials, drug delivery, etc. The macromolecular weights, chemical compositions, and distinctive electronic and molecular structures of these biopolymeric compounds play a role in their protective modes and mechanisms. Chitosan and cellulose are two examples of materials with free hydroxyl and amine groups that can chelate metal ions using their lone pairs of electrons that are easily exploited for coordination bonding at the metal–solution interface. A few inhibitors that were grafted to form more compact structures with polar groups capable of boosting the overall surface energy of the surface were also reviewed. The anticorrosion properties of exudate gums, carboxymethyl and hydroxyethyl cellulose, pectin, starch, pectates, modified chitosans, carrageenan, dextrin/cyclodextrins and alginates were affected by the addition of halide additives [20]

Biopolymer-based coating has been attractive for preparing 3D printed materials, protection of implant materials and industrial applications. Being a biodegradable, cheap and non-toxic polymer, starch is widely used in applications as an inhibitor of corrosion mainly due to low solubility [Figure 5.2] [5].

Starch modification is carried out by reactions through hydroxyl groups of synthetic polymers such as –COOH, anhydride, epoxy, oxazoline and urethane and by free radical polymerization between vinyl monomers and glucose rings [19]. Starch is one of the main sources of energy for animals and humans, including short-term energy storage for plants (Table 5.1).

The semi-crystalline structure of starch granules is irreversibly lost when it is subjected to certain concentrations of shear pressures, heat energy and plasticizers, causing it to change into a continuous matrix. But starch's extraordinary capacity for

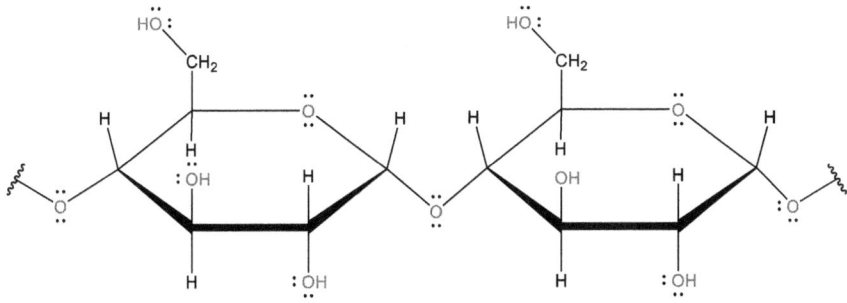

FIGURE 5.2 Starch.

TABLE 5.1
Source of Starch and Their Types

Types of Starch	Sources	References
Wheat	Locust bean gum, bakery yeast, D-glucose and wheat starch	[21]
Corn	Cassava starch, soya flour and corn	[22]
Tapioca	Soya bean flour and corn flour	[23]
Potato	White rice flour, soya flour, rice flour, potato and egg powder	[24]
Cassava	Sorghum flour	[25]

Source: Adapted from [49].

making films, which results in films that are durable, heat-sealable and isotropic, is due to amylose [5]. Starch is utilized in edible films and coatings because it may create translucent, tasteless and colorless layers with characteristics resembling those of synthetic polymers. Additionally, starch is used to thicken foods like gravies, sauces, soups and pie fillings. Foods like pasta and bread can benefit from the addition of starch to improve their consistency and texture. To stop the growth of big ice crystals, stabilizers like starch are added to processed goods like ice cream. Because of its special qualities, starch is used in edible films and coatings [26].

Edible films and coatings made of starch have qualities including flexibility, heat resistance and moisture resistance that are like those of synthetic polymers. Starch-based edible films and coatings are suited for use as food packaging and preservatives due to these characteristics.

Starch has two major components: (1) amylose and (2) amylopectin. Addition of starch helped as the electrolyte during the MAO process, which is expected to decrease structural defects. Modified starch has applications in the food and the non-food industries. In the food industry, starch is used in delayed release. In the non-food industry, it shows biocompatibility and is involved in the interaction of living cells. It is used in the pharmaceutical industry as a binding agent, disintegrating agent and raw material for

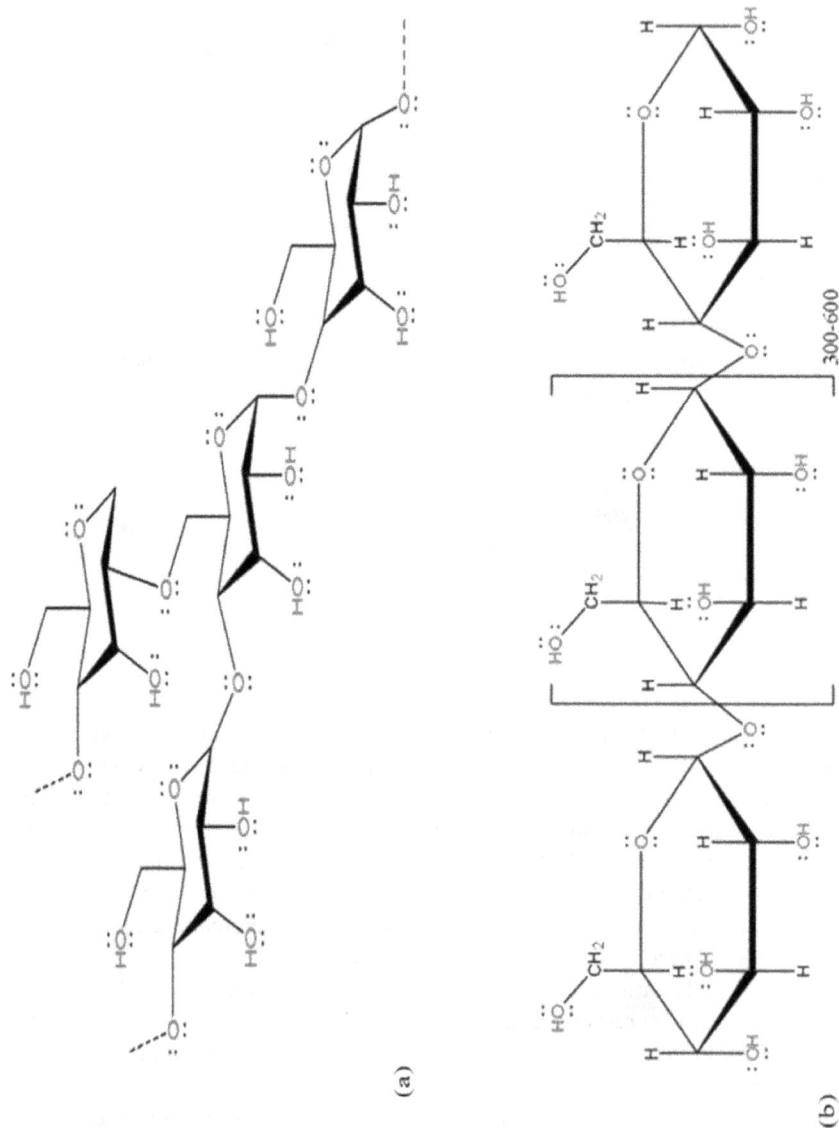

FIGURE 5.3 Chemical structure of (a) amylopectin and (b) amylose molecules of starch.

microspheres because of its excellent adsorption capacity; thus, it is used to improve the dissolution rate of less soluble solvents. However, due to its poor solubility and weak surface adherence, molecular starch is not widely used as a corrosion inhibitor. Its use in neutral and acidic conditions for the inhibition of metals has been reported in several studies based on either chemical or physical modification to increase its anti-corrosion efficacy. As depicted in Figure 5.3a and b, starch has a special molecular structure that allows it to coordinately bind with iron in ferrous substrates by filling any open or partially filled orbitals, which results in corrosion inhibition.

Coating plays a very vital role in the modification of metal surfaces as it protects the surface from being in contact with the atmosphere; in addition, it prevents the degradation of the substrate due to corrosion [1].

Usage of plastic has increased over decades, especially in food packaging, and it has harmful effects on the environment. Furthermore, it is crucial for food products to prevent getting infected by microbes, thus improving shelf life [27]. To cater to this issue, biodegradable films came into the picture for improved protection and environmental preservation as they degrade after disposal (Figure 5.4) [27].

Despite having such great advantages, starch possesses a few drawbacks, including low water solubility, decreased mechanical properties and moisture sensitivity. Researchers have developed a thermoplastic starch matrix packed with nanofillers

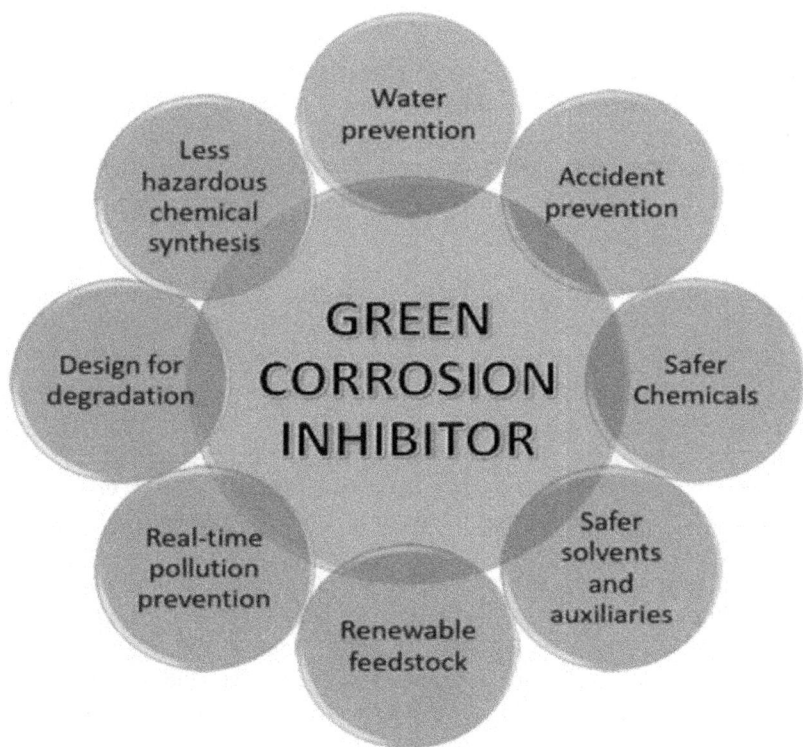

FIGURE 5.4 Corrosion inhibition applications of natural and eco-friendly biopolymer.

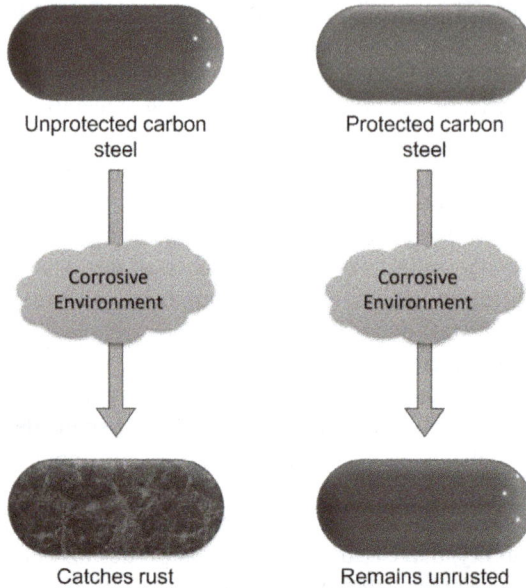

FIGURE 5.5   Corrosion inhibitors for carbon steel.

to address these problems, greatly enhancing the characteristics of the material. Biopolymers improve epoxy coating and slow down mild steel corrosion (Figure 5.5). A blend of starch and pectin is one such instance. The effectiveness of the starch–pectin mixture in preventing mild steel corrosion is reported to be 74%.

5.2   CURRENT APPLICATIONS

Native starch is the starch chain as it was originally found in raw materials. Native starches are rarely employed in the industry due to their disadvantages, including low shear resistance, significant retrogradation and syneresis. Instead, they are frequently altered through physical, chemical or enzymatic processes, or a combination of these, to give them desired functional properties for particular uses. Physically modified starch is a natural product and an extremely safe component [28]. However, the methods that are typically employed to create modified starch are difficult to set up, expensive and time-consuming and frequently involve treatments with dangerous chemicals [29,30].

Numerous industrial processes, including oxidation, esterification, hydroxymethylation, dextrinization and cross-linking, require the modification of native starches (Table 5.2).

The limitations of natural starch characteristics are overcome by these changes, which stabilize the polymers against extreme heating, shear, freezing or storage.

These modified starches are used in a variety of ways in the food industry, most notably in the thickening and emulsifying procedures used in confectionery and bakery products, as well as in the non-food industries like the manufacture

TABLE 5.2
Application of Starches

Type of Starch	Applications	References
Native starches	Textile, paper, food, cardboard and potable alcohol production industry	[31]
Hydrolyzed starches		
Maltodextrins	• Food preparation (bulking agent)	[32]
Glucose syrup	• Pharmaceutical (excipient)	[33]
	• Ingredient in the ice cream industry	[34]
	• Dextrose equivalent of between 40 and 60: food	[39]
	industry	[36]
	• Beverage and confectionery industry	[37]
	• Used directly as a substrate for the manufacture of	[35]
	fermentation products (such as citric acid, lysine,	
	ethanol or glutamic acid)	
	• Processed into other major starch derivatives, such as	
	isoglucose and fructose syrup	
Modified Starches		
Substituted starches (starch	• Textiles	[38]
esters, ethers, cross-linked	• Paper	[39]
starch)	• Water treatment (flocculation)	[40]
Degraded/converted starches	• Oil industry (fluid loss reducer)	[41]
Roast dextrin, oxidized	• Dextrins: adhesives (gummed paper, bag adhesives,	[32]
starch, thin-boiled starch:	bottle labeling)	[42]
Cross-linked starches	• Textiles (textile fabric finishing, printing)	[43]
	• Acid-modified starches: food industry (sweets) and	[33]
	pharmaceuticals	[44]
	• Oxidized starches: food and paper industry (surface	[45]
	sizing, coating)	[46]
	• Textile industry (fabric finishing, warp sizing)	[47]
	• Enzymatically converted starch: paper industry and	[48]
	fermentation industry	[33]
	• Food industry (desserts, bakery products, soups,	
	sauces)	
	• Textile industry (printing)	
	• Adhesives	
	• Pharmaceuticals	

of textiles and paper, adhesive gums, biodegradable materials and sizing agents (Figure 5.6) [50].

Genetic engineering of starch primarily focuses on altering the ratio of amylose to amylopectin, the two main components of starch. These modifications can have a significant impact on starch composition but do not result in substantial changes in molecular properties. Acetylation is a chemical modification that can make the

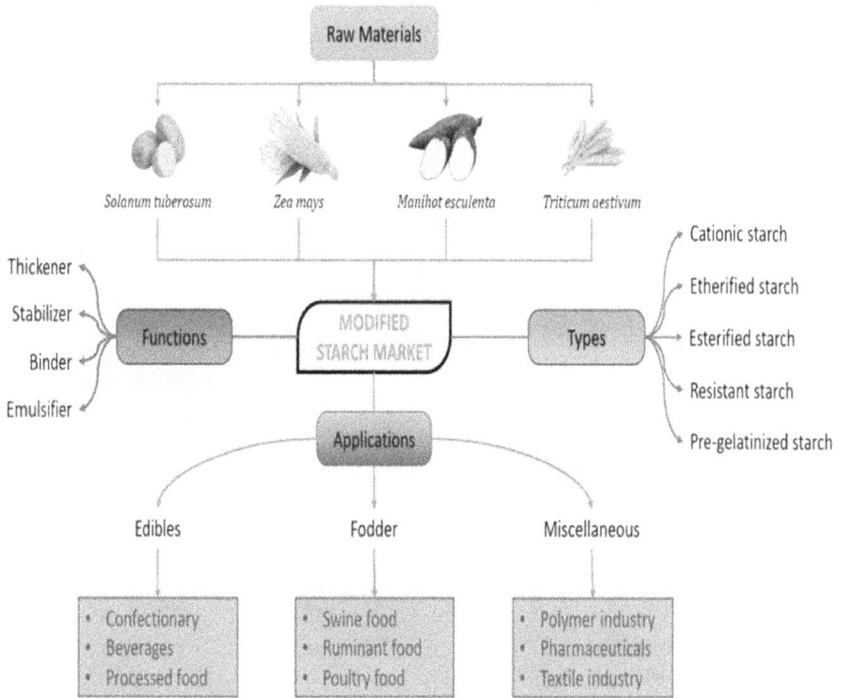

FIGURE 5.6 Modified starch market share by the application. Reproduced from modified starch market size by product 2014–2025 (USD billion). Grand View Research, Inc. [51].

starch more resistant to degradation and dissolution, which is beneficial for applications such as sustained drug release in the upper gastrointestinal tract. Similarly, carboxymethylation is another chemical modification, and by combining these modifications, such as acid-insoluble carboxymethylated starch, with amination, an ionic characteristic can be introduced to the starch [52].

The use of composites is implied by new methods for producing polymers; for instance, starch is added to polypropylene carbonate to make materials that are environmentally friendly and can spontaneously degrade, hence minimizing the impact of the generation of polymeric waste. Additionally, starch exhibits intriguing optical characteristics. For instance, starch granules can produce second-harmonic light [53].

The thermal and hydrophobic characteristics of potato starch (PS) were significantly improved by the grafting of polyorganosiloxane and the subsequent opening of glycosidic rings. The degree of wettability of Al surfaces by PS solution increases as N-[3-(triethoxysilyl) propyl)] hydrolysate is mixed with PS, resulting in a thin, homogeneous solid coating over the Al substrate that is excellent for shielding Al from corrosion [54]. The inhibition process by wheat starch (WS) and the corrosion of mild steel in $0.5\,M\ H_2SO_4$ acid solution are studied, respectively, using weight loss and potentiodynamic polarization measurement techniques. The trend of inhibition efficiency (IE) with temperature, activation energy and heat of adsorption parameters also demonstrates a strong interaction between the WS constituents and the corroding metal surface, suggesting that WS slows the corrosion process by covering the mild steel surface with a chemical adsorption mechanism [55]. This demonstrates the efficiency of a biopolymer blend of starch and pectin as an efficient corrosion inhibitor and demonstrates its capacity to enhance the epoxy resin coating properties on steel during acid corrosion [56].

This research focuses on the prevention of corrosion in C-Mn steel when HCl is present, either as a process fluid or as a product. By modifying maize starch, a bio-copolymer comprising glycerin-grafted starch was created without the toxicological problems associated with traditional synthetic goods. This bio-copolymer was then investigated as a "green" corrosion inhibitor using weight loss procedures and electrochemical techniques [57].

Aluminum corrosion inhibition in 0.3 M HCl was investigated utilizing starch mucor in potassium iodide as an inhibitor at low concentrations. The inhibitor prevented corrosion, and when its concentration was increased, the inhibitor's rate of prevention also increased. When the temperature was increased once the inhibitor (SM in KI) was just slightly incorporated, the efficacy of the inhibition increased. This suggests that the inhibitor (SM in KI) and temperature both affect corrosion control [58].

Dioscorea hispida (Gadong tuber) starch was investigated as a possible green corrosion inhibitor for SAE 1045 carbon steel in 0.6 M NaCl solution. Gadong tubers' raw starch was recovered through the extraction and precipitation procedure before being dried to create fresh starch powder. In order to test the corrosion inhibitor, the insoluble starch powder was then dissolved in a 90% dimethyl sulfoxide solvent. Using potentiodynamic polarization and a corrosive medium containing a 0.6 M NaCl solution, the inhibitive effectiveness of Gadong tuber starch as a green corrosion inhibitor was investigated [59].

The use of new inhibitor starch nanocrystals (SNCs) to inhibit the corrosion of mild steel in 1 M HCl was investigated using potentiodynamic polarization measurements, weight loss techniques and quantum chemical calculations. Density functional theory and quantum chemistry computations help us understand the SNC inhibitory mechanism. The results of the computation of quantum chemical descriptors showed that SNCs exhibit high efficiency in inhibiting corrosion [60].

By grafting acryl amide (AA) with cassava starch (CS), cassava starch–acrylamide graft copolymer (CS-AAGC) was created. It was then tested to understand whether it was an effective inhibitor of 1060 aluminum in $1.0 \, mol \, L/L \, H_3PO_4$ media. Using electrochemical and weight loss techniques, the adsorption behaviors of CS-AAGC and its electrochemical mechanism were studied. Scanning electron microscopy and atomic force microscopy demonstrate that adding CS-AAGC to the media significantly reduces the degree of corrosion and surface roughness of aluminum surfaces [19].

The addition of tapioca starch in this study is intended to increase the AA6061 alloy's resistance to corrosion in seawater. The corrosion behaviors of the AA6061 alloy in saltwater were investigated using gravimetric, potentiodynamic polarization, linear polarization resistance and electrochemical impedance tests. When the AA6061 alloy was electrochemically measured in seawater, it was found that tapioca starch considerably reduces corrosion rates. The way that tapioca starch adheres to metal surfaces was also studied. Tapioca starch precipitates were formed on the metal surface, which slowed the overall corrosion response, according to the examination of scanning electron microscopy and Energy-dispersive X-ray spectroscopy (EDS) [61].

Weight loss and potentiodynamic polarization measurements were used to examine the corrosion inhibition of mild steel in 0.1 M H_2SO_4 in the presence of starch (polysaccharide) in the temperature range of 30°C–60°C. A significant amount of starch significantly slowed down mild steel corrosion. It was also investigated how adding a small amount of cetyl trimethyl ammonium bromide and sodium dodecyl sulfate affected the behaviors of starch in inhibiting corrosion. In the presence of both surfactants, the IE of starch dramatically increased. These starch–surfactant conjugates exhibited 66.21% IE at 300°C with just 200 ppm starch, indicating that they were a mixed type of system (although mostly anodic), according to the results of potentiodynamic polarization. Surfactants seem to affect starch's corrosion-inhibiting behaviors in a synergistic way [62].

The usage of starch in 1 N HCl with 2,6-diphenyl-3-methylpiperidin-4-one (DPMP) using chemical and electrochemical methods for mild steel was investigated. The rate of steel corrosion was shown to decrease as these compounds' concentrations increased; by combining their molecular adsorption at the metal surface, both compounds synergistically reduced steel corrosion. Regardless of the study temperature or the duration of the immersion, the addition of 0.2 mM DPMP significantly improved the corrosion inhibition by starch. The creation of a protective layer, which was supported by Fourier transform infrared spectroscopy, was also attributed by authors as a factor in corrosion inhibition [63].

The effects of physically (activated starch) and chemically (carboxymethylated) altering cassava starch on the corrosion of XC 35 carbon steel in an alkaline 200 mg/l NaCl media were studied using electrochemical impedance spectroscopy. This was confirmed using atomic force microscopy in both the presence and absence of starch. The main reason for corrosion inhibition was found to be inhibitor molecule adsorption produced by starch present at the steel surface. Both examples showed an increase in corrosion inhibition as starch concentration increased; however, carboxymethylated starch functioned less as an inhibitor because its active hydroxyl groups were molecularly substituted [64].

5.3 CASE STUDIES

This study was conducted on the Mediterranean diet lifestyle pattern, which is important to fight against diseases. Resistant starch (RS) is a type of starch which cannot be digested in the small intestine. It increases the number of good bacteria in the gut, which is good for various diseases. It protects red meat from damage, including the decrease in the risk of colonic cancer. RS enters the large intestine, where the microbiota ferments it to produce short-chain fatty acids as the primary end products, which have additional systemic health impacts [65].

In 12 populations from around the world, intakes of starch, non-starch polysaccharides (NSPs), protein and fat have been compared with colon cancer incidence. Starch intake and the likelihood of developing large bowel cancer show significant negative correlations. Even after correcting for fat and protein intakes, the associations between starch, RS and NSPs, and cancer incidence were still statistically significant. The production of fermentation products, such as butyrate, may be the reason for protection against cancer [66]. It is a histone deacetylase inhibitor and a factor in maintaining the healthy function of colorectal mucosa. The enhanced removal of damaged cells and increased repair owing to lower proliferation may be the reason for tumor prevention by resistant starch [67].

Cooked potato starch (CPS) is almost entirely digested in the rat small intestine, whereas raw potato starch (RPS) may evade complete digestion to enter the colon and alter colonic function. In a 6-week feeding study, the effects of RPS and CPS on colonic function—including fecal weight, transmission time and thymidine kinase activity—were compared. According to this research, RPS feeding considerably increases fecal weight while also lengthening overall gastrointestinal transit time [68].

For thousands of patients, hydroxyethyl starch (HES) has been utilized to treat or even prevent hypovolemia. The non-ionic derivative of starch is called HES. In critical care unit patients, HES has become the most often utilized colloid. It has always been a concern that the HES molecule interacts unfavorably with coagulation pathways, impairing blood clotting in the process. Additionally, when HES was administered in particular patient categories, studies in recent years have suggested a substantial risk of acute renal injury and an increase in mortality. The deposition of HES products in different organs rather than their continued circulation is likely the cause of the etiopathogenesis of such deleterious effects [69].

Dewatering operations frequently use geotextile tubes with polyacrylamide flocculants. Sometimes extra flocculant is discharged into the environment during dredging because of fluctuations in solid concentrations, and this could have harmful repercussions. The optimal concentrations of cationic polyacrylamide (CPAM) and cationic starch (C. Starch), an alternative polymer derived from natural sources, were found in this study. Measurements of the residual polymer concentrations were made after treating slurry samples with the recommended dosages of each drug and a 50% overdose. Overdosing C. Starch led to modest residuals (2 ppm), whereas overdosing CPAM led to 17.4 ppm of residual polymer. This strongly suggests that C. Starch should be considered as a replacement for CPAM in dewatering applications [70].

A ZnO lattice was doped with varying amounts of niobium (Nb) (2% and 4%) and a set amount of starch (st). According to the study, ZnO Quantum dots (QDs) improved photocatalytic activity, leading to efficient MB breakdown as well as good antibacterial action. The produced QDs were effective against both Gram-negative and Gram-positive bacteria. The st-ZnO and Nb/st-ZnO nanocomposites from *E. coli* were discovered through molecular docking studies to potentially act as inhibitors of the Dihydrofolate reductase (DHFR) and DNA gyrase enzymes. This study will aid in the creation of binary doped heterostructures for bacterial applications and dye degradation in continuing studies [71].

To examine the effects of starch content in largemouth bass, *Micropterus salmoides*, five diets (D1, D2, D3, D4 and D5) containing 0, 50, 100, 150 and 200 g starch per kg diet were developed. According to the findings, largemouth bass's ability to grow and level of antioxidant protection were both negatively impacted by dietary starch levels above 50 g/kg. The structure and function of the liver may be negatively affected by excessive dietary starch levels, and the number of good gut bacteria may decline [43].

Unmodified starch was investigated as a novel binding polymer for the polymer-enhanced ultrafiltration technique used to remove zinc and lead from aqueous dilute solutions. The effects of transmembrane pressure, pH and metal ion concentration on retention and permeate flux were investigated through experiments. Overall, unmodified starch had superior zinc ion retention over polyethyleneimine, whereas polyethyleneimine demonstrated superior lead ion retention. The pH of the solution was found to have no impact on flux [44].

5.4 FUTURE PREDICTIONS

In terms of microbial counts, it was discovered in the study that all yogurt samples had an increase in total bacteria count after storage. However, the microbial content was significantly reduced when various doses of optimum extracted starch were added. Throughout the whole storage period, no coliform bacteria were found, and the addition of starch had no detrimental effects on the yogurt's flavor, consistency, scent, or general acceptance [72]. This study shows that in the future, starch can be used as an additive which can increase the storage life of food products and the quality of these products would not be hampered. There are numerous papers on the functional modification of starch-based nanoparticles now, but no research has been conducted to determine whether the polar hydroxyl groups of amylose, amylopectin, or SNCs have different functional capabilities. Additionally, once entering the human body, nanoparticles can combine with bodily fluids to create protein crowns [73].

In this work, the inhibitory effectiveness of ferrite nanopowders of Ni, Zn, and Fe in various concentrations of $1\,M\ H_2SO_4$ on the corrosion behaviors of API 5L X80 steel has been examined. It has been discovered that ferrites function as effective steel corrosion inhibitors under specified acidic conditions. The concentrations of ferrite are inhibited by H_2SO_4 in the following order: $0.008 > 0.01 > 0.05 > 0.07 > 1$ M. This corrosive inhibition is sizable and might have a substantial impact on the oil and gas sector [74]. These ferrite-based nanoparticles can be grafted with starch or other biopolymers and can be used as a corrosive inhibitor because, due to their sustainability, economic viability, environmentally friendly approach and wide range of uses, it can be predicted that all the grafted biopolymers will replace both natural and synthetic polymers.

The construction sector is one of the main consumers of metals. Corrosion can occur to metal bars used to construct pillars, frames and other structures. Galvanization is a process used to minimize corrosion, although the use of non-toxic biopolymers has not been reported in this field.

REFERENCES

[1] Christopher, G., Kulandainathan, M.A., & Harichandran, G. (2016). Biopolymers nanocomposite for material protection: Enhancement of corrosion protection using waterborne polyurethane nanocomposite coatings. *Progress in Organic Coatings*, 99, 91–102.
[2] Gräfen, H., Horn, E.-M., Schlecker, H., & Schindler, H. (2011). Corrosion 1. Electrochemical. In *Ullmann's Encyclopedia of Industrial Chemistry*, (Ed.). doi:10.1002/14356007.b01_08.pub2.

[3] Bothi Raja, P., Fadaeinasab, M., Qureshi, A.Z., Rahim, A.A., Osman, H., Litaudon, M., & Awang, K. (2013). Evaluation of green corrosion inhibition by alkaloid extracts of ochrosia oppositifolia and isoreserpiline against mild steel in 1 M HCl medium. *Industrial & Engineering Chemistry Research*, 52(31), 10582–10593.

[4] Charitha, B.P., & Rao, P. (2017). Carbohydrate biopolymer for corrosion control of 6061 Al-alloy and 6061aluminum-15% (v) SiC (P) composite-green approach. *Carbohydrate Polymers*, 168, 337–345.

[5] Muthulakshmi, L., Seghal Kiran, G., Ramakrishna, S., Cheng, K.Y., Ramachandran, R.A., Mathew, M.T., & Pruncu, C.I. (2023). Towards improving the corrosion resistance using a novel eco-friendly bioflocculant polymer produced from Bacillus sp. *Materials Today Communications*, 35, 105438–105438.

[6] Bereket, G., Hür, E., & Öğretir, C. (2002). Quantum chemical studies on some imidazole derivatives as corrosion inhibitors for iron in acidic medium. *Journal of Molecular Structure: THEOCHEM*, 578, 79–88.

[7] Abd El-rehim, S.S., Refaey, S.A.M., Taha, F., Saleh, M.B., & Ahmed, R.A. (2001). Corrosion inhibition of mild steel in acidic medium using 2-amino thiophenol and 2-cyanomethyl benzothiazole. *Journal of Applied Electrochemistry*, 31, 429–435

[8] Tasic, Z.Z., Mihajlovic, M.B.P., & Antonijevi, M.M. (2016). The influence of chloride ions on the anti-corrosion ability of binary inhibitor system of 5-methyl-1H-benzotriazole and potassium sorbate in sulfuric acid solution. *Journal of Molecular Liquids*, 222, 1–7

[9] Herrag, L., Chetouani, A., Elkadiri, S., Hammouti, B., & Aouniti, A. (2008). Pyrazole derivatives as corrosion inhibitors for steel in hydrochloric acid. *Portugaliae Electrochimica Acta*, 26, 211–220.

[10] Elewady, G.Y. (2008). Pyrimidine derivatives as corrosion inhibitors for carbon-steel in 2M hydrochloric acid solution. *International Journal of Electrochemical Science*, 3, 1149–1161

[11] Azzouzi, M.E., Aouniti, A., Tighadouin, S., Elmsellem, H., Radi, S., Hammouti, B., Assyry, A.E., Bentiss, F., & Zarrouk, A. (2016). Some hydrazine derivatives as corrosion inhibitors for mild steel in 1.0M Cl: Weight loss, electrochemical. SEM and theoretical studies. *Journal of Molecular Liquids*, 221, 633–641

[12] Saha, S.K., Dutta, A., Ghosh, P., Sukul, D., & Banerje, P. (2016)Novel Schiff base molecules as efficient corrosion inhibitors for mild steel surface in 1 M HCl medium: Experimental and theoretical approach. *Physical Chemistry Chemical Physics Journal*, 18, 17898–17911

[13] Ashassi-Sorkhabi, H., Majidi, M.R., & Seyyedi, K. (2004). Investigation of inhibition effect of some amino acids against steel corrosion in HCl solution. *Applied Surface Science*, 225, 176–185

[14] Abdel-Gaber, A.M., Masoud, M.S., Khalil, E.A., & Shehata, E.E. (2009). Electrochemical study on the effect of Schiff base and its cobalt complex on the acid corrosion of steel. *Corrosion Science*, 51, 3021–3024.

[15] Singh, P., Singh, A.K., & Singh, V.P. (2013). Synthesis, structural and corrosion inhibition proper ties of some transition metal(II) complexes with o-hydroxyacetophenone-2 thiophenoyl hydrazine. *Polyhedron*, 65, 73–81.,

[16] Mishra, M., Tiwari, K., Singh, A.K., & Singh, V.P. (2015). Versatile coordination behaviour of a multi-dentate Schiff base with manganese(II), copper(II) and zinc(II) ions and their corrosion inhibition study. *Inorganica Chimica Acta*, 425, 36–45.

[17] Manahan, S.E. (1994). *Environmental Science, Technology, and Chemistry, Environmental Chemistry.* CRC Press, New York, 1994

[18] Biswas, A., Das, D., Chung, I.-M., Pal, S., & Nair, U.G. (2019). Biopolymer dextrin and poly (vinyl acetate) based graft copolymer as an efficient corrosion inhibitor for mild steel in hydrochloric acid. *Electrochemical, Surface Morphological and Theoretical Studies*, 275, 867–878.

[19] Deng, S., Li, X., & Du, G. (2021). An efficient corrosion inhibitor of cassava starch graft copolymer for aluminum in phosphoric acid. *Chinese Journal of Chemical Engineering*, 37, 222–231.

[20] Umoren, S.A., & Eduok, U.M. (2016). Application of carbohydrate polymers as corrosion inhibitors for metal substrates in different media: A review, *Carbohydrate Polymers*, 140, 314–341.

[21] Van Riemsdijk, L.E., van der Goot, A.J., Hamer, R.J., & Boom, R.M. (2011). Preparation of gluten-free bread using a meso-structured whey protein particle system. *Journal of Cereal Science*, 53, 355–361. doi:10.1016/j.jcs.2011.02.006.

[22] Miranda, J., Lasa, A., Bustamante, M.A., Churruca, I., & Simon, E. (2014). Nutritional differences between a gluten-free diet and a diet containing equivalent products with gluten. *Plant Foods for Human Nutrition*, 69, 182–187. doi:10.1007/s11130-014-0410-4.

[23] Milde, L.B., Ramallo, L.A., & Puppo, M.C. (2012). Gluten-free bread based on tapioca starch: Texture and sensory studies. *Food and Bioprocess Technology*, 5, 888–896. doi:10.1007/s11947-010-0381-x.

[24] López, A.C.B., Pereira, A.J.G., & Junqueira, R.G. (2004). Flour mixture of rice flour, corn and cassava starch in the production of gluten-free white bread, *Brazilian Archives of Biology and Technology*, 47, 63–70. doi:10.1590/S1516-89132004000100009.

[25] Sabanis, D., Lebesi, D., & Tzia, C. (2009). Effect of dietary fibre enrichment on selected properties of gluten-free bread. *LWT - Food Science and Technology*, 42, 1380–1389. doi:10.1016/j.lwt.2009.03.010.

[26] Majeed, T., Dar, A.H., Pandey, V.K., Dash, K.K., Srivastava, S., Shams, R., G. Jeevarathinam, Singh, P., Echegaray, N., & Ravi Pandiselvam. (2023). Role of additives in starch-based edible films and coating: A review with current knowledge. *Progress in Organic Coatings*, 181, 107597–107597.

[27] Xu, Y.X., Kim, K.M., Hanna, M.A., & Nag, D. (2005). Chitosan-starch composite film: Preparation and characterization. *Industrial Crops and Products*, 21(2), 185–192.

[28] Klein, B., Pinto, V.Z., Vanier, N.L., Zavareze, E. da R., Colussi, R., Evangelho, J.A. do, Gutkoski, L.C., & Dias, A.R.G. (2013). Effect of single and dual heat-moisture treatments on properties of rice, cassava, and pinhao starches. *Carbohydrate Polymers*, 98(2), 1578–1584

[29] Wani, I.A., Jabeen, M., Geelani, H., Masoodi, F.M., Saba, I., & Muzaffar, S. (2014). Effect of gamma irradiation on physicochemical properties of Indian Horse Chestnut (*Aesculus indica* Colebr.) starch. *Food Hydrocolloids*, 35, 253–263.

[30] Sofi, B.A., Wani, I.A., Masoodi, F.A., Saba, I., & Muzaffar, S. (2013). Effect of gamma irradiation on physicochemical properties of broad bean (*Vicia faba* L.) starch. *LWT - Food Science and Technology*, 54(1), 63–72

[31] Shahid ul, I., Shahid, M., & Mohammad, F. (2013). Perspectives for natural product based agents derived from industrial plants in textile applications—a review. *Journal of Cleaner Production*, 57, 2–18.

[32] Herrero, A.M., Carmona, P., Jiménez-Colmenero, F., Ruiz-Capillas, C. (2014). Polysaccharide gels as oil bulking agents: Technological and structural properties. *Food Hydrocolloids*, 36, 374–381.

[33] Daudt, R.M., Külkamp-Guerreiro, I.C., Cladera-Olivera, F., Thys, R.C.S., & Marczak, L.D.F. (2014). Determination of properties of pinhão starch: Analysis of its applicability as pharmaceutical excipient. *Industrial Crops and Products*, 52, 420–429.

[34] Homayouni, A., Azizi, A., Ehsani, M.R., Yarmand, M.S., & Razavi, S.H. (2008). Effect of microencapsulation and resistant starch on the probiotic survival and sensory properties of synbiotic ice cream. *Food Chemistry*, 111(1), 50–55.

[35] Liu, Y., Bhandari, B., & Zhou, W. (2007). Study of glass transition and enthalpy relaxation of mixtures of amorphous sucrose and amorphous tapioca starch syrup solid by differential scanning calorimetry (DSC). *Journal of Food Engineering*, 81(3), 599–610.

[36] Rosicka-Kaczmarek, J., Kwaśniewska-Karolak, I., Nebesny, E., & Miśkiewicz, K. (2013). Influence of variety and year of wheat cultivation on the chemical composition of starch and properties of glucose hydrolysates. *Journal of Cereal Science*, 57(1), 98–106.

[37] Lu, Z., He, F., Shi, Y., Lu, M., & Yu, L. (2010). Fermentative production of L(+)-lactic acid using hydrolyzed acorn starch, persimmon juice and wheat bran hydrolysate as nutrients. *Bioresource Technology*, 101(10), 3642–3648.

[38] Poomipuk, N., Reungsang, A., & Plangklang, P. (2014). Poly-β-hydroxyalkanoates production from cassava starch hydrolysate by Cupriavidus sp. KKU38. *International Journal of Biological Macromolecules*, 65, 51–64.

[39] Becerra, V., & Odermatt, J. (2014). Direct determination of cationic starches in paper samples using analytical pyrolysis. *Journal of Analytical and Applied Pyrolysis*, 105, 348–354.

[40] Letelier-Gordo, C. O., Holdt, S. L., De Francisci, D., Karakashev, D. B., & Angelidaki, I. (2014). Effective harvesting of the microalgae Chlorella protothecoides via bioflocculation with cationic starch. *Bioresource Technology*, 167, 214–218.

[41] Li, M., Liu, G.-L., Chi, Z., Chi, & Z.-M. (2010). Single cell oil production from hydrolysate of cassava starch by marine-derived yeast Rhodotorula mucilaginosa TJY15a. *Biomass and Bioenergy*, 34(1), 101–107.

[42] Setty, C.M., Deshmukh, A.S., & Badiger, A.M. (2014). Hydrolyzed polyacrylamide grafted maize starch based microbeads: Application in pH responsive drug delivery. *International Journal of Biological Macromolecules*, 70, 1–9.

[43] Yang, Z., Yuan, B., Li, H., Yang, Y., Yang, H., Li, A., & Cheng, R. (2014). Amphoteric starch-based flocculants can flocculate different contaminants with even opposite surface charges from water through molecular structure control. *Colloids and Surfaces A: Physicochemical and Engineering Aspects*, 455, 28–35.

[44] Çatal, H., & İbanoğlu, Ş. (2014). Effect of aqueous ozonation on the pasting, flow and gelatinization properties of wheat starch. *LWT - Food Science and Technology*, 59(1), 577–582.

[45] van der Maarel, M.J.E.C., & Leemhuis, H. (2013). Starch modification with microbial alpha-glucanotransferase enzymes. *Carbohydrate Polymers*, 93(1), 116–121.

[46] Gamonpilas, C., Pongjaruvat, W., Methacanon, P., Seetapan, N., Fuongfuchat, A., & Klaikherd, A. (2013). Effects of cross-linked tapioca starches on batter viscosity and oil absorption in deep-fried breaded chicken strips. *Journal of Food Engineering*, 114(2), 262–268.

[47] Lay, C.-H., Kuo, S.-Y., Sen, B., Chen, C.-C., Chang, J.-S., & Lin, C.-Y. (2012). Fermentative biohydrogen production from starch-containing textile wastewater. *International Journal of Hydrogen Energy*, 37(2), 2050–2057.

[48] Czech, Z., Wilpiszewska, K., Tyliszczak, B., Jiang, X., Bai, Y., & Shao, L. (2013). Biodegradable self-adhesive tapes with starch carrier. *International Journal of Adhesion and Adhesives*, 44, 195–199.

[49] Agrosynergie—Description of the Sector (2010). *Evaluation of Common Agricultural Policy Measures Applied to the Starch Sector E*. Commission, Agrosynergie, pp. 17–80.

[50] Santelia, D., & Zeeman, S.C. (2011). Progress in arabidopsis starch research and potential biotechnological applications. *Current Opinion in Biotechnology*, 22(2), 271–280.

[51] Grand View Research, Inc. (2021). Modified Starch Market Size, by Product 2014–2025 (USD Billion). Global Modified Starch Market Size, Share Report, 2019–2025.

[52] Ab'lah, N., Yu, C., Rojsitthisak, P., & Wong, T. (2023). Reinvention of starch for oral drug delivery system design. *International Journal of Biological Macromolecules*, 241, 124506–124506.

[53] Li, M., Liu, P., Zou, W., Yu, L., Xie, F., Pu, H., Liu, H., & Chen, L. (2011). Extrusion processing and characterization of edible starch films with different amylose contents. *Journal of Food Engineering*, 106(1), 95–101.

[54] Toshifumi Sugama, & DuVall, J.E. (1996). Polyorganosiloxane-grafted potato starch coatings for protecting aluminum from corrosion. *Thin Solid Films*, 289(1–2), 39–48.

[55] Al-Moubaraki, A.H., Ganash, A.A., & Al-Malwi, S.D. (2020). Investigation of the corrosion behavior of mild steel/H2SO4 systems. *Moroccan Journal of Chemistry*, 8(1).

[56] Sushmitha, Y., & Rao, P. (2019). Material conservation and surface coating enhancement with starch-pectin biopolymer blend: A way towards green. *Surfaces and Interfaces*, 16, 67–75

[57] Sihem Lahrour, A.B., Bouras, B., Asma Mansri, Lahcene Tannouga, & Marzorati, S. (2019). Glycerin-grafted starch as corrosion inhibitor of c-mn steel in 1 M HCl solution. *Applied Sciences*, 9(21), 4684–4684.

[58] Ezeamaku, U.L., Eze, I., Odimegwu, N., Nwakaudu, A., Okafor, A., Onukwuli, O.D., & Nnanwube, I.A. (2023). Corrosion inhibition of aluminium in 0.3 M HCl using starch mucor in potassium iodide as inhibitor. *Pigment & Resin Technology*. doi: 10.1108/PRT-12-2022-0152.

[59] Othman, N.K., Salleh, E.M., Dasuki, Z., & Lazim, A.M. (2018). Dimethyl sulfoxide-treated starch of dioescorea hispida as a green corrosion inhibitor for low carbon steel in sodium chloride medium. In: Mahmood, A. (Ed.), *Corrosion Inhibitors, Principles and Recent Applications*. Rijeka: IntechOpen.

[60] Thamer, A.N., Kadham, L.H., & Alwaan, I.M. (2022). Preparation and characteristics of starch nanocrystals as corrosion inhibitor. *Experimental and Theoretical Studies*, 34(11), 2854–2864.

[61] Rosliza, R., & Wan Nik, W.B. (2010). Improvement of corrosion resistance of AA6061 alloy by tapioca starch in seawater. *Current Applied Physics*, 10(1), 221–229.

[62] Mobin, M., M. Ijaz Khan, & Parveen, M. (2011). Inhibition of mild steel corrosion in acidic medium using starch and surfactants additives. *Journal of Applied Polymer Science,* 121(3), 1558–1565.

[63] Brindha, T., Mallika, J., & Sathyanarayana, M.V. (2015). Synergistic effect between starch and substituted piperidin-4-one on the corrosion inhibition of mild steel in acidic medium. *Journal of Materials and Environmental Science*, 6, 191–200

[64] Bello, M., Ocho, N., Balsamo, V., López-Carrasquero, F., Coll, S., Monsalve, A., & González, G. (2010). Modified cassava starches as corrosion inhibitors of carbon steel: An electrochemical and morphological approach. *Carbohydrate Polymers*, 82, 561–568

[65] Cione, E., Fazio, A., Curcio, R., Tucci, P., Lauria, G., Cappello, A.R., & Dolce, V. (2021). Resistant starches and non-communicable disease: A focus on mediterranean diet. *Foods*, 10(9), 2062.

[66] Cassidy, A., Bingham, S., & Cummings, J. (1994). Starch intake and colorectal cancer risk: An international comparison. *British Journal of Cancer*, 69(5), 937–942.

[67] Mathers, J.C., Smith, H., & Carter, S.E. (1997). Dose-response effects of raw potato starch on small-intestinal escape, large-bowel fermentation and gut transit time in the rat. *British Journal of Nutrition*, 78(6), 1015–1029.

[68] Calvert, R.J., Otsuka, M., & Satchithanandam, S. (1989). consumption of raw potato starch alters intestinal function and colonic cell proliferation in the rat. *The Journal of Nutrition*, 119(11), 1610–1616.

[69] Kumar, S. (2014). Clinical use of hydroxyethyl starch and serious adverse effects: Need for awareness amongst the medical fraternity. *Medical Journal, Armed Forces India*, 70(3), 209–210.

[70] Duggan, K.L., Morris, M., Bhatia, S.K., Khachan, M.M., & Lewis, K.E. (2019). Effects of cationic polyacrylamide and cationic starch on aquatic life. *Journal of Hazardous, Toxic, and Radioactive Waste*, 23(4), 04019022.

[71] Ikram, M., Shahid, H., Haider, J., Haider, A., Naz, S., Ul-Hamid, A., Shahzadi, I., Naz, M., Nabgan, W., & Ali, S. (2022). Nb/starch-doped ZnO nanostructures for polluted water treatment and antimicrobial applications: Molecular docking analysis. *ACS Omega*, 7(43), 39347–39361

[72] Altemimi, A. (2018). Extraction and optimization of potato starch and its application as a stabilizer in yogurt manufacturing. *Foods*, 7(2), 14

[73] Yu, M., Ji, N., Wang, Y., Dai, L., Xiong, L., & Sun, Q. (2020). Starch-based nanoparticles: Stimuli responsiveness, toxicity, and interactions with food components. *Comprehensive Reviews in Food Science and Food Safety*, 20(1), 1075–1100.

[74] Usman, C., Bhola, R., Mittal, V., & Mishra, B. (2014). Ni0. 5Zn0. 5Fe2O4 as a potential corrosion inhibitor for API 5L X80 steel in acidic environment. *International Journal of Electrochemical Science*, 9, 4478–4492.

6 Cellulose and Cellulose Derivatives as Corrosion Inhibitors

Omar Dagdag and Hansang Kim
Gachon University

Rajesh Haldhar
Yeungnam University

Elyor Berdimurodov
National University of Uzbekistan

ABBREVIATIONS

FTIR	Fourier transform infrared
TGA	Thermal gravimetric analysis
DLS	Dynamic light scattering
SEM-EDS	Scanning electron microscopy and energy dispersive X-ray
PDP	Potentiodynamic polarization
EIS	Electrochemical impedance spectroscopy
CMC	Carboxymethyl cellulose
MCC	Microcrystalline cellulose
HEC	Hydroxyethylcellulose
EC	Ethylcellulose
MS	Mild steel
LPR	Linear polarization resistance
WL	Weight loss
CS	Carbon steel
EFM	Electrochemical frequency modulation
Cdl	Double layer capacitance
AHEC	Aminated hydroxyethyl cellulose
XRD	X-ray diffraction
SDS	Sodium dodecyl sulfate

DOI: 10.1201/9781003400059-6

6.1   INTRODUCTION

To inhibit corrosion, polymers are preferred over organic chemicals because of their high number of adsorption sites, capacity to form complexes with metallic ions, affordability, and environmental friendliness [1]. By adhering to the metal surface, these compounds create a barrier film that protects the metal from corrosion. Additionally, polymers are preferable substitutes for both inorganic and organic inhibitors. Research on corrosion inhibition has focused on using biopolymers as green chemicals due to their non-toxic, environmentally beneficial, and biodegradable corrosion inhibition characteristics, whereas synthetic polymers are non-biodegradable. Many studies have discussed how biopolymers are used to prevent corrosion [2,3].

Cellulose is one of the most affordable and widely accessible biopolymers found in nature [4]. Under acidic conditions, cellulose can be utilized as a composite to prevent MS deterioration [5]. Because of their chemical reactivity, cellulose derivatives can be used as corrosion inhibitors [6].

Hemp, cotton, and wood contain significant amounts of cellulose (up to 57%, 90%, and 40%–45%, respectively), a natural polymer that is rich in carbohydrates. In cellulose, $\beta \rightarrow 1$–4-glycosidic binding is used to join glucose units together (Figure 6.1) [7]. The molecular structure of cellulose changes to $(C_5H_{10}O_5)_n$ [8]. Anselme Payen, a French scientist [9], discovered cellulose, while Hermann Staudinger established its polymer structure in 1920 [9]. Kobayashi and Shoda achieved the first chemical synthesis of cellulose in 1992 [9]. The restricted solubility of cellulose in polar liquids prevents its use in a wide range of biological and commercial applications [10]. Hence, several investigations have been conducted to increase its solubility in these solutions. One such effort involves the use of exogenous additives such as ionic solutions. Another popular method is chemical modification by adding polar substituents to the molecular structure of cellulose. The two most significant cellulose derivatives, HEC and CMC, have been used extensively in a variety of industrial applications, including biological, medicinal, and anticorrosive ones [11]. In polar liquids, HEC and CMC show relatively high solubility.

Although chemically unaltered cellulose has a range of electron-rich polar functional groups, such as adsorption centers, its solubility limits its potential use as a metal corrosion inhibitor. Hence, the majority of cellulose's applications in corrosion inhibition are in the form of anticorrosive composites [2]. Additionally, corrosion experts and engineers strive to use cellulose as an inhibitor of aqueous-phase corrosion through correct functionalization due to its high availability and environmental friendliness. HEC and CMC are used as aqueous-phase inhibitors in several electrolyte experiments for this reason [11].

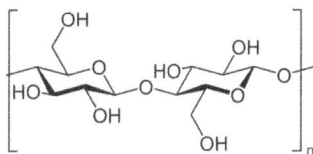

FIGURE 6.1   Chemical structure of cellulose. Reprinted with permission from [11]. © 2022 Elsevier Publications.

6.2 USE OF CELLULOSE AND ITS DERIVATIVES TO PREVENT CORROSION

As a sodium salt, cellulose gum, often known as CMC, is readily available. Although it has the same structural characteristics as regular cellulose, its cellulosic -gluco-pyranosyl moiety has reactive carboxymethyl groups connected to hydroxyl groups (Figure 6.2).

CMC, which is derived from molecular cellulose, is one of the most prevalent water-insoluble polysaccharides used in industries where exudate gums are substituted because they are superior binders, thickeners, and stabilizers for food and pharmaceuticals [12].

Alkali-assisted cellulose/chloroacetic acid reaction is a common method used to synthesize CMC. The protective property of CMC on steel in acid-induced corrosion prevention is attributable to the potential physisorption of protonated CMC via molecular interaction with the negatively charged mild electrode by means of weak adsorption by hydrated ions of electrolyte anions. CMC protonation takes place most frequently in the carbonyl group, typically generating polycations. For MS in various acid solutions, the anticorrosive characteristics of CMC have been extensively investigated.

Using WL, hydrogen evolution, and thermometric methods, Solomon et al. [13] evaluated the inhibitory and adsorption behavior of CMC, using the chemical structure of the repeat unit shown in Figure 6.3a as acquired in the concentration ranging from 0.1 to 0.5 g/l for MS in H_2SO_4 solution at 30°C–60°C. The inhibitory efficacy

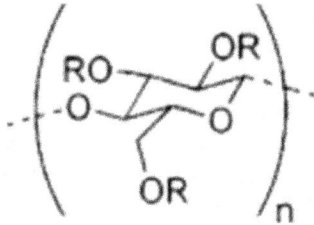

FIGURE 6.2 Molecular structure of carboxymethyl cellulose [R=H or CH2COOH] and hydroxyethylcellulose [R=H or CH2CH2OH].

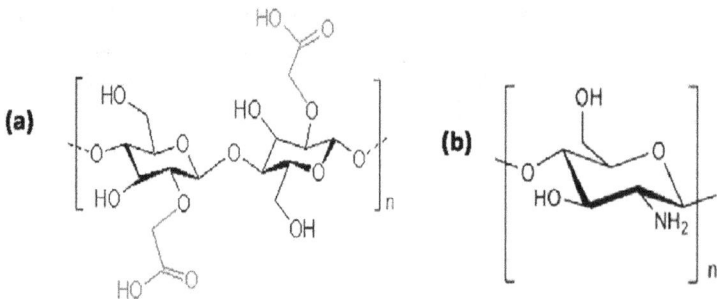

FIGURE 6.3 Chemical structure of (a) CMC and (b) chitosan repeat unit. Reprinted with permission from [12,13]. © 2010 Elsevier Publications.

was found to be increased with an increase in CMC concentration but decreased with an increase in temperature, suggesting a physical adsorption mechanism.

In another study, Solomon et al. [14] assessed the corrosion-preventative properties of two single-component, environmentally friendly inhibitor substances, chitosan and CMC. Figure 6.3b depicts the chemical compositions of both these substances.

Using EIS and PDP techniques, as well as surface morphology description of corroded steel samples with and without inhibitors, the performance of these two natural polymers, chitosan and CMC, as single-component corrosion inhibitors on API 5L X60 pipeline steel was determined in comparison with commercial inhibitor formulations. The findings showed that there is a significant variation between the abilities of both inhibitors to inhibit API 5L X60 steel and commercial inhibitor formulations. As inhibitor concentrations increased, so did the effectiveness of the inhibition. It was proposed that the abilities of all inhibitors to control corrosion were significantly affected by the length of their immersion. Additionally, it was shown that an increase in temperature reduces the inhibitory efficiency. PDP results showed that all inhibitors exhibit mixed-type inhibition.

In line with Hasanin et al. [15], the aim of the present work is to investigate the synergistic effects of two naturally occurring, environmentally friendly, non-toxic, biodegradable, and economically viable corrosion inhibitors for copper corrosion in a 3.5% NaCl solution which were synthesized in situ using various cellulosic materials and niacin. The chemical makeup of niacin and the repeating units of MCC, EC, and CMC are shown in Figure 6.4.

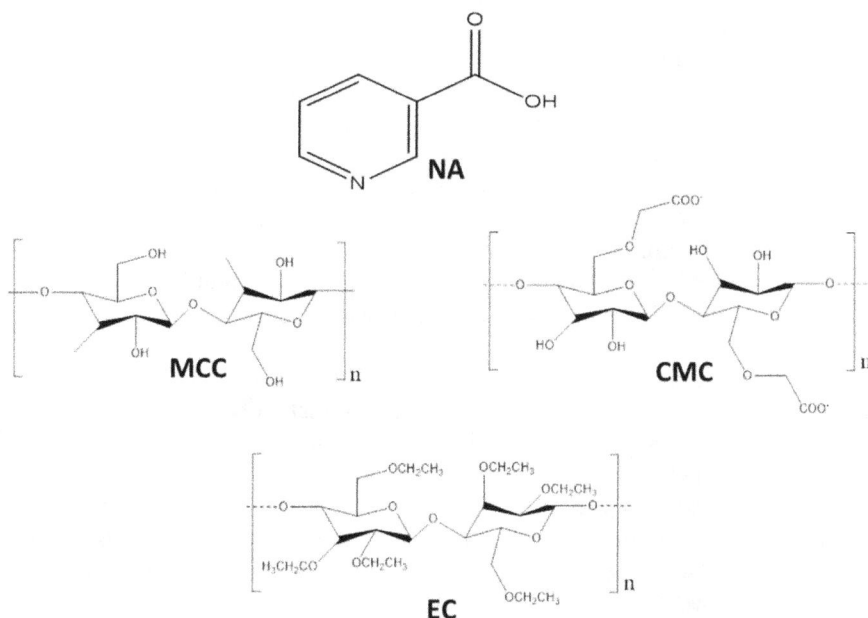

FIGURE 6.4 The chemical structure of niacin and repeat units of MCC, EC, and CMC. Reprinted with permission from [14]. © 2016 Elsevier Publications.

Cellulose nanocomposites were characterized using FTIR, TGA, DLS, and SEM-EDS. Using PDP and EIS methods, the anticorrosive capabilities of these nanocomposites for copper in the 3.5% NaCl solution were assessed. Using SEM and EDS, the surface morphology of uninhibited and inhibited composites was investigated. PDP measurements demonstrated the mixed-type inhibitory action of cellulose-based inhibitors. When used as a green corrosion inhibitor for copper in the 3.5% NaCl solution, the ethyl cellulose–niacin composite outperformed, with an efficiency of 94.7%, both MCC–niacin composite and CMC–niacin composite, whose inhibition efficiencies were 33.2% and 83.4%, respectively.

Bayol et al. [16] reported similar results in 1 M HCl using chemical (WL method) and electrochemical (PDP, LPR, and EIS) approaches. They reported that the inhibition of MS corrosion was concentration dependent, and CMC was identified as a mixed-type inhibitor.

Using electrochemical and chemical techniques, Rajeswari et al. [17] evaluated the corrosion inhibition of hydroxypropyl cellulose, gellan gum, and glucose for cast iron in 1 M HCl. They reported that both temperature and CMC concentration affect corrosion inhibition. Cast iron was better protected at higher inhibitor concentrations, lower temperatures, and longer immersion times. The inhibitory properties of CMC were found to be of mixed type, and its adsorption followed the Langmuir adsorption isotherm. Thermodynamic and kinetic characteristics suggested a physical adsorption process.

Recently, electrochemical investigations of aluminum and aluminum silicon alloys demonstrated the corrosion inhibition of methyl cellulose in 0.1 M NaOH [15]. PDP results showed that this chemical system's inhibitory activity mostly affected the anodic process, indicating that methyl cellulose was an anodic-type inhibitor. The adsorption of this substance at the metal surface via a variety of locations and the physical displacement of corrosive molecules across the metal–solution contact were attributable to the mechanisms by which it inhibited corrosion. In addition, the alloys' corrosion decreased with methyl cellulose concentration but not with temperature. To further clarify the process of molecule adsorption, kinetic/thermodynamic parameters were evaluated to show that cellulose's physisorption at the metal surface followed the Langmuir adsorption isotherm.

In medicine, HEC, which is also generated from cellulose, is utilized for gastrointestinal fluid medication dissolution and is largely used as a thickening, binding, and gelling/stabilizing substance.

Figure 6.2 shows the molecular structure of HEC. Its structural characteristics are identical to those of CMC, with the exception that carboxymethyl groups are replaced with hydroxyethyl groups while remaining in the same location on the cellulose glucopyranyl moiety. These distinct functional groups (OH and COOH) on the cellulose backbone of this carbohydrate polymer and its high molecular size, which leads to wider coverage of the metal surface and inhibits corrosive ions and molecules, are the reasons why cellulose is a successful corrosion inhibitor under various conditions.

El-Haddad et al. [19] reported a corrosion inhibition efficacy (%η) in the range of 97% for 5 mM HEC for 1018 grade CS in 3.5 wt.% NaCl solution. Using EIS, PDP, and EFM, the anticorrosion capabilities of this carbohydrate polymer were investigated.

Under these experimental conditions, electrochemical polarization results showed that HEC was a mixed-type inhibitor, and SEM-EDS analysis demonstrated that corrosion inhibition was attributable to molecule adsorption on the steel surface.

Arukalam et al. [16] conducted a similar study using WL, EIS, and PDP methods for MS corrosion in an aerated 0.5 M H_2SO_4 solution. They reported that both temperature and HEC content increased corrosion inhibition. PDP results showed that HEC is a mixed-type inhibitor and that its adsorption on the steel surface gradually reduced C_{dl}.

Sangeetha et al. [17] synthesized AHEC and evaluated it using FTIR and chemical and electrochemical tests to determine its corrosion inhibition on MS in 1 M HCl. Figure 6.5 shows the synthesis mechanism of AHEC.

As the concentration of AHEC increased, the effectiveness of corrosion inhibition increased. At room temperature, an IE of up to 93% was achieved for 900 ppm of AHEC. Hence, AHEC showed a higher IE than other cellulose derivatives that have been attempted thus far. As the temperature increased from 303 K to 323 K, the corrosion inhibition capacity decreased. According to PDP results, AHEC functions as a mixed-type inhibitor. It was also found that the inhibitor's adsorption on the MS

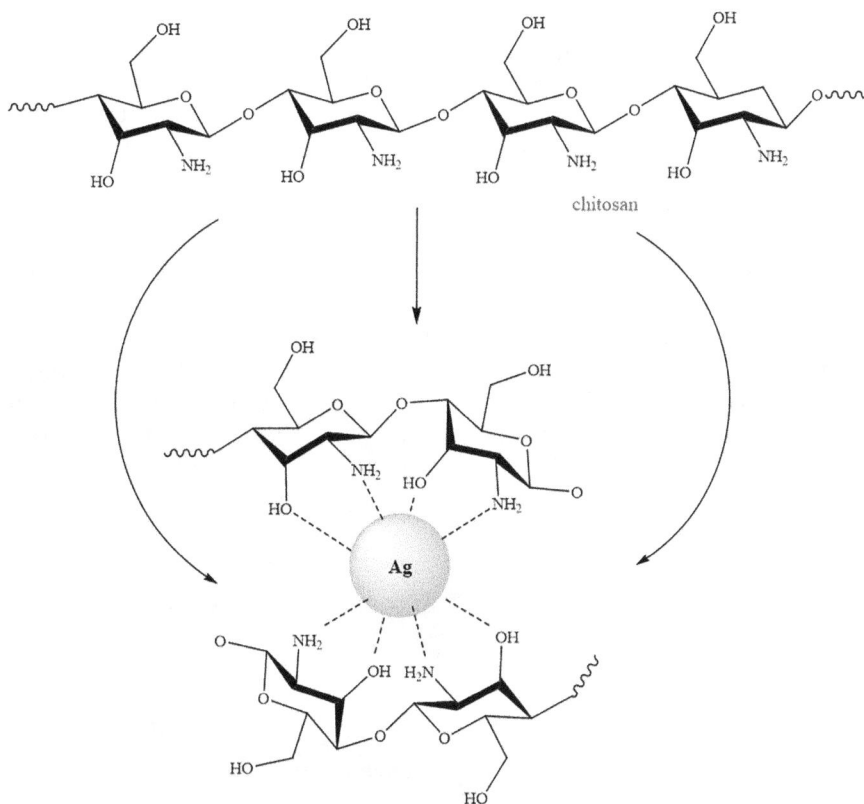

FIGURE 6.5   Synthesis of AHEC. Reprinted with permission from [17]. © 2021 Elsevier Publications.

surface follows the Frumkin adsorption isotherm and is based on a physical process. Impedance graphs showed that the Rct increased as the concentration of the inhibitor increased. SEM and AFM studies demonstrated that the inhibitor formed a protective coating on the MS surface.

Al Kiey et al. [18] synthesized cellulose tetrazole (CTZ) and reported it as a safe corrosion inhibitor for CS in 1 M HCl solution. Figure 6.6 shows the synthesis mechanism of CTZ.

FTIR spectral methods, SEM, XRD, and TGA analyses were used to characterize the synthesized CTZ. CS corrosion in 1 M HCl was further examined for the inhibitory effect of CTZ using WL, PDP, and EIS methods. A combination of SEM-EDS was used to describe the surface of CS both with and without CTZ inhibitors. UV spectroscopy was used to investigate corrosive solutions. The observed spectra supported the adsorption of these compounds on the iron surface, resulting in the formation of a Fe–inhibitor adsorption layer. Electrochemical tests revealed that CTZ has a maximum effectiveness of 94.2% at 100 ppm when used as a high-performance, environmentally friendly CS inhibitor in 1 M HCl medium. According to PDP measurements, the development of CTZ adsorption films, which followed the Langmuir adsorption process, is responsible for positive outcomes. The effective corrosion inhibition performance was attributable to the production of CTZ adsorption films, which occurred after the Langmuir adsorption isotherm. SEM and roughness pictures showed that CTZ can successfully prevent acid assault by forming an adsorbed coating on the CS surface. In electrochemical tests, CTZ was found to be a mixed-type corrosion inhibitor that functions mostly as a cathodic inhibitor.

While investigating the inhibition potential of HEC for MS corrosion in 1 M HCl using chemical, surface analysis, and electrochemical approaches, Mobin and Rizvi [19] found that the inhibition potential of HEC may be synergistically increased using external additives. A WL investigation revealed that the concentration of HEC is necessary for its inhibitory capacity and an increase in HEC concentration increases its protection potential. The highest IE of 85.8% was observed for HEC at 500 ppm. However, the inhibitory potential of HEC (at 500 ppm) was found to be synergistically increased to 90.77%, 91.07%, and 91.62%, respectively, in the presence

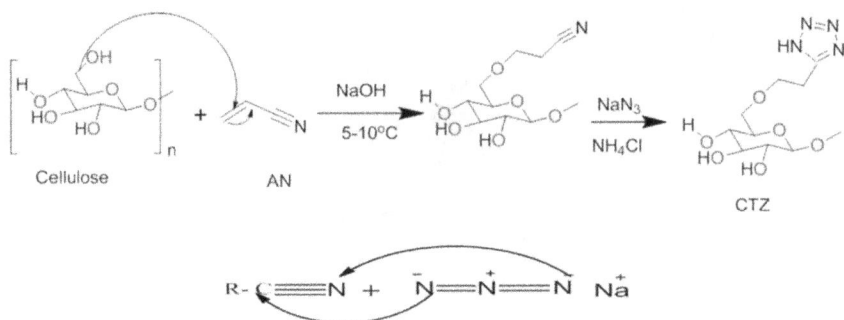

FIGURE 6.6 Plausible mechanisms of CTZ synthesis. Reprinted with permission from [18]. © 2021 Elsevier Publications.

of 5 ppm concentrations of SDS, PCP, and XT. With or without SDS, CPC, or XT, an increase in HEC concentration resulted in a corresponding increase in metallic surface covering.

As shown in PDP investigations, HEC had a significant impact on both anodic and cathodic Tafel curves and constants. HEC acts as a mixed-type corrosion inhibitor at all of the investigated doses, both with and without SDS, CPC, and XT. According to EIS results, HEC acts as an interface-type inhibitor both in the presence and in the absence of SDS, CPC, and XT. SEM and AFM investigations showed that the adsorption manner of corrosion inhibition was strengthened. It was also shown that XT and HEC together significantly enhance the surface morphology of SEM as well as AFM pictures. Figure 6.7 shows AFM images of MS surfaces that have been polished and corroded in 1 M HCl. Surface roughness was found to decrease in the presence of HEC as well as HEC + TX.

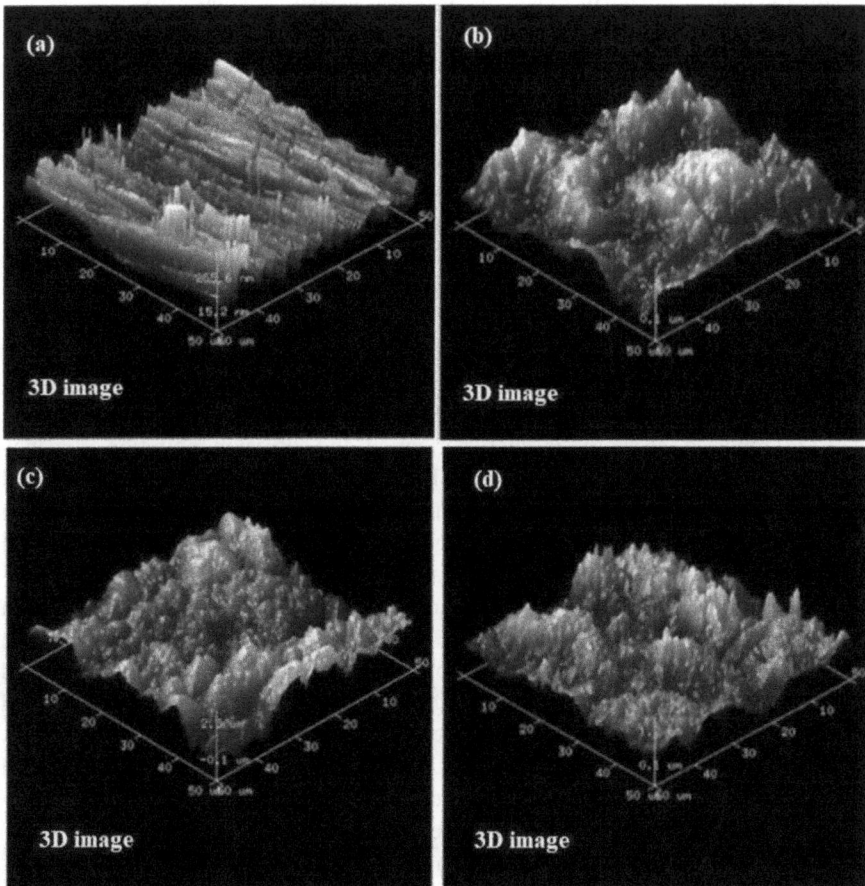

FIGURE 6.7   AFM images of polished MS surface (a), corroded in 1 M HCl without (b) and with HEC (500 ppm) (c) and HEC (500 ppm) + TX (5 ppm) (d). Reprinted with permission from [11]. © 2022 Elsevier Publications.

Nwanonenyi revealed the corrosion inhibition potential of hydroxypropyl cellulose (HPC) for aluminum in 0.5 M HCl and 2 M H_2SO_4 solutions [20]. To prove HPC's anticorrosive potential, several theoretical and experimental investigations have been conducted using DFT. According to PDP results, at 5 g/L concentrations of 0.5 M HCl and 2 M H_2SO_4, HPC shows the highest IE of 84.29% and 95.92%, respectively. HPC is effective in both electrolytes by negatively influencing anodic and cathodic Tafel reactions. In addition, corrosion current densities decreased significantly, which shows that HPC blocks the active areas causing corrosion. Furthermore, HPC adsorption on aluminum surfaces in 0.5 M HCl and 2 M H_2SO_4 follows the Langmuir adsorption isotherm model. Another study that examined the corrosion inhibition capacity of hydroxypropyl methylcellulose (HPMC) with HEC reported a similar conclusion. Both HEC and HPMC function as corrosion inhibitors mostly of the cathodic type. At a concentration of 2,000 mg/L, HEC and HPMC show the highest IE of 83.25% and 84.68%, respectively.

Other cellulose derivatives, including cellulose acetate [21] and CMC [22], are utilized as efficient corrosion inhibitors, according to a literature review. The majority of these chemicals function as mixed-type corrosion inhibitors, with some studies indicating a cathodic-type preponderance [23]. These substances prevent metallic corrosion by delaying cathodic and anodic reactions by adhering to the active areas of the metallic surface. The Langmuir adsorption isotherm was mostly followed when these chemicals were adsorbed onto metallic surfaces. These chemicals' behavior at the interface and adsorption was also investigated using EIS investigations. Increased Rct and charge transfer resistance values observed in most experiments showed that these compounds act as interface-type corrosion inhibitors. This finding is further supported by the ability of these compounds to reduce the width of Nyquist curves. Surface examinations (SEM and AFM) and computational analyses provide substantial evidence for the adsorption mechanism of these chemicals' ability to suppress corrosion.

Using chemical, electrochemical, and surface morphological studies, Solomon et al. [24] reported the synthesis, characterization, and corrosion inhibition efficiency of CMC along with Ag nanoparticle (AgNPs) composite (CMC/AgNPs) for St37 steel in 15% H_2SO_4. Their findings demonstrated that CMC/AgNPs prevented corrosion by adsorption, with the adsorption being strengthened by SEM, EDS, AFM, and FTIR techniques. SEM analysis demonstrated that the presence of CMC/AgNPs considerably improves the metal's surface shape, validating the adsorption mechanism of corrosion inhibitors. The makeup of the St37 steel EDS spectra recorded both with and without CMC/AgNPs after 25 hours of immersion was changed, which provided additional support for this observation. Other studies on CS and MS in acidic and NaCl electrolytes have extensively reported the corrosion prevention characteristics of cellulose derivatives. Additionally, they are utilized to prevent corrosion in Al and Cu. Furthermore, how halide ions might enhance the protective properties of cellulose derivatives for aluminum has also been extensively explored.

Using the WL technique, Manimaran et al. [25] assessed the CMC–Zn^{2+} system's ability to inhibit corrosion on CS in the formation of water either with or without Zn^{2+}. Their findings demonstrated that the combination of 250 ppm CMC with 50 ppm Zn^{2+} has a corrosion prevention rate as high as 98%. The creation of a protective coating among the Fe^{2+}–CMC complex and $Zn(OH)_2$ is attributable to the mechanism by which CMC inhibits corrosion, and the CMC–Zn^{2+} system has a synergistic effect. Additionally, to counteract the low solubility of cellulose, it is possible to modify it or use cellulose derivatives.

Farhadian et al. [26] synthesized a chemically altered form of hydroxyethyl cellulose (CHEC) and assessed its effectiveness in preventing corrosion in low CS. They showed that CHEC can successfully retain its anticorrosion activity over the temperature range of 20°C–80°C, and at that temperature, the corrosion inhibition rate may reach 93%. According to a mechanism study, CHEC chemically adsorbs on the surface of the corroded metal, which prevents the anode's metal dissolution and the cathode's hydrogen evolution reaction. This adsorption pattern is consistent with the Langmuir adsorption isotherm model.

6.3 CONCLUSION

Cellulose is a macromolecular polysaccharide containing glucose. The unit of glucose is finally produced during cellulose acid hydrolysis. Cellulose is a polymer made up of just one dehydrated glucose. Why do cellulose and glucose have such different physical and chemical characteristics? This is exactly the same as starch's poor water solubility, which is attributable to the intramolecular hydrogen bond. The fiber molecule has a linear shape and around 3,000–5,000 units of glucose due to the β-1, 4-glycosidic link. Because of many hydrogen and van der Waals bonds that hold macromolecules to a long, linear chain, they are insoluble in water. Numerous O atoms in cellulose can establish coordination bonds with metal atoms to shield the metal from corrosive conditions. However, as was already indicated, natural cellulose has a high degree of polymerization and low water solubility, which has a significant impact on its usage as a corrosion inhibitor.

Therefore, little information about the ability of cellulose to prevent metal corrosion is reported. To increase its solubility and limit corrosion, cellulose may be chemically modified with a variety of groups. Researchers are thus interested in identifying the role of functionalized cellulose in corrosion inhibition.

REFERENCES

[1] A. Yurt, V. Bütün, and B. Duran, "Effect of the molecular weight and structure of some novel water-soluble triblock copolymers on the electrochemical behaviour of mild steel," *Materials Chemistry and Physics*, vol. 105, pp. 114–121, 2007.

[2] S. A. Umoren and U. M. Eduok, "Application of carbohydrate polymers as corrosion inhibitors for metal substrates in different media: A review," *Carbohydrate Polymers*, vol. 140, pp. 314–341, 2016.

[3] M. Abdelraof, M. S. Hasanin, M. M. Farag, and H. Y. Ahmed, "Green synthesis of bacterial cellulose/bioactive glass nanocomposites: Effect of glass nanoparticles on cellulose yield, biocompatibility and antimicrobial activity," *International Journal of Biological Macromolecules*, vol. 138, pp. 975–985, 2019.

[4] A. Abou Hammad, M. Abd El-Aziz, M. Hasanin, and S. Kamel, "A novel electromagnetic biodegradable nanocomposite based on cellulose, polyaniline, and cobalt ferrite nanoparticles," *Carbohydrate Polymers*, vol. 216, pp. 54–62, 2019.

[5] A. Yabuki, T. Shiraiwa, and I. W. Fathona, "pH-controlled self-healing polymer coatings with cellulose nanofibers providing an effective release of corrosion inhibitor," *Corrosion Science*, vol. 103, pp. 117–123, 2016.

[6] M. V. Fiori-Bimbi, P. E. Alvarez, H. Vaca, and C. A. Gervasi, "Corrosion inhibition of mild steel in HCL solution by pectin," *Corrosion Science*, vol. 92, pp. 192–199, 2015.

[7] A. Akaracharanya, T. Taprig, J. Sitdhipol, and S. Tanasupawat, "Characterization of cellulase producing Bacillus and Paenibacillus strains from Thai soils," *Journal of Applied Pharmaceutical Science*, vol. 4, pp. 6–11, 2014.

[8] S. Gawande, "Cellulose: A natural polymer on the earth," *International Journal of Polymer Science & Engineering*, vol. 3, pp. 32–37, 2017.

[9] S. Kobayashi, K. Kashiwa, J. Shimada, T. Kawasaki, and S. I. Shoda, "Enzymatic polymerization: The first in vitro synthesis of cellulose via nonbiosynthetic path catalyzed by cellulase," in *Makromolekulare Chemie: Macromolecular Symposia*, Basel: Hüthig & Wepf Verlag, vol. 54, pp. 509–518, 1992.

[10] B. Medronho, A. Romano, M. G. Miguel, L. Stigsson, and B. Lindman, "Rationalizing cellulose (in) solubility: Reviewing basic physicochemical aspects and role of hydrophobic interactions," *Cellulose*, vol. 19, pp. 581–587, 2012.

[11] C. Verma, M. Quraishi, and K. Y. Rhee, "Aqueous phase polymeric corrosion inhibitors: Recent advancements and future opportunities," *Journal of Molecular Liquids*, vol. 348, p. 118387, 2022.

[12] M. Solomon, S. Umoren, I. Udosoro, and A. Udoh, "Inhibitive and adsorption behaviour of carboxymethyl cellulose on mild steel corrosion in sulphuric acid solution," *Corrosion science*, vol. 52, pp. 1317–1325, 2010.

[13] S. A. Umoren, A. A. AlAhmary, Z. M. Gasem, and M. M. Solomon, "Evaluation of chitosan and carboxymethyl cellulose as ecofriendly corrosion inhibitors for steel," *International Journal of Biological Macromolecules*, vol. 117, pp. 1017–1028, 2018.

[14] M. S. Hasanin and S. A. Al Kiey, "Environmentally benign corrosion inhibitors based on cellulose niacin nano-composite for corrosion of copper in sodium chloride solutions," *International Journal of Biological Macromolecules*, vol. 161, pp. 345–354, 2020.

[15] S. Eid, M. Abdallah, E. Kamar, and A. El-Etre, "Corrosion inhibition of aluminum and aluminum silicon alloys in sodium hydroxide solutions by methyl cellulose," *Journal of Materials and Environmental Science,* vol. 6, pp. 892–901, 2015.

[16] I. Arukalam, I. Madufor, O. Ogbobe, and E. Oguzie, "Inhibition of mild steel corrosion in sulfuric acid medium by hydroxyethyl cellulose," *Chemical Engineering Communications*, vol. 202, pp. 112–122, 2015.

[17] Y. Sangeetha, S. Meenakshi, and C. S. Sundaram, "Corrosion inhibition of aminated hydroxyl ethyl cellulose on mild steel in acidic condition," *Carbohydrate Polymers*, vol. 150, pp. 13–20, 2016.

[18] S. A. Al Kiey, M. S. Hasanin, and S. Dacrory, "Potential anticorrosive performance of green and sustainable inhibitor based on cellulose derivatives for carbon steel," *Journal of Molecular Liquids*, vol. 338, p. 116604, 2021.

[19] M. Mobin and M. Rizvi, "Adsorption and corrosion inhibition behavior of hydroxyethyl cellulose and synergistic surfactants additives for carbon steel in 1M HCl," *Carbohydrate Polymers*, vol. 156, pp. 202–214, 2017.

[20] S. Nwanonenyi, H. Obasi, and I. Eze, "Hydroxypropyl cellulose as an efficient corrosion inhibitor for aluminium in acidic environments: Experimental and theoretical approach," *Chemistry Africa*, vol. 2, pp. 471–482, 2019.

[21] K. Andarany, A. Sagir, A. Ahmad, S. Deni, and W. Gunawan, "Cellulose acetate layer effect toward aluminium corrosion rate in hydrochloric acid media," *IOP Conference Series: Materials Science and Engineering*, vol. 237, p. 012042, 2017.

[22] O. Egbuhuzor, I. Madufor, S. Nwanonenyi, and J. Bokolo, "Adsorption behavior and corrosion rate model of sodium carboxymethyl cellulose (NA-CMC) polymer on aluminium in hcl solution," *Nigerian Journal of Technology*, vol. 39, pp. 369–378, 2020.

[23] M.-M. Li, Q.-J. Xu, J. Han, H. Yun, and Y. Min, "Inhibition action and adsorption behavior of green inhibitor sodium carboxymethyl cellulose on copper," *International Journal of Electrochemical Science*, vol. 10, pp. 9028–9041, 2015.

[24] M. M. Solomon, H. Gerengi, and S. A. Umoren, "Carboxymethyl cellulose/silver nanoparticles composite: Synthesis, characterization and application as a benign corrosion inhibitor for St37 steel in 15% H2SO4 medium," *ACS Applied Materials & Interfaces*, vol. 9, pp. 6376–6389, 2017.

[25] N. Manimaran, S. Rajendran, M. Manivannan, J. A. Thangakani, and A. S. Prabha, "Corrosion inhibition by carboxymethyl cellulose," *European Chemical Bulletin*, vol. 2, pp. 494–498, 2013.

[26] A. Farhadian, S. A. Kashani, A. Rahimi, E. E. Oguzie, A. A. Javidparvar, S. C. Nwanonenyi, et al., "Modified hydroxyethyl cellulose as a highly efficient eco-friendly inhibitor for suppression of mild steel corrosion in a 15% HCl solution at elevated temperatures," *Journal of Molecular Liquids*, vol. 338, p. 116607, 2021.

7 Chitosan and its Derivatives as Corrosion Inhibitors

Humira Assad and Abhinay Thakur
Lovely Professional University

Ashish Kumar
Government of Bihar

ABBREVIATIONS

BP	Biopolymer
CI	Corrosion inhibitor
CS	Chitosan
EIS	Electrochemical impedance spectroscopic
GO	Graphene oxide
LAI	Langmuir adsorption isotherm
MCRs	Multicomponent reactions
MS	Mild steel
NP	Natural polymers
OAM	Oleic acid-modified
PDP	Potentiodynamic polarization
PFG	Polar functional group
PILs	Polymeric ionic liquids
PM	Polymer matrix
QCC	Quantum chemical calculation
SA	Steel alloy
SB	Schiff base

7.1 INTRODUCTION

One of the ultimate wishes of humans is to lead a trouble-free existence, yet this is not always attainable. For instance, with respect to metals, their extreme melting and boiling points, strong conductivity, bendiness, ductility, high densities, prodigious tensile potency, and distinctive sheen are just a few of their fascinating characteristics [1]. If metals can retain these characteristics, they could improve human existence due to their applications in sectors such as electrochemical engineering and structural construction [2]. Corrosion is the phenomenon in which metals engage with some

 DOI: 10.1201/9781003400059-7

elements that are constantly present in the environment. It causes several serious problems, including material deterioration, aesthetic degradation, and environmental pollution. Corrosion affects the entire world, with an estimated yearly expense of $2.5 trillion, or 3.4% of the global GDP [3]. Corrosion control is therefore crucial for the environment, economy, society, and aesthetics [4]. According to estimates, adopting the best practices for metal corrosion avoidance could reduce yearly corrosion expenses worldwide by 15%–35% [5]. To counteract the corrosion threat, a variety of tactics have been employed, such as design, material selection, electrochemical methods (anodic and cathodic protection), coverings, and environmental modification (the use of CIs) [6]. Corrosion control techniques that involve CIs are frequently used. Heteroatoms such as nitrogen, sulfur, and oxygen are found in organic molecules known as inhibitors [7]. These atoms, which are rich in electrons, can easily create a coating on the metal by forming coordinate bonds with it. The metal cannot further dissolve because of this protective shield [8]. Some of the synthetic organic molecules are reportedly harmful, whereas others require expensive ingredients. Additional toxicity could develop both in the preparation process and during application. These poisonous substances may negatively affect both the environment and human health. Since human safety and environmental protection take priority over corrosion inhibition, using hazardous chemicals to combat corrosion is restricted by severe environmental regulations. Owing to the increasing demands of "green chemistry" throughout the world, researchers have recently shifted their attention to the use of CIs that are ecologically safe for the environment [9]. Utilizing MCRs, chemical reactions catalyzed by energy-efficient microwave and ultrasound irradiations, plant compounds, surfactants, ionic liquids, and BPs, among others, have made significant strides in the direction of "green and sustainable corrosion inhibition" over the past two decades [10,11]. Although there is a wealth of research supporting the usage of naturally occurring plant compounds as CIs, there are very few practical uses for them. This is because it is challenging to identify the component that inhibits corrosion and to suggest a possible corrosion inhibition mechanism [12]. Some of the surfactants are also shown to have negative skin effects after prolonged exposure. The use of BPs as green CIs has now been reported in an increasing amount of literature [13]. Giant molecules called BPs have a higher number of heteroatoms, which makes it easier for them to adsorb to metal surfaces as shown in Figure 7.1.

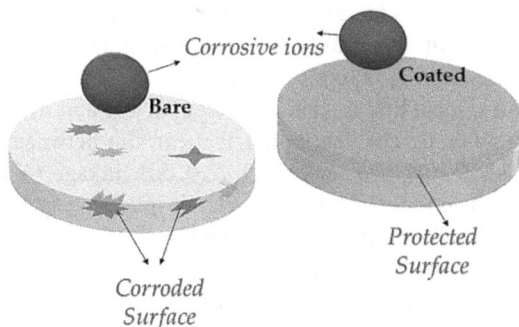

FIGURE 7.1 Schematic illustration of BP protection of metals against corrosion.

These substances meet the requirements to be used as effective ecological CIs for metals because of their easy accessibility, environmental impact, non-toxicity, biodegradability, and chemical stability. In this context, a wide range of NPs, including alginate, sodium CS, pectin, carboxy-methyl, and hydroxy-ethyl cellulose, function as anticorrosive substances to protect metals and their alloys against disintegration for numerous electrolytic settings such as HCl, H_2SO_4, and NaCl [14,15]. CS, one of the available BPs, stands out for having an excellent capacity to inhibit corrosion. It is distinguished by the presence of oxygen (of the functional moieties' alcohol and ether) and nitrogen (of the functional moiety amine) atoms in its backbone chain. These locations are well known for serving as efficient adsorption centers on metallic substrates [16]. Based on this, we sought to clarify in this chapter the advantages of using CS and its derivatives as persistent substances for reducing the corrosion of metallic components in different hostile settings.

7.2   ORIGIN AND STRUCTURE OF CS

CS has the potential to be a groundbreaking material in the next-generation coating. D-glucosamine and N-acetyl-D-glucosamine combine by a β-1,4-glycosidic bond to form the polysaccharide known as CS [17]. It is the second most prevalent NP on the Earth and is easily synthesized through partial de-acetylation of chitin, a natural substance found in the shells of numerous crustaceans, including crabs and shrimps [18]. The chitin-to-CS alteration addresses certain disadvantages of chitin, such as its limited H_2O solubility, which restricts its use and dissemination. The goal of de-acetylating chitin is to release amino groups from acetyl groups and give the polymer a cationic character [19]. Depending on the type of product desired, the de-acetylation technique is conducted either at room temperature (homogeneous de-acetylation) or at a higher temperature (heterogeneous de-acetylation). However, for industrial uses, the latter is preferred. De-acetylation reactions may occasionally be conducted in the presence of thiophenol, which scavenges oxygen, or under an N_2 environment, which prevents chain breakdown that invariably results from peeling reactions under highly alkaline conditions [20]. CS's antibacterial effect is thought to be attributable to its cationic character, which interacts with negative charges on microbial cell membranes. CS also has other qualities such as non-toxicity, biocompatibility, and biodegradability, as well as the ability to cure itself [21]. Its self-restorative capabilities have been explored and used in anticorrosive coatings. Epoxy typically has CS incorporated into it to enhance its natural capacity for self-healing [22]. The self-healing nature of CS-based polymers can be improved using numerous techniques, including dynamic covalent bonds and hydrogen bonding. In the affected parts, dynamic covalent bonds that can flip between or swap between many molecules can shatter and reestablish [23]. SB linkage (-CH=N-), a type of active covalent bond, is one such instance. It can be produced by joining an aldehyde group (such as glutaraldehyde) from one polymer to the NH_2 unit of CS. In this case, the delicate balance of the SB linkage determines the self-healing abilities of the CS-based polymer. CS becomes a polycation in an acidic environment. Here, the CS-based polymer's self-healing capacity is dependent on robust hydrogen-bonding linkages. CS is regarded as an excellent matrix for anticorrosion treatment owing to its polycationic character under acid environments. The $-NH_3^+$ units bonding with the OH^- produced by corrosion's pH increase can neutralize the pH of the defective area. In addition, the water that has been penetrated makes CS chains more mobile, which enhances their self-healing capacity [24]. There is a vast array of alternative uses for CS. Despite the qualities and range of uses stated above, CS has rarely been used. It has some drawbacks, including weak mechanical strength [25], which makes it unsuitable for many applications. To address these limitations, CS is mostly employed in composite materials. CS coating on metal surfaces can enhance its capabilities as an anticorrosion barrier through surface treatment with poly(maleic anhydride-alt-1-octadecene) and poly(ethylene-alt-maleic anhydride) [26]. CS serves as a physical barrier in this situation, slowing down the propagation of aggressive ions. Therefore, CS is an excellent choice for a polymeric matrix in a completely natural covering for oxidization prevention. Not only does it permit antibacterial activity and matrix for anticorrosive compounds, but it preferably guarantees the self-restorative characteristic for a prolonged lifetime covering. Furthermore, it can operate as a proactive shield against the transmission of aggressive ions through specific surface functionalization. The use of chemically altered CS in biological and industrial applications, such as the prevention of corrosion and the elimination of metallic contaminants from industrial influents, has recently become common [27]. Due to its polymeric composition and ability to adsorb on metal surfaces, CS shows a fair amount of anticorrosive activity in its pure form. This protection is therefore exceptional. Different polar substituents found in CS affect both the material's solubility and its capacity to cooperate with metal surfaces. CS naturally has a variety of e^-- opulent polar situates or functional moieties, such as hydroxyl ($-OH$), amino ($-NH_2$), hydroxy-methyl ($-CH_2OH$), cyclic and acyclic ether ($-O-$), and amide ($-NHCOCH_3$), by which it can simply adsorb on the outer layer of a metallic object as shown in Figure 7.2. CS exhibits both physical and chemical adsorption. Typically, the contribution of an unshared electron pair by nitrogen and oxygen heteroatoms allows for chemisorption to take place (coordination bonding). Physisorption, also known as electrostatic adsorption, takes place when charged CS units interact/bond with a charged metallic substrate via electrostatic (ionic) connections. A few polar CS substituents, such as $-OH$ and $-NH_2$ units, go through protonation in aqueous solutions and transform into their positively charged forms. Nevertheless, the buildup of counterions of electrolyte particles at the cationic metal outer layer causes the metal surface to become negatively charged. For several metal–electrolyte systems, CS and its derivatives have been utilized as CIs. Although pure CS has robust chelating properties, its uses in corrosion inhibition and elimination of metal contaminants from manufacturing influents are limited due to its insolubility in aqueous mixtures [28]. Consequently, efforts are being made to solubilize them through chemical modification and co-solvent addition. Chemical changes that increase aqueous-phase solubility are followed because they increase both the solubility and the capacity of the modified material to attach to metal surfaces [29]. It has been widely documented that chemically modified CS exhibits a relatively high level of protective efficiency in comparison with pure CS [30].

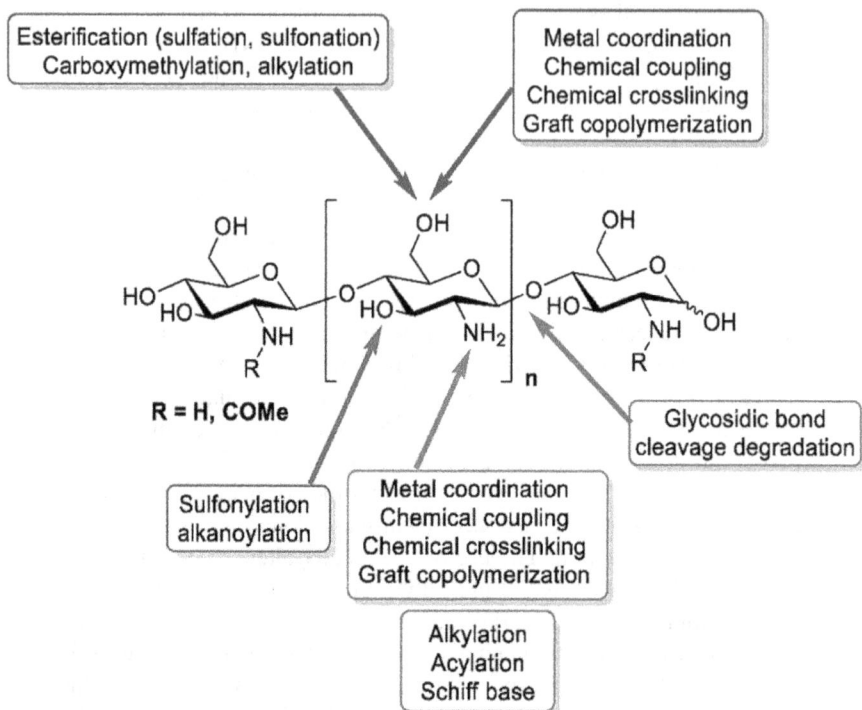

FIGURE 7.2 Structure of CS and its functional groups capable of chemical modification [31]. Reproduced with permission from MDPI. Copyright 2021.

7.3 CHITOSAN AND ITS DERIVATIVES AS SUSTAINABLE CIS

CS is a naturally existing polymer that satisfies the criteria for being categorized as a "green corrosion inhibitor," making it an affordable substitute for often employed inhibitors in industrial applications. One of the essential requirements that determine a CI's use is its solubility in the target corrosive media. CS and its derivatives are used as efficient CIs for metals and alloys for a variety of electrolytic solutions, such as HCl, H_2SO_4, and NaCl owing to their ecologically benevolent and green characteristics. CS is widely used to prevent steel from being corroded by media as listed below:

- Pure CS,
- Functionalized CS, and
- Composite CS.

Composite and fictitious forms can also be categorized into different types. The following section clarifies the essential details of CS-based inhibitors, with a quick review of recent findings.

FIGURE 7.3 Microstructure of CS [33]. Reproduced with permission from MDPI. Copyright 2022.

7.3.1 PURE FORM OF CS

Because pure CS is a non-toxic, biodegradable, and biocompatible NP, it is frequently used as a BP for a range of commercial and biological purposes. The microstructure of CS is shown in Figure 7.3. The protonation of the amine group in CS causes it to partially dissolve in aqueous, low-pH solutions. Many variables, including the degree of de-acetylation, pH, temperature, and molecular weight, affect the solubility of CS in aqueous acid solutions [32]. When a protonation level of more than 50% is attained, CS can be dissolved.

As shown in the literature, CS forms in HCl solutions when the concentration of acid is higher than 1 M [34]. According to Pardo-Castaño and Bolaños [35], temperature has an adverse impact on the dissolution of medium-molecular-weight (MW) CS in CH_3COOH solutions. Moreover, Panda et al. [32] reported that when the MW of native CS is increased, the water solubility of three compositions of p-coumaric acid-modified CS is decreased. Chang et al. reported a similar tendency, who found that CS with an MW > 29.2 kDa has low solubility [36]. In the literature, there are only a few studies that discuss using pure CS to block steel substrates [37]. Yet, CS is utilized as an inhibitory coating or additive to protect different metals and alloys in diverse settings, as discussed below.

Due to their high mechanical strength and affordability, SAs, particularly MS and carbon steel, are the most frequently utilized materials in buildings. Yet, when exposed to acidic liquids, they are extremely vulnerable to corrosive degradation. In hydrochloric acid, CS is used as a CI for SA [37]. CS mostly succeeds in these solutions by adhering to the metallic surface. PFGs such as -OH and $-NH_2$ serve as adsorption hubs in adsorption processes. Using electrochemical and weight loss studies, Rabizadeh et al. investigated the impact of CS on the corrosion behavior of MS in 0.1 M HCl solution [38]. Their findings showed that as the dosage of CS increased from 0.3 to 1.8 mM, the percentage of metal corrosion governed by weight loss data decreased. Corrosion current density (Icorr), which was determined by analyzing polarization curves, for the HCl solution was 156.54 $\mu A/cm^2$, but it was dramatically reduced to 73.9 and 11.2 µA/cm^2 for 0.3 and 1.8 mM levels of CS, respectively. Moreover, Gupta and colleagues [39] examined the ability of CS to prevent corrosion in MS when revealed to 1 M sulfamic acid, which is relevant to the sugar industry. They reported that CS acts as an efficient anticorrosive substance and that it becomes effective by building an outer-layer defensive film by its adsorption that follows the LAI. It was discovered through chemical and electrochemical tests that an increase in CS dosage results in an improvement in its potential to inhibit corrosion. CS displays the highest efficiency of 73.8% at 200 ppm. Although CS exhibits the best protectiveness at 200 ppm + 5 ppm KI, which is 91.68%, the inhibitory potency of CS is synergistically boosted by the inclusion of 5 ppm KI. CS's effectiveness at inhibiting the oxidization of MS in seawater has also been explored relative to its molecular weight. Comparing the effectiveness of CS and its simplest derivative, namely carboxy-methyl CS, as CIs in wastewater, CS shows a lower inhibitory efficacy (23%) than carboxy-methyl CS (38%) [40]. This conclusion is supported by the values obtained in charge transfer resistance (Rct) measurements for CS and carboxy-methyl CS, which are 190.6 and 502.9 Ω cm^2, respectively. According to Feng et al., [41], a constricted adherence to the metallic outer layer is attributable to the structure of carboxy-methyl CS that has more -OH PFGs as an adsorption hub.

Besides, a lot of research has been carried out on CS as a CI for Cu [42] and aluminum alloys [43]. These research findings suggest that CS has a high potential for inhibiting corrosion and that it is an efficient CI. Using several investigations, El-Haddad [44] examined the anticorrosive capability of CS for Cu in 0.5 M HCl media. In these investigations, CS showed the maximum potential (95%) at a concentration of 8 ×10^6 M. CS induces a considerable decrease in the rates of I$_{corr}$ in corrosive media. This finding shows that CS prevents Cu deterioration by adhering to the surface of the metal following the LAI. Furthermore, Abd-El-Nabey et al. [45] used EIS and PDP approaches to demonstrate the inhibitory activities of CS toward aluminum in 0.1 M HCl. Their findings revealed that CS has an optimal inhibitory effectiveness of 90% at a dosage of 0.028 g/L.

7.3.2 FUNCTIONALIZED CS

The functionalization of CS-based substances as CIs is the prevailing trend, followed by their application. This innovative method intends to improve the solubilization of these biocompounds in nearly corrosive conditions as well as their capacity to adsorb and adhere to metal surfaces. This way, more PFGs are joined to the backbone of the CS molecule. The R$_{3-X}$NH$_X$ unit, which is the functional site, is frequently used in chemical alterations in the CS BP. Hence, several CS-based derivatives with unique structural makeups have been synthesized and used to slow down or stop metal corrosion in a variety of hostile conditions, as shown in Figure 7.4. Most of these compounds are applied as inhibiting additives as opposed to inhibiting coatings. When compared with the standard CS form, the inhibitory efficacy of even the simplest CS derivative (carboxy-methyl CS) is improved, increasing from 23% to 38% for steel in wastewater solutions [40]. According to their molecular makeup, functionalized

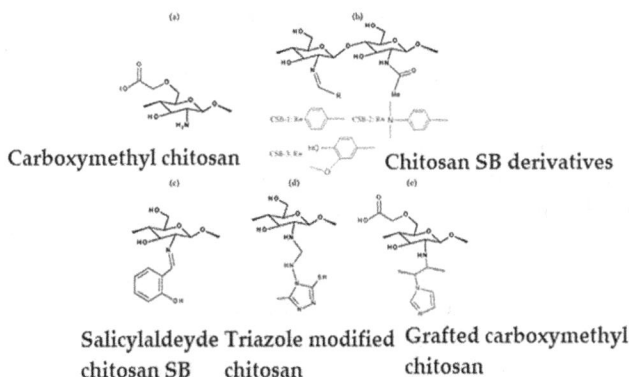

FIGURE 7.4 Molecular structure of some CS derivatives used as CIs [46]. Reproduced with permission from IntechOpen. Copyright 2021.

CS-based inhibiting additives can be divided into many groups, including CS SB, CS polymeric salts, CS surfactants, PEG cross-linked CS, grafted CS, etc. In this chapter, we are limited to using a single example of each functionalized CS set to illustrate its inhibitory effect.

The class of organic chemicals known as SBs is crucial and beneficial for a broad array of commercial and biological implementations [47]. Because of the existence of imine linkage, or CH=N–, SBs have garnered a lot of attention for use in the domain of corrosion inhibition over the past 10 years. According to observations, they are effective CIs for various metallic substances, especially under acidic conditions [48]. Recent research by Ansari et al. investigated salicylaldehyde–CS SBs (150 ppm) for J55 steel type in a 3.5% NaCl solution saturated with CO_2 at increased temperature. The advantages of the produced functionalized CS include its environmental friendliness, safety, straightforwardness, and affordable synthesis of employed SBs, and improvements in inhibitor solubility relative to unaltered CS. The findings further demonstrate that the salicylaldehyde–CS SB is an effective anticorrosion compound for the oil and gas sectors due to the aforementioned benefits [49].

Because of their distinct chemical and physical characteristics, the development of PILs has garnered a lot of interest. Corrosion inhibition capabilities of CS PILs have been assessed. In a recent study, BHC and PHC were synthesized and used to inhibit P110 steel in a 3.5% NaCl solution saturated with CO_2 at 80°C. This study was further extended to the exploration of CS quaternary ammonium salts. At 100 ppm of PHC and BHC, respectively, the findings demonstrated that the inhibition efficiency was 91.6% and 93.3%. Further evidence that the inhibitors were efficiently adsorbed on the outer layer of the steel came from the EDS and contact angle data. Therefore, it can be concluded from the literature that CS polymeric salts demonstrate inhibitory efficiency higher than 90% at a concentration of 300 ppm or lower, thus establishing them as affordable inhibitors [50].

However, the occurrence of both hydrophilic and hydrophobic regions on each unit of a surfactant makes it a surface-active reagent. The propensity of these chemical substances to orientate at the metal–solution interface makes them highly useful as CIs in the petrochemical industry. To comprehend the association between the effectiveness of corrosion inhibition and the surface-active characteristics of the components, a variety of hydrophobically enhanced CS derivatives were assessed. For example, a series of seven hydrophobically modified CS surfactants were created, and their anticorrosion ability for carbon steel in an acid medium was tested. These surfactant-functionalized CS derivatives showed higher inhibitory efficiencies between 93% and 74% (at 250 ppm) compared with pure CS [39]. Atta et al. synthesized amphiphilic CS nanoparticles and used them in a 1.0 M HCl solution to actively prevent steel from corroding [51]. To synthesize hydrophobic CS, nanoparticles were created by amidating CS with unsaturated fatty acids such as oleic and linoleic acids. Methylene chloride was used as an emulsifying solvent to convert the obtained surfactants into nanoparticles, which were then cross-linked with sodium tripolyphosphate. At 50 and 250 ppm of CSOA–MPEG nanoparticles and CSLA–MPEG nanoparticles, the values of inhibition efficiency extrapolated from the polarization data were 92.1% and 97.3%, and 90.9% and 93.5%, respectively. These findings showed that the proposed CS-based nanoparticles are efficient CIs even at low concentrations [51]. Another conclusion that could be drawn is that inhibitors that are close to nanoscale in size may exhibit higher potentials than those that are of micron size.

Recently, it has been found that PEG-cross-linked CS (CS–PEG) can prevent MS from oxidizing. Formaldehyde was used to synthesize this inhibitor, and CH_2O_2 was used as a solvent to disperse CS. Some of the free $R_{3-X}NH_X$ groups in CS interacted with the formaldehyde throughout the process to produce SB intermediates. Then, two PEG-OH groups at its two terminals independently interacted with two distinct intermediates to produce cross-linking. Using electrochemical and weight loss tests, the produced CS–PEG was found to be an MS CI in 1.0 M HCl [52] and 1.0 M sulfamic acid [53] solutions. To fully comprehend the proposed inhibitor's hypothesized adsorption mechanism, QCCs were also employed. At the optimum concentration of 200 ppm, 93.9% corrosion IE was attained. In contrast to PEG and CS alone, CS–PEG displayed higher E_{HOMO} and lower E_{LUMO} values, as well as lower ΔE, according to DFT calculations. This confirmed the sturdier adsorption of CS–PEG. This led to the conclusion that the addition of PEG to the CS backbone has a surprising synergistic effect on the ability of CS to restrict growth.

In polymer chemistry, the term "grafting" is often used to denote the attachment of polymer chains to an outer layer. According to a literature review, grafted CS is frequently used as CIs in a variety of systems [54]. A careful examination of the results of these investigations shows that grafted CS exhibits higher protective effectiveness than non-grafted CS [55]. To develop a CI for API X70 steel in a 1 M HCl solution, Eduok and colleagues [56] produced and tested poly-N-vinyl imidazole)-grafted carboxy-methyl CS (CMCh-g-PVI). They showed that the diameter of the Nyquist curves increases when CMCh-g-PVI is present, indicating that it acts as an interface-type CI. The character of the anodic and cathodic curves was significantly modified by the inclusion of CMCh-g-PVI, according to PDP research. The protective efficacy of Ch, CMCh, and CMCh-g-PVI was consistent with the order of surface smoothness.

7.3.3 COMPOSITE CS

CS composites have mostly been used as inhibitory coatings to shield decomposition in a broad array of settings. As reported in several studies, composite coatings can be created using CS to provide protective systems for metal substrates. The capacity of polymer coatings to defend against corrosion may be improved by strengthening their mechanical qualities, adhesiveness, and protective effectiveness. By introducing organic and inorganic fillers into the PM, this goal might be accomplished. Compared with micron-sized additives, nanoscale fillers infer better impediment qualities in polymer coatings. In general, a representation of composite generation is presented in Figure 7.5.

It has been extensively reported that CS and its derivatives, which contain different inorganic/organic elements, act as inhibitory systems to preserve SAs. Instead of being added as inhibitory additives, these composites have primarily been used as inhibitory coatings. Nanocomposite coatings made of BP CS have been studied as a potential defense against copper corrosion [57]. Silica nanoparticles with a CS matrix with 2-mercaptobenzothiazole have been used as makeup coatings. CS coatings become more effective in protecting against corrosion after the addition of both inorganic and organic nanoparticles, which is a significant step toward creating long-lasting corrosion protection coatings for various metals. Steel has been reported to be well protected by CS/nano-ZnO (CS/ZnO) composite in a variety of hostile situations. Rasheed et al. investigated the strong biocorrosion prevention of the CS/ZnO composite for S150 carbon steel against mixed-culture SRB [58]. The R_{ct} values were 3.2 and 2.8 times higher than those of the control after 21 and 28 days of incubation under the influence of the CI, respectively, showing active corrosion prevention with the highest IE of 74%. These findings demonstrate that the inclusion of ZnO nanoparticles significantly enhances the characteristics of the CS coating. Moreover, carbon steel was electrodeposited with a CS-Zn-Ni (CS/Zn-Ni) composite to increase its anticorrosion resistance brought on by SRB [59]. The EDS data revealed that the coating's Ni content was influenced by the amount of CS present.

CS is used in a low concentration of 0.2 g/L since high dosages (0.6 and 1.0 g/L) decrease the quantity of Ni and lead to low corrosion resistance. CS/Zn-Ni coatings show more potent SRB inhibitory characteristics than Zn-Ni covering. Another study [60] investigated and compared the resistance to oxidation of carbon steel utilizing

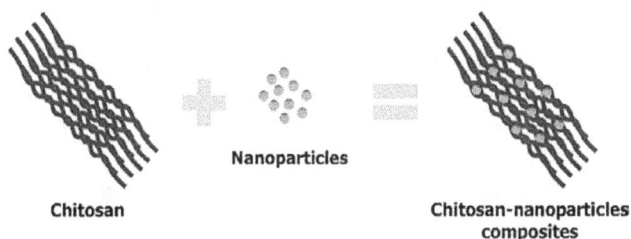

Nanoparticles

Chitosan

Chitosan-nanoparticles composites

FIGURE 7.5 Illustration of the process for the formation of CS–nanoparticle composites [46]. Reproduced with permission from IntechOpen. Copyright 2021.

two distinct CS approaches, namely pure CS covering and OAM CS–GO composite coating. This study reported that in contrast to pure CS coating, the corrosion inhibition of the OAM CS–GO composite coating increased by a factor of 100. Hence, the corrosion prevention of carbon steel is improved using the OAM CS–GO composite. Thus, the usage of CS–nanoparticle composites improves the corrosion fortification of several surfaces, including steel and copper. According to Bahari et al. [57], adding nanoparticles reduces the swelling of CS coatings, and cross-linked CS coatings outperform non-cross-linked ones in terms of reducing corrosion of the Cu surface. John et al. [61] concluded that MS corrosion was mitigated by nanostructured CS/ZnO nanoparticle layers centered on chemical stability and coating oxidation control. Therefore, the structure of the coating (including the PM, cross-linking, and adhesion) affects the process of controlling corrosion.

7.4 CONCLUSION

The lifespan of certain metal surfaces can be increased using effective inhibitors and organic coatings. CS is a substance that has a strong ability to prevent metal corrosion when introduced into corrosive environments. This chapter clarified the usefulness of CS and its derivatives, both individually and in composite forms, for the prevention of corrosion in different kinds of metals. Several studies have reported that pure CS generally acts as a mild inhibitor. However, the decreased water solubility of this BP restricts its use to a wide range of aggressive environments, such as nearly neutral solutions. Yet, such a vulnerability can be circumvented and the preventive capabilities of CS can be enhanced through various functionalization techniques. Functionalization is an effective method for significantly enhancing the corrosion inhibition potential of CS. The inhibitory efficiencies of CS composites and nanofillers, such as metal nanoparticles, are higher than those of pure CS. This enhancement might be the result of direct interactions between these nanoparticles and the metallic surface. The CS inhibitors employed as additives show improved abrasion resistance with immersion time, providing longer protection. Nevertheless, the active corrosion prevention of CS-based coatings is more suited for short-term fortification. From these findings, it can be inferred that CS and its derivatives are environmentally friendly and long-lasting compounds that are perfect replacements for conventional toxic polymeric materials utilized in anticorrosive applications. Nonetheless, the search for new CS derivatives and composites with better corrosion inhibition characteristics is still ongoing.

REFERENCES

[1] Assad, H. and A. Kumar, Understanding functional group effect on corrosion inhibition efficiency of selected organic compounds. *Journal of Molecular Liquids*, 2021. **344**, 117755.
[2] Umoren, S. and M. Solomon, Recent developments on the use of polymers as CIs-a review. *The Open Materials Science Journal*, 2014. **8**(1), 39–54.
[3] Hou, B., et al., The cost of corrosion in China. *NPJ Materials Degradation*, 2017. **1**(1), 4.
[4] Patni, N., S. Agarwal, and P. Shah, Greener approach towards corrosion inhibition. *Chinese Journal of Engineering*, 2013. **2013**, 1–10.

[5] Koch, G.H., et al., *Corrosion Cost and Preventive Strategies in the United States.* Federal Highway Administration, Washington, D.C., 2002.

[6] Umoren, S.A. and M.M. Solomon, Protective polymeric films for industrial substrates: A critical review on past and recent applications with conducting polymers and polymer composites/nanocomposites. *Progress in Materials Science*, 2019. **104**, 380–450.

[7] Assad, H., R. Ganjoo, and S. Sharma, A theoretical insight to understand the structures and dynamics of thiazole derivatives. *Journal of Physics: Conference Series.* 2022. **2267**, 012063.

[8] Stefaniak, A.B., et al., Dissolution of the metal sensitizers Ni, Be, Cr in artificial sweat to improve estimates of dermal bioaccessibility. *Environmental Science: Processes & Impacts*, 2014. **16**(2), 341–351.

[9] Thakur, A., et al., *Coordination Polymers as CIs, in Functionalized Nanomaterials for Corrosion Mitigation: Synthesis, Characterization, and Applications.* 2022, ACS Publications, Washington, DC, pp. 231–254.

[10] Ganjoo, R., et al., Experimental and theoretical study of Sodium Cocoyl Glycinate as corrosion inhibitor for mild steel in hydrochloric acid medium. *Journal of Molecular Liquids*, 2022. **364**, 119988.

[11] Thakur, A., et al., Computational and experimental studies on the efficiency of Sonchus arvensis as green corrosion inhibitor for mild steel in 0.5 M HCl solution. *Materials Today: Proceedings*, 2022. **66**, 609–621.

[12] Sushmitha, Y. and P. Rao, Material conservation and surface coating enhancement with starch-pectin biopolymer blend: A way towards green. *Surfaces and Interfaces*, 2019. **16**, 67–75.

[13] Faraj, L. and G.M. Khan, Application of natural product extracts as green CIs for metals and alloys in acid pickling processes-A. *International Journal of Electrochemical Science*, 2015. **10**, 6120–6134.

[14] Umoren, S.A., et al., Exploration of natural polymers for use as green CIs for AZ31 magnesium alloy in saline environment. *Carbohydrate Polymers*, 2020. **230**, 115466.

[15] Umoren, S.A. and U.M. Eduok, Application of carbohydrate polymers as CIs for metal substrates in different media: A review. *Carbohydrate Polymers*, 2016. **140**, 314–341.

[16] Zeng, Y., et al., An eco-friendly nitrogen doped carbon coating derived from CS macromolecule with enhanced corrosion inhibition on aluminum alloy. Surface and Coatings Technology, 2022. **445**, 128709.

[17] Islam, S., M.R. Bhuiyan, and M. Islam, Chitin and CS: Structure, properties and applications in biomedical engineering. *Journal of Polymers and the Environment*, 2017. **25**, 854–866.

[18] Zeng, J.-B., et al., Chitin whiskers: An overview. Biomacromolecules, 2012. **13**(1), 1–11.

[19] Li, M., et al., Simple preparation of aminothiourea-modified CS as corrosion inhibitor and heavy metal ion adsorbent. *Journal of Colloid and Interface Science*, 2014. **417**, 131–136.

[20] Badawy, M.E. and E.I. Rabea, A biopolymer CS and its derivatives as promising antimicrobial agents against plant pathogens and their applications in crop protection. *International Journal of Carbohydrate Chemistry*, 2011. 2011, 1–29.

[21] Wahba, M.I., Enhancement of the mechanical properties of CS. *Journal of Biomaterials Science, Polymer Edition*, 2020. **31**(3), 350–375.

[22] Njoku, D.I., et al., Understanding the anticorrosive protective mechanisms of modified epoxy coatings with improved barrier, active and self-healing functionalities: EIS and spectroscopic techniques. *Scientific Reports*, 2017. **7**(1), 15597.

[23] Chakma, P. and D. Konkolewicz, Dynamic covalent bonds in polymeric materials. *Angewandte Chemie International Edition*, 2019. **58**(29), 9682–9695.

[24] Ding, F., et al., Recent advances in CS-based self-healing materials. *Research on Chemical Intermediates*, 2018. **44**, 4827–4840.

[25] Prateepchanachai, S., et al., Mechanical properties improvement of CS films via the use of plasticizer, charge modifying agent and film solution homogenization. *Carbohydrate Polymers*, 2017. **174**, 253–261.

[26] Carneiro, J., et al., Functionalized CS-based coatings for active corrosion protection. Surface and Coatings Technology, 2013. **226**, 51–59.

[27] Loganathan, P., et al., Progress, challenges, and opportunities in enhancing NOM flocculation using chemically modified CS: A review towards future development. *Environmental Science:* Water Research & Technology, 2020. **6**(1), 45–61.

[28] del Carpio-Perochena, A., et al., Chelating and antibacterial properties of CS nanoparticles on dentin. *Restorative Dentistry & Endodontics*, 2015. **40**(3), 195–201.

[29] Ashassi-Sorkhabi, H. and A. Kazempour, CS, its derivatives and composites with superior potentials for the corrosion protection of steel alloys: A comprehensive review. *Carbohydrate Polymers*, 2020. **237**, 116110.

[30] Verma, C., et al., Corrosion inhibition potential of CS based Schiff bases: Design, performance and applications. *International Journal of Biological Macromolecules*, 2021. **184**, 135–143.

[31] Aranaz, I., et al., CS: An overview of its properties and applications. *Polymers*, 2021. **13**(19), 3256.

[32] Panda, P.K., et al., Modification of different molecular weights of CS by p-Coumaric acid: Preparation, characterization and effect of molecular weight on its water solubility and antioxidant property. *International Journal of Biological Macromolecules*, 2019. **136**, 661–667.

[33] Song, X., et al., Electrophoretic deposition and corrosion resistance of TiB2-CS coating on 6Cr13 martensitic stainless steel. Coatings, 2022. **12**(11), 1662.

[34] Rinaudc, M., G. Pavlov, and J. Desbrieres, Solubilization of CS in strong acid medium. *International Journal of Polymer Analysis and Characterization*, 1999. **5**(3), 267–276.

[35] Pardo-Castaño, C. and G. Bolaños, Solubility of CS in aqueous acetic acid and pressurized carbon dioxide-water: Experimental equilibrium and solubilization kinetics. *The Journal of Supercritical Fluids*, 2019. **151**, 63–74.

[36] Chang, S.-H., et al., pH Effects on solubility, zeta potential, and correlation between antibacterial activity and molecular weight of CS. *Carbohydrate Polymers*, 2015. **134**, 74–81.

[37] Fayomi, O., et al., Effect of water-soluble CS on the electrochemical corrosion behaviour of mild steel. Chemical Data Collections, 2018. **17**, 321–326.

[38] Rabizadeh, T. and S. Khameneh Asl, CS as a green inhibitor for mild steel corrosion: Thermodynamic and electrochemical evaluations. *Materials and Corrosion*, 2019. **70**(4), 738–748.

[39] Gupta, N.K., et al., CS: A macromolecule as green corrosion inhibitor for mild steel in sulfamic acid useful for sugar industry. *International Journal of Biological Macromolecules*, 2018. **106**, 704–711.

[40] Sun, H., et al., Enhanced removal of heavy metals from electroplating wastewater through electrocoagulation using carboxymethyl CS as corrosion inhibitor for steel anode. *Environmental Science:* Water Research & Technology, 2018. **4**(8), 1105–1113.

[41] Feng, P., et al., Synergistic protective effect of carboxymethyl CS and cathodic protection of X70 pipeline steel in seawater. RSC Advances, 2017. **7**(6), 3419–3427.

[42] Brou, Y., et al., CS biopolymer effect on copper corrosion in 3.5 wt.% NaCl solution: Electrochemical and quantum chemical studies. *International Journal of Corrosion and Scale Inhibition*, 2020. **9**(1), 182–200.

[43] Fayomi, O., I. Akande, and A. Popoola, Corrosion protection effect of CS on the performance characteristics of A6063 alloy. *Journal of Bio-and Tribo-Corrosion*, 2018. **4**, 1–6.

[44] El-Haddad, M.N., CS as a green inhibitor for copper corrosion in acidic medium. *International Journal of Biological Macromolecules*, 2013. **55**, 142–149.

[45] Abd-El-Nabey, B., et al., Anticorrosive properties of CS for the acid corrosion of aluminium. *Portugaliae Electrochimica Acta*, 2015. **33**(4), 231–239.

[46] El Ibrahimi, B., et al., The application of CS-based compounds against metallic corrosion. *Chitin and CS-Physicochemical Properties and Industrial Applications*, 2021.

[47] Hamed, A.A., et al., Synthesis, characterization and antimicrobial activity of a novel CS Schiff bases based on heterocyclic moieties. *International Journal of Biological Macromolecules*, 2020. **153**, 492–501.

[48] Haque, J., et al., Microwave-induced synthesis of CS Schiff bases and their application as novel and green CIs: Experimental and theoretical approach. *ACS Omega*, 2018. **3**(5), 5654–5668.

[49] Ansari, K., et al., CS Schiff base: An environmentally benign biological macromolecule as a new corrosion inhibitor for oil & gas industries. *International Journal of Biological Macromolecules*, 2020. **144**, 305–315.

[50] Cui, G., et al., CS oligosaccharide derivatives as green CIs for P110 steel in a carbon-dioxide-saturated chloride solution. *Carbohydrate Polymers*, 2019. **203**, 386–395.

[51] Atta, A.M., et al., Synthesis of nonionic amphiphilic CS nanoparticles for active corrosion protection of steel. *Journal of Molecular Liquids*, 2015. **211**, 315–323.

[52] Srivastava, V., et al., PEG-functionalized CS: A biological macromolecule as a novel corrosion inhibitor. *ChemistrySelect*, 2018. **3**(7), 1990–1998.

[53] Chauhan, D.S., et al., Thiosemicarbazide and thiocarbohydrazide functionalized CS as ecofriendly CIs for carbon steel in hydrochloric acid solution. *International Journal of Biological Macromolecules*, 2018. **107**, 1747–1757.

[54] Liu, J., et al., Synthesis, characterization, bioactivity and potential application of phenolic acid grafted CS: A review. *Carbohydrate Polymers*, 2017. **174**, 999–1017.

[55] Kong, P., et al., Corrosion by polyaniline/salicylaldehyde modified CS in hydrochloric acid solution. *International Journal of Electrochemical Science*, 2019. **14**, 9774–9784.

[56] Eduok, U., E. Ohaeri, and J. Szpunar, Electrochemical and surface analyses of X70 steel corrosion in simulated acid pickling medium: Effect of poly (N-vinyl imidazole) grafted carboxymethyl CS additive. *Electrochimica Acta*, 2018. **278**, 302–312.

[57] Bahari, H.S., et al., CS nanocomposite coatings with enhanced corrosion inhibition effects for copper. *International Journal of Biological Macromolecules*, 2020. **162**, 1566–1577.

[58] Rasheed, P.A., et al., Controlling the biocorrosion of sulfate-reducing bacteria (SRB) on carbon steel using ZnO/CS nanocomposite as an eco-friendly biocide. Corrosion Science, 2019. **148**, 397–406.

[59] Zhai, X., et al., Microbial corrosion resistance and antibacterial property of electrode-posited Zn-Ni-CS coatings. Molecules, 2019. **24**(10), 1974.

[60] Fayyad, E.M., et al., Oleic acid-grafted CS/graphene oxide composite coating for corrosion protection of carbon steel. *Carbohydrate Polymers*, 2016. **151**, 871–878.

[61] John, S., et al., Enhancement of corrosion protection of mild steel by CS/ZnO nanoparticle composite membranes. *Progress in Organic Coatings*, 2015. **84**, 28–34.

8 Collagen Biopolymers in Sustainable Corrosion Inhibition
Recent Trends in Sustainability and Efficiency

Elyor Berdimurodov
National University of Uzbekistan

Khasan Berdimuradov
Tashkent Institute of Chemical Technology

Ilyos Eliboev, Lazizbek Azimov, Yusufboy Rajabov, Jaykhun Mamatov, Abduvali Kholikov, and Khamdam Akbarov
National University of Uzbekistan

Oybek Mikhliev
Karshi Engineering Economics Institute

8.1 INTRODUCTION

Metallic corrosion is a widespread and expensive issue that affects a variety of industries, infrastructure, and daily life. It is a natural process that takes place when metals, especially those that include iron, combine with their surroundings and produce oxides and other compounds. This process may cause the mechanical characteristics of the metal to deteriorate, the loss of structural integrity, and ultimately failure. According to a study by the National Association of Corrosion Engineers (NACE), the annual cost of corrosion worldwide is projected to be over $2.5 trillion, or around 3.4% of the GDP. This statistic emphasizes the critical need for efficient and long-lasting corrosion prevention techniques to lessen the negative effects of metallic corrosion on the economy, environment, and public safety [1,2].

DOI: 10.1201/9781003400059-8

Many industries, including oil and gas, transportation, construction, and manufacturing, struggle with corrosion. These problems can result in machinery breakdowns, production halts, environmental contamination, and safety risks. In order to counteract corrosion, a variety of tactics have been followed, including the use of corrosion inhibitors, protective coatings, and material selection. When introduced to a corrosive environment, corrosion inhibitors can slow down the pace at which a metal deteriorates. Depending on how they work, these inhibitors might be categorized as anodic, cathodic, or mixed inhibitors. However, the utilization of harmful compounds in the usage of corrosion inhibitors raises questions regarding their long-term viability and environmental effects [3,4].

Protective coatings act as a physical barrier between the metal surface and the corrosive environment, and thus, corrosion is prevented or delayed. These coatings can be organic or inorganic; however, organic coatings are more frequently utilized because of how simple and adaptable they are to implement. However, standard organic coatings, such as epoxy and polyurethane, are made from nonrenewable petroleum-based resources, which pollute the environment. These coatings' negative effects on the environment are further exacerbated by the fact that they frequently produce volatile organic compounds (VOCs) and other dangerous chemicals during manufacture and removal [5,6].

Due to these difficulties, there has been an increase in interest in synthesizing eco-friendly corrosion prevention techniques that utilize less harmful chemicals and rely less on nonrenewable resources. Utilizing biopolymers, which originate from renewable biological sources and have biodegradable qualities, is one possible strategy. Due to its special qualities and extensive availability, collagen has emerged as a particularly appealing alternative among several biopolymers for applications involving sustainable corrosion prevention [7,8].

Collagen forms the highest proportion of protein content in the animal world, which makes up around 30% of the total protein in mammals. It is a fibrous, structural protein that gives tendons, bones, and other tissues including skin strength and flexibility. Fish scales, animal skins, bones, and other sources of collagen may be used to extract collagen, which can then be transformed into gels, films, and fibers. Collagen is a perfect material for corrosion protection applications because of its biocompatibility, biodegradability, and capacity to form films [9,10].

Collagen biopolymers for sustainable corrosion prevention have advanced significantly in recent years as a result of research into diverse formulations, manufacturing methods, and applications. These initiatives have resulted in the synthesis of novel collagen-based materials such as hydrogels, coatings, and composites that provide corrosion protection that is both efficient and ecologically beneficial. The performance and functionality of collagen-based materials have also been shown to be improved by the incorporation of various additives, including nanoparticles, corrosion inhibitors, and other biopolymers, leading to multifunctional coatings with self-healing, antimicrobial, and antifouling properties [11–13].

With an emphasis on their effectiveness and environmental advantages, this chapter seeks to offer a thorough review of current developments in the development of collagen biopolymers for long-term corrosion prevention. Discussion topics will include the sustainability and effectiveness of these materials in reducing corrosion; the characteristics and uses of collagen biopolymers; and the most recent developments in collagen-based corrosion prevention materials. It is intended that this analysis would highlight the potential of collagen biopolymers as a workable substitute for conventional corrosion protection techniques, opening the door for more studies and innovations in this developing sector [14–16].

8.2 COLLAGEN BIOPOLYMERS: PROPERTIES AND APPLICATIONS

A naturally occurring protein called collagen is essential for maintaining the structural integrity of different animal tissues. The animal kingdom's most prevalent protein has drawn a lot of interest due to its potential uses in a variety of areas, including sustainable corrosion prevention. This section will examine the characteristics and uses of collagen biopolymers, emphasizing their applicability for anticorrosion and other relevant uses [17].

8.2.1 PROPERTIES OF COLLAGEN BIOPOLYMERS

Three polypeptide chains that are linked to form a triple-helical structure make up the fibrous protein known as collagen. Collagen is a perfect material for load-bearing applications thanks to its distinctive structure, which also gives it remarkable mechanical strength and flexibility. Collagen biopolymers have a number of essential characteristics that make them excellent for corrosion prevention and other uses, as follows [18,19]:

1. **Biocompatibility**: Since collagen is a natural part of animal tissues, it demonstrates good biocompatibility. Collagen-based materials can be employed in applications requiring interaction with living cells or tissues, such as tissue engineering and medical devices, thanks to this feature.
2. **Biodegradability**: Collagen is a natural protein that can be broken down into simpler components by enzymes and microbes. Compared to petroleum-based polymers, which may linger in the environment for a long time, this feature makes collagen-based products more ecologically benign.
3. **Capability to build films**: Collagen can cling and form stable films on a variety of surfaces, including metals. This characteristic is highly useful for corrosion prevention applications because it can inhibit or delay the contact between the metal surface and the corrosive environment by forming a protective coating [20,21].
4. **Hydrophilicity**: Collagen has a high affinity for water, making it a hydrophilic substance. Collagen-based materials can absorb and hold onto moisture thanks to this feature, which is helpful in some situations including medicine administration and wound healing. However, as moisture may encourage corrosion, this feature can also result in difficulties in corrosion defense applications. Researchers frequently modify collagen to increase its water resistance or mix it with other substances that have hydrophobic qualities to address this problem.
5. **Chemical reactivity**: Collagen's molecular structure contains functional groups, such as amine and carboxyl groups, which enable it to interact with different chemicals and cross-linking agents. This reactivity may be used to improve the performance of collagen-based materials in certain applications by altering their characteristics or forming covalent connections with other polymers or additives.

8.2.2   APPLICATIONS OF COLLAGEN BIOPOLYMERS

Collagen biopolymers are used in a variety of fields, from materials science to biomedicine, as a result of their special characteristics. A few significant uses for collagen biopolymers are as follows (Figure 8.1) [22]:

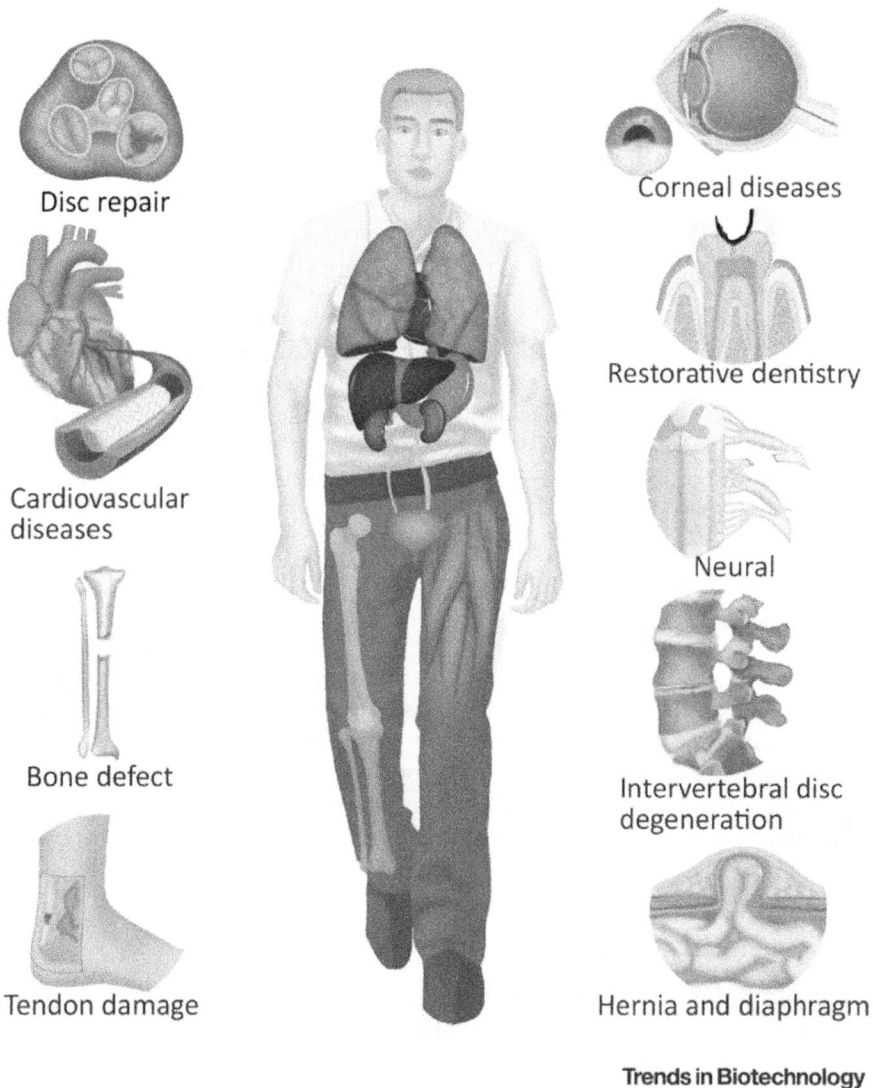

Disc repair

Cardiovascular diseases

Bone defect

Tendon damage

Corneal diseases

Restorative dentistry

Neural

Intervertebral disc degeneration

Hernia and diaphragm

Trends in Biotechnology

FIGURE 8.1   Collagen cross-linking: biomedical and pharmaceutical applications [22].

Protection against corrosion: To offer a barrier against corrosive environments, collagen-based coatings and films can be applied to metal surfaces. Collagen's capacity for film formation ensures strong adhesion to metal surfaces, and its biodegradability and biocompatibility make it an environmentally friendly substitute for traditional corrosion protection techniques.

Biomedical uses: Collagen biopolymers have been widely used in a variety of biomedical applications, including drug delivery systems, tissue engineering scaffolds, and wound dressings. This is because these materials are biocompatible and biodegradable. Collagen is an excellent material for controlled drug release because of its hydrophilic nature, which also enables it to absorb and release bioactive molecules.

Food packaging applications: Collagen-based films and coatings can be used as an alternative to petroleum-based plastics. Being biodegradable with the ability to reduce the environmental impact of packaging waste, these materials can act as a barrier against moisture, oxygen, and microbial contamination.

Cosmetics and personal care products: These products frequently contain collagen because collagen supports skin hydration, elasticity, and general health. Additionally, collagen-based materials can be utilized as carriers for these products' active ingredients, resulting in a controlled release and enhanced product performance.

In conclusion, collagen biopolymers have a special set of qualities that make them appropriate for a variety of uses, including sustainable corrosion protection. Collagen is a promising candidate for devising environmentally friendly and successful corrosion protection techniques due to its biocompatibility, biodegradability, and capacity for film formation, as well as its ability to be altered and combined with other materials. Collagen biopolymers are expected to have more uses as research into their potential advances, which will help different industries create more environmentally friendly products and technologies [23,24].

8.3 RECENT TRENDS IN COLLAGEN-BASED CORROSION PROTECTION

The abundance of collagen, a naturally occurring protein, in animal tissues has made it a promising candidate for long-term corrosion protection. Collagen biopolymers have special qualities like biocompatibility, biodegradability, and film-forming capacity that make them suitable for a variety of uses, including corrosion protection. In this chapter, recent developments in material development, adaptations, and applications are highlighted along with trends in collagen-based corrosion protection [25,26].

8.3.1 COLLAGEN MODIFICATIONS FOR ENHANCED PERFORMANCE

Collagen's hydrophilic nature, which can promote corrosion by soaking up moisture, is one of the main obstacles to using it to prevent corrosion. Researchers have created a number of modification techniques to enhance the mechanical and water resistance of collagen-based materials in order to address this limitation [27–29]. Some of these methods consist of:

1. **Cross-linking**: By creating covalent bonds between collagen molecules, the process of cross-linking creates a structure that is more stable and water resistant. For applications involving corrosion protection, collagen-based materials' properties have been improved by the use of a variety of cross-linking agents, including glutaraldehyde, genipin, and carbodiimide. Due to their promising properties, magnesium (Mg) alloys have received extensive research as potential biodegradable implant materials in clinical orthopedics. However, their development has been severely constrained by physiological conditions such as the uncontrollable rate of corrosion. The corrosion resistance and biocompatibility of the AZ31 Mg alloy substrate are improved by synthesizing a polycaprolactone/collagen/cerium-hydroxyapatite (PCL/Col/Ce-HA) fiber coating using the electrospinning technique. According to the scanning electron microscopy analysis, the PCL/Col/Ce-HA fiber coating with 5 wt.% of Ce-HA displays a dense surface with a consistent fiber size of 396 nm, offering a reliable physical barrier against corrosive ions. During prolonged immersion times, the coating's increased adhesion strength keeps it on the Mg substrate and ensures its durability (Figure 8.2) [30].
2. **Grafting**: Grafting involves joining functional groups or other polymers to the collagen's structural support, which can increase the water resistance, mechanical strength, and metal-surface adhesion of collagen. For instance, to improve collagen's corrosion protection capabilities, researchers have grafted hydrophobic polymers such as polystyrene and poly(methyl methacrylate) onto it.
3. **Blending**: Collagen is blended with other biopolymers or artificial polymers to produce hybrid materials with enhanced properties. For instance, combining collagen with chitosan, alginate, or polyurethane can produce materials with improved mechanical strength, corrosion protection, and water resistance [31–33].

To solve the corrosion problem of Mg, a novel bilayered coating of silane, titanium dioxide, and collagen has been created by reducing the rate of biodegradation and enhancing biointegration (Figure 8.3). A suitable environment for the in situ deposition of modified hydroxyapatite (HA), which is known for its biocompatibility and capacity to stimulate bone growth, is provided by the first layer of silane–TiO_2. This layer serves as a protective layer, slowing the rate of corrosion of magnesium alloys and enhancing their mechanical qualities. The adhesion of the second layer, fibrillar collagen, which is essential for cell growth and proliferation, can also be improved by the silane–TiO_2 layer. Due to its biocompatibility and capacity to encourage cell adhesion, proliferation, and differentiation, collagen, a natural protein present in the extracellular matrix of various tissues, has been extensively used in tissue

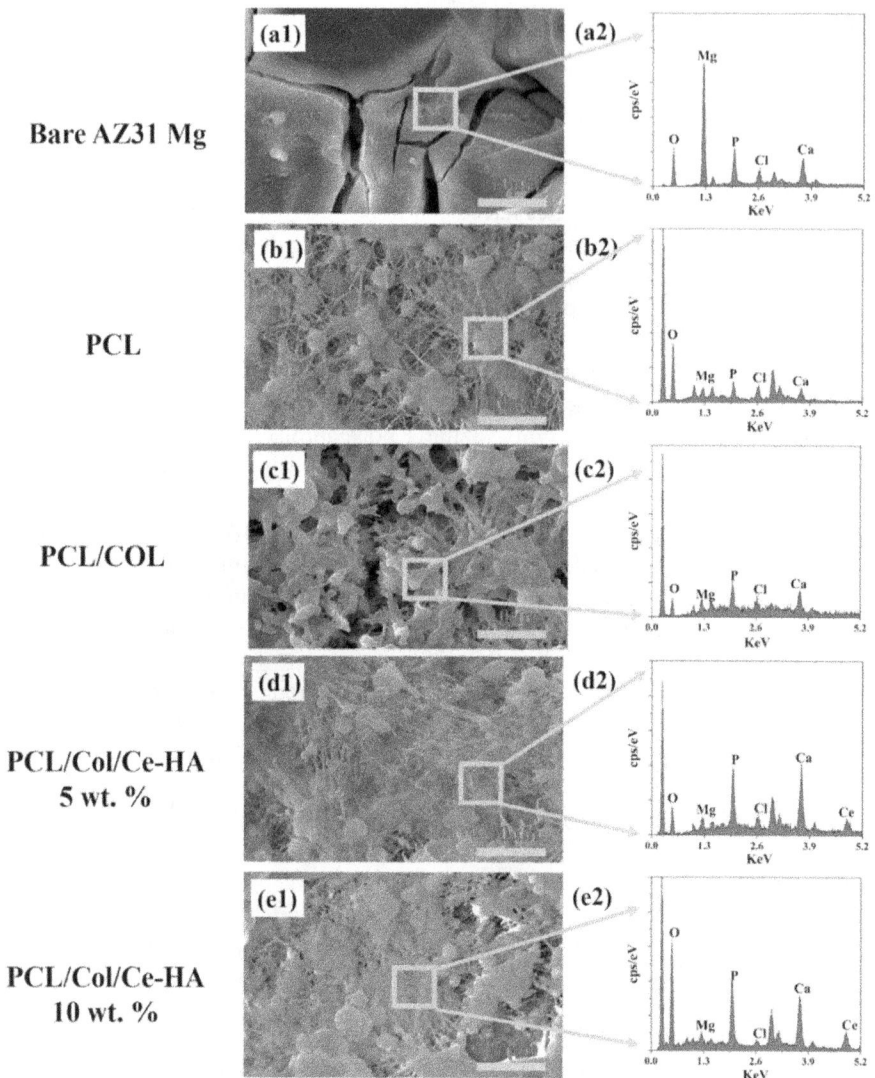

FIGURE 8.2   SEM images and EDS spectra of (a) bare AZ31 Mg alloy, (b) PCL, (c) PCL/Col, (d) PCL/Col/Ce-HA 5 wt.% and (e) PCL/Col/Ce-HA 10 wt.% fiber coatings after 336 h of immersion in SBF solution [30].

engineering. The fibrillar collagen layer in this coating enhances cell growth and proliferation on the silane–TiO$_2$ surface, resulting in improved Mg alloy biointegration. Over an extended period of time, the corrosion mechanisms of Mg alloys coated with this bilayered silane–TiO$_2$/collagen coating were clarified in an organic medium. The results revealed a significant decrease in corrosion rate, indicating that this novel coating could successfully regulate the biodegradation of magnesium alloys while encouraging their biointegration [34].

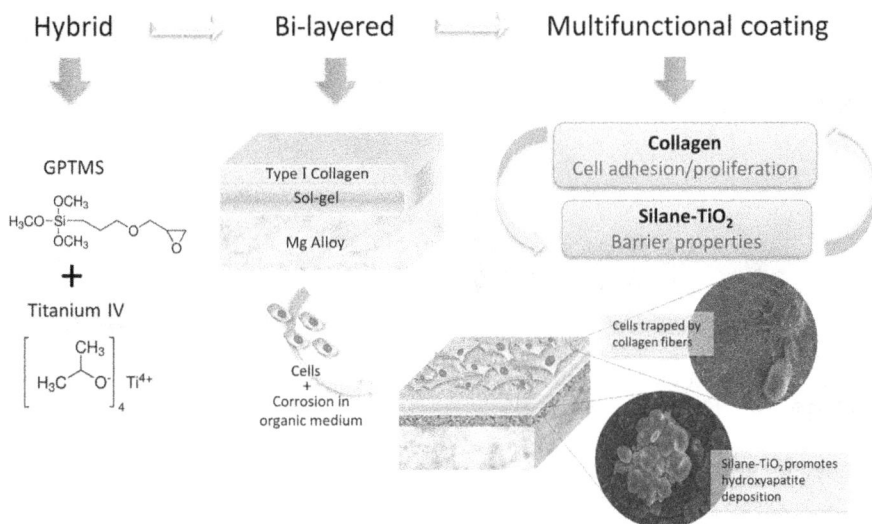

FIGURE 8.3 Bilayered silane–TiO$_2$/collagen coating to control biodegradation and biointegration of Mg alloys [34].

8.3.2 DEVELOPMENT OF COLLAGEN-BASED COMPOSITES AND NANOCOMPOSITES

Due to their ability to incorporate different additives, such as nanoparticles, corrosion inhibitors, and other biopolymers, collagen-based composites and nanocomposites have emerged as promising trends in corrosion protection [35–37]. Examples of nanocomposites and collagen-based composites synthesized for corrosion protection include:

1. **Collagen/clay nanocomposites**: Collagen films' barrier properties against moisture, oxygen, and corrosive species can be improved by adding clay nanoparticles. As clay nanoparticles can migrate to the damaged areas and restore the protective film upon exposure to moisture, these nanocomposites can also display self-healing qualities.

2. **Collagen/silver nanoparticles composites**: Silver nanoparticles can improve the conductivity and barrier performance of collagen films in addition to having antimicrobial properties. Collagen/silver nanoparticle composites have been created by researchers to effectively protect metal surfaces from corrosion and microbial contamination.

3. **Collagen/corrosion inhibitor composites**: By releasing the inhibitor as needed, corrosion inhibitors like benzotriazole that are incorporated into collagen films can offer active corrosion protection. These composites can assist in extending the coated metal's lifespan and decreasing the frequency of maintenance.

Electrochemical impedance spectroscopy was used to investigate the corrosion performance of the collagen/silica sol–gel (CSOL) sample (Figure 8.3) in a 3.5% NaCl

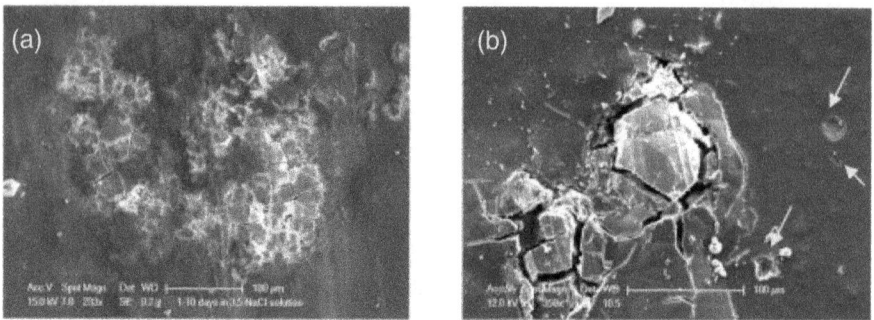

FIGURE 8.4 SEM image of SOL coated AA2024 sample after immersion in 3.5% NaCl solution for (a) 10 days and (b) 14 days [38].

solution for up to five weeks of exposure. The line at slope $= -1$, in the low-frequency range, suggests that the CSOL coating's impedance is under capacitive control from the beginning of immersion. This response suggests that there is some corrosion resistance in this coating. The corrosion process of a solgel-coated material is shown in Figure 8.4. Pits appearing after 10 days of immersion indicate the start of corrosion. The increasing number of pits and their color change from white to dark brown suggest the breakdown of the coating and the formation of different corrosion products. Energy-dispersive X-ray spectroscopy (EDS) mapping shows the coating is completely ruptured where pitting has occurred, due to the "jacking stress" effect from the volume expansion of the corrosion products [38].

8.3.3 SMART AND RESPONSIVE COLLAGEN COATINGS FOR CORROSION PROTECTION

The creation of intelligent and responsive coatings that can react to environmental cues like changes in pH, temperature, or the presence of particular ions is another emerging trend in collagen-based corrosion protection. When necessary, these coatings can release corrosion inhibitors or healing agents, providing active corrosion protection and extending the coated metal's lifespan [39–41]. Several instances of perceptive and adaptable collagen coatings include:

1. **pH-responsive collagen coatings**: Researchers have created pH-responsive collagen coatings that, in response to pH changes, can release corrosion inhibitors like benzotriazole. In environments with changing pH levels, like those found in industrial or marine applications, these coatings can offer active corrosion protection.
2. **Temperature-responsive collagen coatings**: Collagen coatings that are temperature responsive may release anticorrosion or anti-inflammatory substances in response to temperature changes. These coatings can be especially helpful in applications such as automotive or aerospace components, where temperature variations can lead to thermal stresses and encourage corrosion.

3. **Ion-responsive collagen coatings**: In the presence of particular ions, such as chloride ions, which are frequently present in corrosive environments, ion-responsive collagen coatings can release corrosion inhibitors or healing agents. These coatings can aid in shielding metals from localized corrosion that can cause early failure, such as pitting and crevice corrosion.

The production of advanced materials with improved performance, adjustments to improve water resistance and mechanical properties, and the development of intelligent and responsive coatings have been the recent trends in collagen-based corrosion protection. As they can help lessen the environmental impact of corrosion protection methods while delivering effective and long-lasting performance, these developments hold great promise for the future of sustainable corrosion protection. Collagen-based materials are expected to become more significant as this field of study advances, contributing to the designing of effective and environmentally friendly corrosion protection techniques [42–44].

8.3.4 BIOINSPIRED AND BIOMIMETIC APPROACHES IN COLLAGEN-BASED CORROSION PROTECTION

Approaches that are bioinspired and biomimetic have become a promising trend as researchers continue to investigate the potential of collagen-based materials for corrosion protection. These methods seek to create novel and long-lasting corrosion protection strategies by imitating the structures, traits, and operations of biological systems. Collagen-based corrosion protection techniques that are bioinspired and biomimetic include the following [45,46]:

1. **Hierarchical structures**: In biological tissues like skin and bone, collagen fibers are arranged in a hierarchical fashion to provide mechanical strength, toughness, and flexibility. To improve the performance of collagen-based coatings and films in applications involving corrosion protection, researchers are looking into how to mimic these hierarchical structures in these materials.
2. **Self-healing mechanisms**: When they are injured, some biological systems, like skin, have the capacity to repair themselves. Collagen-based corrosion protection materials may eventually include self-healing mechanisms, according to researchers. For instance, incorporating corrosion inhibitors or healing agents into collagen films can enable the release of these substances upon damage, aiding the self-healing process.
3. **Adaptive and responsive properties**: The presence of certain ions, changes in pH, or variations in temperature can cause biological systems to adapt and change. Researchers are focused on synthesizing intelligent and responsive collagen coatings that can release corrosion inhibitors or healing agents when necessary, providing active corrosion protection, by modeling these adaptive and responsive properties.

The Field emission scanning electron microscopy (FE-SEM) images of the HA/CTS/COL/h-BN coatings after 12 weeks of immersion in r-SBF at 37°C are shown

FIGURE 8.5 High-magnification FE-SEM images: (a) hydroxyapatite, (b) collagen, and (c) h-BN after immersion to r-SBF for 12 weeks [47].

in Figure 8.5. The images show the development of pores on the coating surfaces following immersion that are at least 20 microns in size. It is clear from a closer look at 5K magnification that these pores are made of globular aggregations. Numerous similar studies have noted globular aggregations, and it is well known that they are biomimetic apatites. In the biomineralization process, biomimetic apatites are essential because they act as a template for the growth and development of new bone tissue. These apatites suggest that the HA/CTS/COL/h-BN coatings have the potential to encourage osseointegration and bone regeneration. It is also important to note that the development of pores on coating surfaces indicates that the substance has a porous structure. Because it allows for the infiltration of cells, nutrients, and growth factors that facilitate the formation of new bone tissue and its integration with the surrounding bone, porosity is a crucial component for bone tissue engineering applications. Since it has been noted that pore sizes in this range are ideal for promoting bone ingrowth and vascularization, the presence of pores with sizes of 20 μm and greater is particularly significant [47].

8.3.5 EMERGING APPLICATIONS OF COLLAGEN-BASED CORROSION PROTECTION MATERIALS

New and innovative applications are being investigated as the field of collagen-based corrosion protection develops, making use of the special qualities of these materials. These applications include, among others:

1. **Biomedical devices and implants**: Collagen is a desirable material for corrosion protection in biomedical devices and implants, including orthopedic implants, dental implants, and cardiovascular stents, due to its biocompatibility and biodegradability. The lifespan of metal implants can be extended, and the risk of corrosion-related complications like inflammation and tissue damage can be decreased with collagen-based coatings.
2. **Marine and underwater structures**: Due to the salt water, chloride ions, and microorganisms that are present in the marine environment, it is highly corrosive. For marine and underwater structures like ships, offshore platforms, and underwater pipelines, collagen-based corrosion protection materials can offer an environmentally friendly substitute to traditional coatings.

Collagen coatings with intelligence and responsiveness may perform better in these applications by releasing anticorrosion inhibitors or anti-inflammatory compounds in response to changes in the marine environment.

3. **Automotive and aerospace components**: Materials that are lightweight, strong, and resistant to adverse environmental conditions like temperature changes, moisture, and corrosive chemicals are required in the automotive and aerospace industries. For a variety of components, including engine parts, fuel tanks, and structural components, collagen-based materials can offer corrosion protection while also reducing the environmental impact of traditional corrosion protection techniques.

4. **Infrastructure and construction**: Collagen-based corrosion protection materials can be used in a variety of infrastructure and building projects, including pipelines, bridges, and steel bar reinforcement. These materials can aid in preventing corrosion, extending the lifespan of these structures, and reducing maintenance expenses. Their functionality in these applications may be further improved by the development of self-healing and intelligent collagen coatings.

5. **Energy and power generation**: Solutions for corrosion protection are needed in the energy and power generation industries for a variety of components, including wind turbines, solar panels, and nuclear power plants. These components can be protected against corrosion using collagen-based materials, which can improve their durability and lower maintenance costs.

The development of long-lasting and environmentally friendly corrosion control strategies has a lot of potential with collagen-based corrosion protection materials. The performance and functionality of these materials have significantly improved as a result of recent trends in the field, such as material modifications, composites and nanocomposites, intelligent and responsive coatings, and bioinspired approaches. To address the remaining issues and further boost the scalability, economy, and long-term performance of collagen-based corrosion protection materials, however, more studies and innovations are required. Collagen-based materials are anticipated to play a significant role in the future of sustainable corrosion protection and contribute to the global efforts to reduce the environmental impact of corrosion and its associated costs as these difficulties are overcome.

8.4 SUSTAINABILITY AND EFFICIENCY OF COLLAGEN BIOPOLYMERS IN CORROSION PROTECTION

Various industries, infrastructure, and the environment are affected by the widespread global problem of corrosion, which causes significant financial losses and environmental damage. Corrosion mitigation has traditionally been accomplished using techniques like metallic coatings, organic coatings, and inhibitors. These techniques, however, frequently have detrimental effects on the environment, such as the release of toxic substances and the production of hazardous waste. As a result, there is an increasing amount of interest in the synthesis of environmentally friendly and long-lasting corrosion protection materials, like collagen biopolymers.

The most prevalent protein in the animal kingdom, collagen is crucial for the structural integrity of connective tissues like skin, bones, and tendons. Collagen has become a promising material for corrosion protection applications because of its special qualities, including biocompatibility, biodegradability, and adaptability. This chapter discusses recent developments, difficulties, and potential future applications of collagen biopolymers in corrosion protection.

8.4.1 Sustainability of Collagen Biopolymers

Comparing collagen biopolymers to conventional materials, there are several environmental advantages to using them for corrosion protection:

1. **Renewable and abundant source**: Animal byproducts, which are plentiful and frequently regarded as waste products, are natural and renewable sources of collagen. In comparison to petroleum-based polymers and non-renewable metallic coatings, collagen is therefore a sustainable material.
2. **Biodegradability**: Because collagen biopolymers are biodegradable, microorganisms in the environment can break them down into simpler compounds. This characteristic reduces the buildup of waste and the release of toxic substances that are related to traditional corrosion protection techniques.
3. **Low environmental impact**: Collagen biopolymers typically have less environmental impact during production, processing, and disposal than conventional materials. Collagen extraction and processing, for instance, do not call for the use of dangerous chemicals or the production of hazardous waste.

8.4.2 Efficiency of Collagen Biopolymers in Corrosion Protection

In numerous corrosion protection applications, including coatings, films, and composites, collagen biopolymers have demonstrated promising results. Collagen-based materials now perform and function much better thanks to recent developments in material modifications, nanocomposites, and bioinspired methods, making them more effective at preventing corrosion [16,48]:

1. **Material modifications**: Collagen can be changed chemically and physically to improve its mechanical characteristics, stability, and corrosion protection performance. Examples include cross-linking, grafting, and blending with other polymers. The adhesion of collagen coatings to metal substrates, which is essential for efficient corrosion protection, can also be improved by these modifications.
2. **Nanocomposites**: Collagen matrices' corrosion protection abilities can be greatly enhanced by adding nanoparticles like metal oxides, carbon nanotubes, and graphene. Collagen coatings' barrier abilities can be improved by nanoparticles, preventing the diffusion of corrosive species to the metal

surface. Additionally, by discharging corrosion inhibitors or scavenging reactive oxygen species, nanoparticles can offer active corrosion protection.

3. **Bioinspired approaches**: Bioinspired approaches aim to create novel and long-lasting corrosion protection solutions by imitating the structures, characteristics, and functions of biological systems. Hierarchical structures, self-healing mechanisms, and adaptive and responsive properties are a few examples of bioinspired strategies in collagen-based corrosion protection.

8.4.3 CHALLENGES AND FUTURE PERSPECTIVES

Although collagen-based corrosion protection has come a long way, there are still a number of problems to be solved:

1. **Scalability and cost-effectiveness**: Collagen biopolymers need to be produced and processed on a large scale for corrosion protection applications in order to compete with traditional materials. To create scalable and affordable processes for the creation, modification, and processing of collagen-based materials, more research and innovations are needed.

2. **Durability and long-term performance**: It is necessary to further research and optimize the durability and long-term performance of collagen-based corrosion protection materials, especially in challenging environments. This entails comprehending the mechanisms by which these materials degrade and devising tactics to increase their resistance to environmental elements like UV radiation, temperature changes, and microbial attack.

3. **Standardization and performance evaluation**: For collagen-based corrosion protection materials to be widely used and commercialized, standardized evaluation techniques must be created. In order to compare the performance of these materials with traditional corrosion protection techniques, performance criteria, testing procedures, and benchmarks must be established.

8.5 CONCLUSION

Collagen presents a promising substitute for traditional corrosion protection techniques, which frequently have detrimental environmental effects. Collagen is a naturally abundant and renewable material. Collagen has special qualities like biocompatibility, biodegradability, and adaptability that make it a desirable material for coatings, films, and composites, among other uses.

This chapter covered the advantages of collagen biopolymers for the environment, recent developments in material modification, nanocomposites, and bioinspired methods, as well as the difficulties and prospects for the future of the field. Additionally, it looked into potential new uses for collagen-based corrosion protection materials in a variety of fields, including biomedical devices, marine structures, automotive and aerospace components, infrastructure and building, and energy and power generation.

It is anticipated that collagen-based materials will be essential in creating long-lasting and environmentally friendly corrosion protection techniques as research and innovation in this area continue. Scalability, cost-effectiveness, and long-term performance issues must be resolved if these materials are to be widely used and made commercially available. Furthermore, the adoption of collagen-based corrosion protection materials across a variety of industries will be facilitated by the development of standardized methods for assessing their effectiveness.

In conclusion, the growing demand for collagen biopolymers for corrosion protection emphasizes the significance of synthesizing environmentally friendly and sustainable materials to prevent corrosion and reduce the associated financial and environmental costs. The knowledge gained from this chapter will be an invaluable foundation for future research and innovation in the search for long-term corrosion protection solutions as the field continues to develop.

REFERENCES

[1] Bahgat Radwan, A., et al., Electrospun highly corrosion-resistant polystyrene-nickel oxide superhydrophobic nanocomposite coating. *Journal of Applied Electrochemistry*, 2021. **51**, 1605–1618.

[2] Berdimurodov, E., et al., Green β-cyclodextrin-based corrosion inhibitors: Recent developments, innovations and future opportunities. *Carbohydrate Polymers*, 2022. **292**, 119719.

[3] Berdimurodov, E., et al., MOFs-based corrosion inhibitors. In: V. S. Saji (Ed.), *Supramolecular Chemistry in Corrosion and Biofouling Protection*. 2021, CRC Press, London, pp. 287–305.

[4] Berdimurodov, E., et al., Inhibition properties of 4,5-dihydroxy-4,5-di-p-tolylimidazolidine-2-thione for use on carbon steel in an aggressive alkaline medium with chloride ions: Thermodynamic, electrochemical, surface and theoretical analyses. *Journal of Molecular Liquids*, 2021. **327**, 114813.

[5] Berdimurodov, E., et al., Experimental and theoretical assessment of new and eco-friendly thioglycoluril derivative as an effective corrosion inhibitor of St2 steel in the aggressive hydrochloric acid with sulfate ions. *Journal of Molecular Liquids*, 2021. **335**, 116168.

[6] Berdimurodov, E., et al., A gossypol derivative as an efficient corrosion inhibitor for St2 steel in 1 M HCl + 1 M KCl: An experimental and theoretical investigation. *Journal of Molecular Liquids*, 2021. **328**, 115475.

[7] Berdimurodov, E., et al., Novel bromide-cucurbit[7]uril supramolecular ionic liquid as a green corrosion inhibitor for the oil and gas industry. *Journal of Electroanalytical Chemistry*, 2021. **901**, 115794.

[8] Berdimurodov, E., et al., Novel cucurbit[6]uril-based [3]rotaxane supramolecular ionic liquid as a green and excellent corrosion inhibitor for the chemical industry. *Colloids and Surfaces A: Physicochemical and Engineering Aspects*, 2022. **633**, 127837.

[9] Xu, S., et al., Super high-quality SEM/FIB imaging of dentine structures without collagen fiber loss through a metal staining process. Scientific Reports, 2022. **12**(1), 2369.

[10] Park, C.H., H.R. Pant, and C.S. Kim, Effect on corrosion behavior of collagen film/fiber coated AZ31 magnesium alloy. *Digest Journal of Nanomaterials & Biostructures (DJNB)*, 2013. **8**(3), 1227–1234.

[11] Berdimurodov, E., et al., Novel gossypol-indole modification as a green corrosion inhibitor for low-carbon steel in aggressive alkaline-saline solution. *Colloids and Surfaces A: Physicochemical and Engineering Aspects*, 2022. **637**, 128207.

[12] Berdimurodov, E., et al., Novel glycoluril pharmaceutically active compound as a green corrosion inhibitor for the oil and gas industry. *Journal of Electroanalytical Chemistry*, 2022. **907**, 116055.
[13] Dagdag, O., et al., Functionalized nanomaterials for corrosion mitigation: synthesis, characterization and applications. In: C. Verma, C. M. Hussain, and M. A. Quraishi (Eds.), *Functionalized Nanomaterials for Corrosion Mitigation: Synthesis, Characterization, and Applications*. 2022, ACS Publications, Washington, D.C., pp. 67–85.
[14] Berdimurodov, E., et al., New and green corrosion inhibitor based on new imidazole derivate for carbon steel in 1 M Hcl medium: Experimental and theoretical analyses. *International Journal of Engineering Research in Africa*, 2022. **58**, 11–44.
[15] Berdimurodov, E., et al., Ionic liquids as green and sustainable corrosion inhibitors I. In: L. Guo and C. Verma (Eds.), *Eco-Friendly Corrosion Inhibitors Principles, Designing and Applications*. 2022, Elsevier, Amsterdam, pp. 331–390.
[16] Cai, T., et al., Flexible cellulose/collagen/graphene oxide based triboelectric nanogenerator for self-powered cathodic protection. Materials Letters, 2022. **306**, 130904.
[17] Tozar, A., İ.H. Karahan, and Y. Yücel, Optimization of the electrophoretic deposition parameters for biocomposite hydroxyapatite/chitosan/collagen/h-BN coatings on Ti6Al4V biomedical implants. *Metallurgical and Materials Transactions A*, 2019. **50**, 1009–1020.
[18] Martinesi, M., et al., Biocompatibility studies of low temperature nitrided and collagen-I coated AISI 316L austenitic stainless steel. *Journal of Materials Science: Materials in Medicine*, 2013. **24**, 1501–1513.
[19] Zanca, C., et al., Behavior of calcium phosphate-chitosan-collagen composite coating on AISI 304 for orthopedic applications. *Polymers*, 2022. **14**(23), 5108.
[20] Berdimurodov, E., et al., Thioglycoluril derivative as a new and effective corrosion inhibitor for low carbon steel in a 1 M HCl medium: Experimental and theoretical investigation. *Journal of Molecular Structure*, 2021. **1234**, 130165.
[21] Berdimurodov, E., et al., New anti-corrosion inhibitor (3ar,6ar)-3a,6a-di-p-tolyltetrahydroimidazo[4,5-d]imidazole-2,5(1 h,3h)-dithione for carbon steel in 1 M HCl medium: gravimetric, electrochemical, surface and quantum chemical analyses. *Arabian Journal of Chemistry*, 2020. **13**, 7504–7523.
[22] Gu, L., et al., Novel biomedical applications of crosslinked collagen. *Trends in Biotechnology*, 2019. **37**(5), 464–491.
[23] Dagdag, O., et al., Graphene and graphene oxide as nanostructured corrosion inhibitors. *Carbon Allotropes: Nanostructured Anti-Corrosive Materials*. 2022, p. 133.
[24] Dagdag, O., et al., Recent progress in epoxy resins as corrosion inhibitors: design and performance. *Journal of Adhesion Science and Technology*, 2022. **37**, 1–22.
[25] Xie, X., et al., Synergistic bacteria killing through photodynamic and physical actions of graphene oxide/Ag/collagen coating. *ACS Applied Materials & Interfaces*, 2017. **9**(31), 26417–26428.
[26] de Damborenea, J., et al., Corrosion-erosion of TiN-PVD coatings in collagen and cellulose meat casing. *Surface and Coatings Technology*, 2007. **201**(12), 5751–5757.
[27] Berdimurodov, E., et al., 8-Hydroxyquinoline is key to the development of corrosion inhibitors: An advanced review. *Inorganic Chemistry Communications*, 2022. **144**, 109839.
[28] Berdimurodov, E., et al., The recent development of carbon dots as powerful green corrosion inhibitors: A prospective review. *Journal of Molecular Liquids*, 2021. **349**, 118124.
[29] Dagdag, O., et al., Epoxy coating as effective anti-corrosive polymeric material for aluminum alloys: Formulation, electrochemical and computational approaches. *Journal of Molecular Liquids*, 2021. **346**, 117886.

[30] Chen, Z., et al., Electrospinning polycaprolactone/collagen fiber coatings for enhancing the corrosion resistance and biocompatibility of AZ31 Mg alloys. *Colloids and Surfaces A: Physicochemical and Engineering Aspects*, 2023. **662**, 131041.

[31] Dagdag, O., et al., Synthesis, physicochemical properties, theoretical and electrochemical studies of Tetraglycidyl methylenedianiline. *Journal of Molecular Structure*, 2022. 133508.

[32] Dagdag, O., et al., Rheological and simulation for macromolecular matrix epoxy bifunctional aromatic amines. *Polymer Bulletin*, 2021. **79**, 7571–7587.

[33] Dewangan, Y., et al., N-hydroxypyrazine-2-carboxamide as a new and green corrosion inhibitor for mild steel in acidic medium: experimental, surface morphological and theoretical approach. *Journal of Adhesion Science and Technology*, 2022. **36**, 1–21.

[34] Córdoba, L.C., et al., Bi-layered silane-TiO_2/collagen coating to control biodegradation and biointegration of Mg alloys. *Materials Science and Engineering: C*, 2019. **94**, 126–138.

[35] Haldhar, R., et al., *Corrosion Inhibitors: Industrial Applications and Commercialization*, in *Sustainable Corrosion Inhibitors II: Synthesis, Design, and Practical Applications*. 2021, American Chemical Society, Washington, DC, pp. 10–219.

[36] Kaur, J., et al., Euphorbia prostrata as an eco-friendly corrosion inhibitor for steel: electrochemical and DFT studies. *Chemical Papers*, 2022. **77**, 1–20.

[37] Kumar, A.A., V.D. Kumar, and E. Berdimurodov, Recent trends in noble-metals based composite materials for supercapacitors: A comprehensive and development review. *Journal of the Indian Chemical Society*, 2022. **100**, 100817.

[38] Gobara, M., et al., Corrosion behaviour of AA2024 coated with an acid-soluble collagen/hybrid silica sol-gel matrix. *Progress in Organic Coatings*, 2015. **89**, 57–66.

[39] Rbaa, M., et al., Development process for eco-friendly corrosion inhibitors. In: *Eco-Friendly Corrosion Inhibitors*. 2022, Elsevier, Amsterdam, pp. 27–42.

[40] Rbaa, M., et al., Synthesis of new halogenated compounds based on 8-hydroxyquinoline derivatives for the inhibition of acid corrosion: Theoretical and experimental investigations. *Materials Today Communications*, 2022. **33**, 104654.

[41] Shahmoradi, A.R., et al., Theoretical and surface/electrochemical investigations of walnut fruit green husk extract as effective inhibitor for mild-steel corrosion in 1M HCl electrolyte. *Journal of Molecular Liquids*, 2021. **338**, 116550.

[42] Verma, D.K., et al., Ionic liquids as green and smart lubricant application: an overview. *Ionics*, 2022. **28**, 1–10.

[43] Verma, D.K., et al., N-hydroxybenzothioamide derivatives as green and efficient corrosion inhibitors for mild steel: Experimental, DFT and MC simulation approach. *Journal of Molecular Structure*, 2021. **1241**, 130648.

[44] Zhu, M., et al., Insights into the newly synthesized N-doped carbon dots for Q235 steel corrosion retardation in acidizing media: A detailed multidimensional *study. Journal of Colloid and Interface Science*, 2022. **608**, 2039–2049.

[45] Alarcon, E.I., et al., The biocompatibility and antibacterial properties of collagen-stabilized, photochemically prepared silver nanoparticles. *Biomaterials*, 2012. **33**(19), 4947–4956.

[46] Córdoba, L.C., et al., Hybrid coatings with collagen and chitosan for improved bioactivity of Mg alloys. *Surface and Coatings Technology*, 2018. **341**, 103–113.

[47] Tozar, A. and İ.H. Karahan, A comprehensive study on electrophoretic deposition of a novel type of collagen and hexagonal boron nitride reinforced hydroxyapatite/chitosan biocomposite coating. *Applied Surface Science*, 2018. **452**, 322–336.

[48] Yu, M., et al., Controlled release of naringin in metal-organic framework-loaded mineralized collagen coating to simultaneously enhance osseointegration and antibacterial activity. *ACS Applied Materials & Interfaces*, 2017. **9**(23), 19698–19705.

9 Hyaluronic Acid as Corrosion Inhibitors

Sonish Rashid and Shahab A. A. Nami
Aligarh Muslim University

9.1 INTRODUCTION

Corrosion is a major concern in various industries, including oil and gas, automotive, aerospace, and marine sectors. It leads to significant economic losses and poses safety risks. To mitigate corrosion, the use of corrosion inhibitors has been widely adopted [1–3]. Corrosion inhibitors are substances that, when added to a corrosive substance, reduce or prevent the corrosion process by either forming a protective film on the metal surface or altering the electrochemical reactions.

Hyaluronic acid (HA) is a naturally occurring biopolymer that has gained attention not only in the fields of medicine and cosmetics but also in corrosion protection. HA is a linear polysaccharide composed of repeating units of D-glucuronic acid (GlcA) and N-acetyl-D-glucosamine (GlcNAc) (Figure 9.1). It is known for its excellent biocompatibility, biodegradability, and high water-binding capacity. These properties make HA an attractive candidate for various applications, including corrosion inhibition [4–10].

9.1.1 HYALURONIC ACID: A BRIEF OVERVIEW

HA, also known as hyaluronan, is a naturally occurring polysaccharide that is widely distributed throughout the human body as well as in many other organisms. HA is widely distributed in various tissues and fluids, such as the skin, joints, eyes, and connective tissues. It is a major component of the extracellular matrix and plays a crucial role in various biological processes, such as tissue hydration, lubrication, and cell signaling. HA is a linear, high-molecular-weight polymer composed of repeating

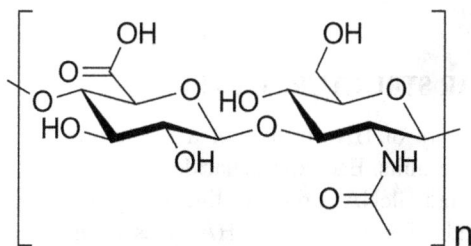

FIGURE 9.1 Chemical structure of HA polymer.

DOI: 10.1201/9781003400059-9

disaccharide units of GlcA and GlcNAc. It possesses several desirable properties, including biocompatibility, biodegradability, non-toxicity, and water solubility, making it an attractive candidate for various biomedical and industrial applications. HA has gained significant attention and recognition in both the biomedical and cosmetic fields due to its remarkable properties [11].

9.2 CORROSION INHIBITION MECHANISMS

9.2.1 FILM FORMATION

HA can form a protective film on the metal surface, acting as a physical barrier between the metal and the corrosive environment. This film prevents direct contact between the metal and corrosive species, thus reducing the corrosion rate.

9.2.2 ADSORPTION

HA can adsorb onto the metal surface, leading to the formation of a monolayer or multilayer coverage. The adsorption of HA molecules alters the electrochemical reactions occurring on the metal surface, subsequently inhibiting the corrosion process. The adsorption may occur through interactions between the functional groups of HA and the metal surface.

9.2.3 COMPLEXATION

HA has many functional groups, such as carboxyl (–COOH) and hydroxyl (–OH), which can form complexes with metal ions. This complexation process can reduce the availability of metal ions for corrosion reactions, inhibiting the corrosion process.

9.2.4 PASSIVATION

HA may facilitate the formation of a passive oxide layer on the metal surface. This oxide layer acts as a protective barrier against further corrosion by reducing the metal's reactivity with the corrosive environment.

These mechanisms are not mutually exclusive and can occur simultaneously or in combination. The specific mechanism of HA as a corrosion inhibitor may depend on factors such as pH, concentration, temperature, and the composition of the corrosive medium (Figure 9.2).

9.3 MOLECULAR STRUCTURE OF HA

The molecular structure of HA consists of a linear polysaccharide composed of repeating disaccharide units. Each disaccharide unit consists of GlcA and GlcNAc residues. The GlcA and GlcNAc units are linked together by alternating β-1,4 and β-1,3 glycosidic bonds. The structure of HA gives it unique properties, including high water-binding capacity and excellent biocompatibility.

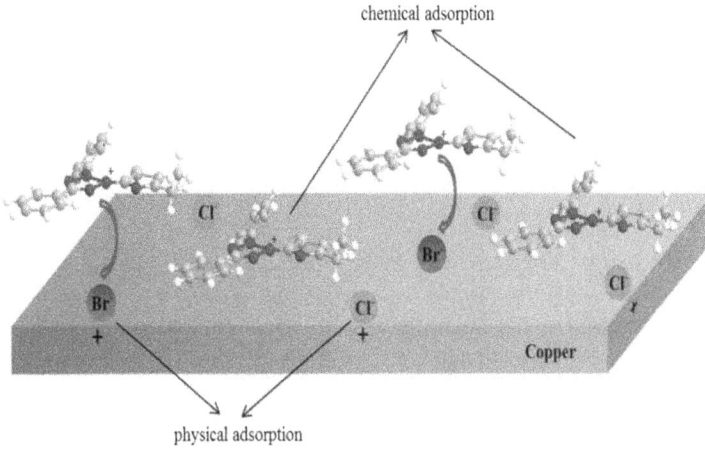

FIGURE 9.2   Different mechanisms of adsorption on Cu-surface.

9.3.1 ADSORPTION MECHANISM

The adsorption mechanism of HA as a corrosion inhibitor can involve inter-actions between the functional groups of HA and the metal surface. Although the specific adsorption mechanism of HA as a corrosion inhibitor is not available in the literature to the best of our knowledge, but one study pertaining to the adsorption behavior of HA on hydroxyapatite, which is a mineral component of bones, has been reported. The study gives an insight into the adsorption mechanism of HA on a solid surface. It explores the influence of solution pH, ionic strength, and HA concentration on the adsorption process. The authors discussed the possible mechanisms involved in HA adsorption, including electrostatic interactions, hydrogen bonding, and hydrophobic interactions. The findings highlight the importance of pH and ionic strength in modulating the adsorption behavior of HA.

9.4   EXPERIMENTAL STUDIES ON HA AS CORROSION INHIBITORS

Several studies have been reported on the corrosion inhibition performance of HA on different metallic substrates. These studies have utilized various experimental techniques, including electrochemical measurements, surface analysis, and corrosion rate calculations. For example, in one of the studies on carbon steel corrosion inhibition, HA was found to exhibit excellent inhibitory effects, reducing the corrosion rate and increasing the polarization resistance. The adsorption of HA on the metal surface was found to follow the Langmuir adsorption isotherm, suggesting a monolayer coverage of the inhibitor. Similar positive results have been reported for other metals such as aluminum, copper, and zinc. It has been demonstrated that the concentration of HA and the pH of the environment can significantly influence its effectiveness as a corrosion inhibitor.

The effect of HA concentration and pH on its corrosion inhibition properties may be summarized as:

9.4.1 CONCENTRATION OF HA

The concentration of HA in the corrosion inhibition system has a direct impact on its effectiveness. Generally, higher concentrations of HA lead to better corrosion inhibition performance. This is because a higher concentration provides a greater number of HA molecules available for adsorption onto the metal surface, resulting in a thicker and more protective barrier layer.

However, there is typically an optimum concentration range for HA, beyond which the corrosion inhibition efficiency may reach a plateau or even decrease. This can occur due to the formation of an excessive or loosely bound HA layer that is less effective in preventing the diffusion of corrosive species.

9.4.2 pH OF THE ENVIRONMENT

The pH of the corrosive environment also plays a crucial role in the corrosion inhibition behavior of HA. The adsorption and protective properties of HA are influenced by the pH-dependent ionization of its carboxyl and hydroxyl groups. The pH affects the net charge of HA molecules and the metal surface, which in turn affects the electrostatic interactions and adsorption behavior.

In general, HA exhibits better corrosion inhibition performance in neutral to slightly acidic pH conditions. This is because, at lower pH values, the carboxyl groups of HA are protonated, leading to more positively charged HA molecules. The electrostatic attraction between the positively charged HA and the negatively charged metal surface enhances the adsorption and, thereby, the formation of a protective barrier.

At very high or very low pH values, the corrosion inhibition efficiency of HA may decrease. At high pH, the deprotonation of carboxyl groups in HA reduces the positive charge, leading to weaker adsorption onto the metal surface. At low pH, the high concentration of hydrogen ions can compete with HA for adsorption sites, thus reducing its effectiveness.

It is important to note that the optimal concentration of HA and the pH range for effective corrosion inhibition may vary depending on the specific metal substrate, corrosive environment, and experimental conditions. Therefore, systematic studies and optimization are necessary to determine the most suitable concentration and pH conditions for achieving maximum corrosion inhibition efficiency using HA.

By carefully adjusting the concentration of HA and controlling the pH of the environment, it is possible to enhance the corrosion inhibition performance of HA and develop effective corrosion protection systems for various industrial applications (Figure 9.3).

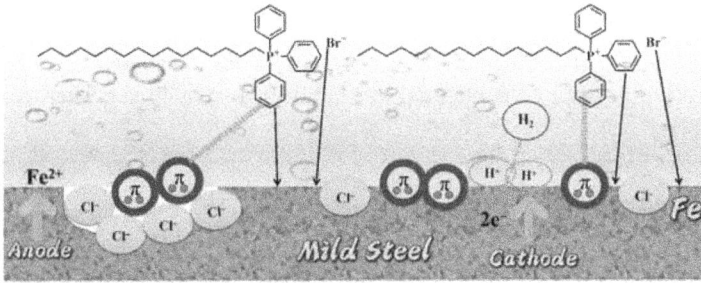

FIGURE 9.3 Adsorption mechanism of HPP on the surface of mild steel in an acidic medium.

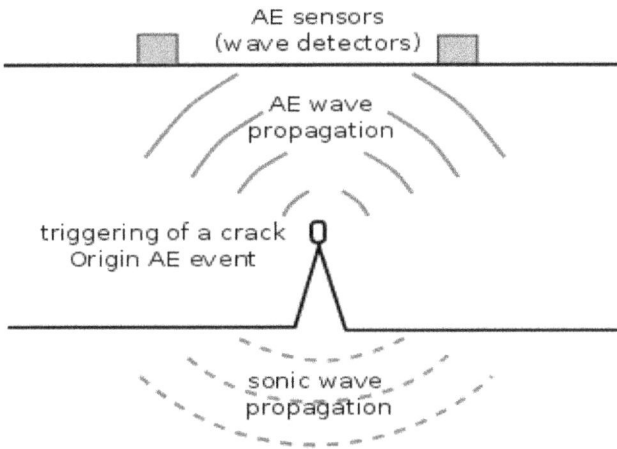

FIGURE 9.4 Pitting corrosion.

9.5 CORROSION TESTING TECHNIQUES

Several corrosion testing techniques are commonly used to evaluate the corrosion behavior of materials. Here is a list of a few widely employed techniques, along with a reference for further exploration (Figure 9.4):

9.5.1 ELECTROCHEMICAL TECHNIQUES

Electrochemical techniques are widely used for corrosion testing. They include:

a. Potentiodynamic Polarization:
   This technique involves measuring the current as a function of the applied potential to determine the corrosion potential and corrosion rate.
b. Electrochemical Impedance Spectroscopy (EIS):
   EIS measures the impedance of a corroding system as a function of frequency. It provides information about the corrosion rate, polarization resistance, and characteristics of the corrosion interface.

9.5.2 Salt Spray Test

The salt spray test, also known as the salt fog test, is used as an accelerated corrosion test. It involves exposing the material to a controlled spray of saltwater to assess its corrosion resistance.

9.5.3 Immersion Testing

Immersion testing involves immersing the material in a corrosive solution for a specific duration of time to evaluate its corrosion behavior. Weight loss measurements and visual inspection are commonly employed in this technique.

9.5.4 Cyclic Corrosion Test

The cyclic corrosion test simulates the cyclic exposure of materials to different corrosive environments, including humidity, temperature, and chemical exposure. It aims to mimic real-time conditions to assess the durability of materials.

9.5.5 Scanning Electron Microscopy (SEM) and Energy-Dispersive X-ray Spectroscopy (EDS)

SEM and EDS are used to examine the surface morphology, composition, and elemental distribution of the corroded samples, providing valuable information on the corrosion mechanisms.

Similarly, the corrosion inhibition efficiency of HA on mild steel in a hydrochloric acid solution was also studied. Here, weight loss measurements, potentiodynamic polarization, and EIS were carried out to evaluate the corrosion inhibition performance of HA. The results demonstrate that HA acts as an effective corrosion inhibitor, reducing the corrosion rate and forming a protective film on the metal surface.

The study also explores the corrosion inhibition efficiency of HA and chitosan on carbon steel in acidic media. Corrosion tests, including potentiodynamic polarization

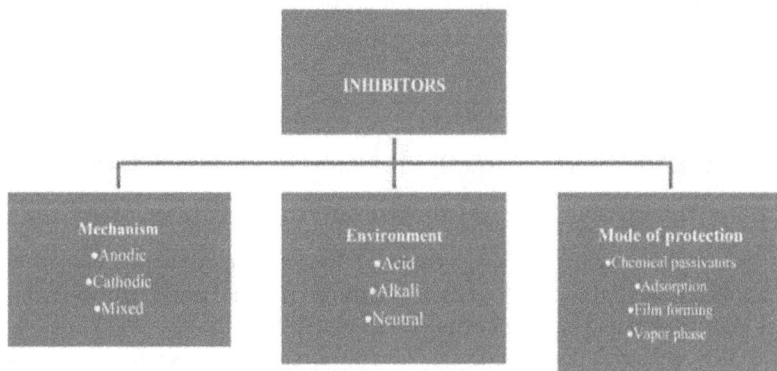

FIGURE 9.5 Block depicting the importance of HA as a corrosion inhibitor.

and EIS, were done to evaluate the inhibition performance of HA. The results indicate that HA exhibits considerable corrosion inhibition properties by forming a protective film on the metal surface (Figure 9.5).

9.6 ADVANTAGES AND LIMITATIONS OF HA AS CORROSION INHIBITOR

9.6.1 ADVANTAGES OF HA AS A CORROSION INHIBITOR

1. **Biocompatibility**: HA is a naturally occurring biopolymer found in the human body, making it highly biocompatible. Its use as a corrosion inhibitor is advantageous for applications in industries such as biomedicine and healthcare, where compatibility with biological systems is essential.
2. **Environmental benignness**: HA is considered a "green" corrosion inhibitor due to its biodegradability and non-toxic nature. Its use as a corrosion inhibitor aligns with sustainable practices and environmental regulations.
3. **Film formation**: HA can form a protective film on the metal surface, acting as a physical barrier against corrosive species. The film provides long-lasting protection and restricts the direct contact between the metal and the corrosive environment.
4. **Self-healing properties**: HA exhibits self-healing properties, enabling it to repair and regenerate the protective film on the metal surface if damaged during the corrosion process. This self-healing capability enhances the long-term corrosion protection provided by HA.
5. **Adsorption ability**: HA has many functional groups, such as carboxyl and hydroxyl groups, which facilitate its adsorption onto the metal surface. The adsorbed HA molecules modify the metal surface properties and inhibit corrosion reactions, reducing the corrosion rate.

These advantages make HA a promising corrosion inhibitor for various practical applications.

9.6.2 LIMITATIONS OF HA AS A CORROSION INHIBITOR

1. **Solubility and stability**: HA is soluble in water and certain solvents, which may limit its application in non-aqueous or organic corrosive environments. Additionally, the stability of HA under varying pH, temperature, and corrosive conditions can influence its long-term effectiveness as a corrosion inhibitor.
2. **Concentration dependency**: The corrosion inhibition efficiency of HA can be concentration-dependent. Higher concentrations of HA may lead to better corrosion protection; however, this may result in increased amount of HA, thus enhancing cost requirements.
3. **Adsorption competition**: HA may face competition for adsorption sites on the metal surface from other species present in the corrosive environment.

This competition can affect the adsorption behavior and efficiency of HA as a corrosion inhibitor.

4. **Surface coverage**: Achieving a complete and uniform surface coverage of HA on the metal surface can be challenging, while incomplete coverage may leave certain areas susceptible to corrosion attack.

5. **Limited corrosion mechanism coverage**: The corrosion inhibition mechanism of HA may be more effective against specific types of corrosion mechanisms, such as general corrosion or localized corrosion, and may not provide the same level of protection against other types of corrosion mechanisms.

9.7 APPLICATIONS OF HA AS A COROSION INHIBITOR

See Figure 9.6.

9.7.1 BIOMEDICAL AND HEALTHCARE INDUSTRIES

HA's biocompatibility and non-toxic nature make it suitable for corrosion protection in biomedical and healthcare applications. It can be employed to protect metallic implants, medical devices, and equipment from corrosive environments, ensuring their long-term performance and biocompatibility. It also finds application in cosmetic industries too.

9.7.2 MARINE AND OFFSHORE STRUCTURES

The marine environment is highly corrosive due to the presence of higher concentrations of salt in seawater, accompanied by humidity. HA can be utilized as a corrosion

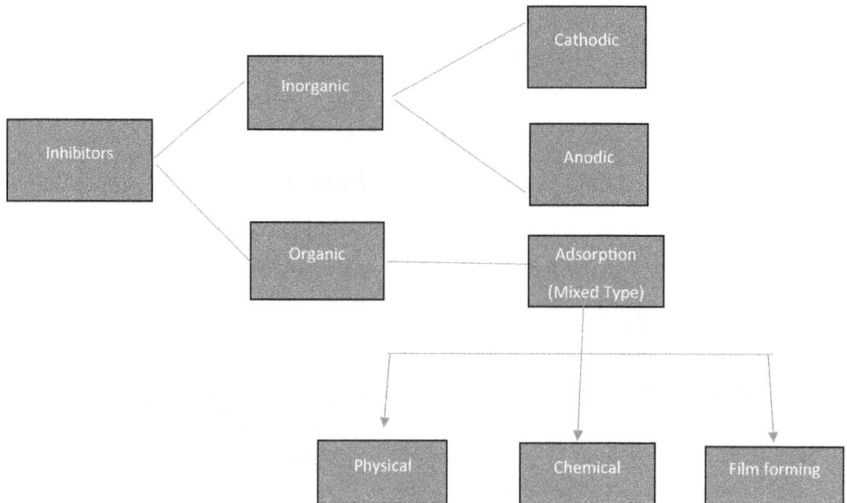

FIGURE 9.6   Corrosion inhibitor coatings.

inhibitor to protect marine and offshore structures, such as ships, offshore platforms, pipelines, and underwater equipment, from corrosion damage.

9.7.3 INDUSTRIAL AND CHEMICAL PROCESSING

HA may find applications as a corrosion inhibitor in various industrial sectors, including chemical processing, oil and gas, and manufacturing. It can be employed to protect metallic equipment, storage tanks, pipelines, and processing units from corrosive chemicals and harsh environments.

9.7.4 INFRASTRUCTURE AND CONSTRUCTION

HA has the potential to be used as a corrosion inhibitor in infrastructure and construction applications. It can help protect metallic components of bridges, buildings, and infrastructure systems from corrosion caused by exposure to environmental factors like moisture, humidity, and air pollutants.

9.7.5 AUTOMOTIVE AND TRANSPORTATION

HA may be applied as a corrosion inhibitor in the automotive and transportation industries to protect metal parts and components of vehicles, including chassis, frames, engines, and exhaust systems, from corrosion due to exposure to road salts, frictional heat, moisture, and other environmental conditions.

9.8 SYNERGISTIC EFFECTS

Synergistic effects refer to the enhanced performance achieved when combining multiple corrosion inhibitors or materials. When HA is combined with other corrosion inhibitors, such as organic compounds or inorganic substances or particles, it can lead to synergistic effects that result in improved corrosion inhibition properties [12].

For instance, HA can be combined with traditional organic corrosion inhibitors to create a composite inhibitor system. The combined action of HA and the traditional inhibitor can provide better adsorption, barrier formation, and corrosion inhibition compared to using either inhibitor alone. Synergistic effects can occur due to complementary mechanisms of action, improved adsorption affinity, or cooperative interactions between the different inhibitors. Hence, synergistic effects obtained by combining HA with other corrosion inhibitors and the use of composite materials incorporating HA can lead to enhanced corrosion protection. These approaches allow for the customization of corrosion inhibition properties, improved performance, and the development of more advanced corrosion-resistant materials for various industrial applications.

The following are some common approaches for combining HA with other corrosion inhibitors.

9.8.1 ORGANIC CORROSION INHIBITORS

HA can be combined with traditional organic corrosion inhibitors, such as benzotri-azoles, imidazolines, or amines. The combination of HA with organic inhibitors can result in improved adsorption onto the metal surface, increased barrier formation, and enhanced corrosion inhibition properties. The presence of HA can enhance the stability and effectiveness of the organic inhibitor, leading to a synergistic effect in corrosion protection.

9.8.2 INORGANIC CORROSION INHIBITORS

HA can also be combined with inorganic corrosion inhibitors, such as chromates, phosphates, or silicates. Inorganic inhibitors often provide excellent passivation properties and can form protective films on the metal surface. The addition of HA can enhance the adhesion and stability of the inorganic film, improve the coverage of the metal surface, and provide additional corrosion inhibition mechanisms. The combination of HA with inorganic inhibitors can lead to synergistic effects in terms of corrosion protection and film stability.

9.8.3 NANOPARTICLES

HA can be combined with nanoparticles, such as metal nanoparticles (e.g., silver and gold) or metal oxide nanoparticles (e.g., zinc oxide and cerium oxide). These nanoparticles can act as active corrosion inhibitors, providing localized protection and reducing the corrosion rate. The addition of HA can improve the dispersion and stability of the nanoparticles, facilitate their adsorption onto the metal surface, and enhance their overall corrosion inhibition properties.

9.9 CHALLENGES AND FUTURE PERSPECTIVE OF HA AS CORROSION INHIBITOR

HA is primarily known for its applications in the fields of medicine, cosmetics, and biomaterials rather than corrosion inhibition. Hence, research focusing specifically on the challenges and future perspectives of HA as a corrosion inhibitor is limited. However, it is worth noting that the use of natural polymers, including HA, as corrosion inhibitors is an emerging area of research. Some general challenges and future perspectives may include:

1. **Optimization of formulation**: Developing optimized formulations of HA-based corrosion inhibitors, including adjusting pH, concentration, and other additives, to enhance their stability and effectiveness in different corrosive environments.
2. **Understanding mechanisms**: Investigating the adsorption mechanisms, film formation, and interaction of HA with metal surfaces to gain a deeper understanding of its corrosion inhibition mechanisms.
3. **Compatibility with other materials**: Assessing the compatibility of HA-based corrosion inhibitors with different coating materials and paints to develop effective synergistic corrosion protection systems.

4. **Long-term performance**: Evaluating the long-term performance and durability of HA-based corrosion inhibitors under extended exposure to corrosive environments to ensure their sustained effectiveness.
5. **Multifunctional applications**: Exploring the potential of incorporating additional functionalities into HA-based corrosion inhibitors, such as self-healing properties, drug release capabilities, or stimuli-responsive behavior, to enhance their overall performance and application versatility.

9.10   CONCLUSION

HA exhibits significant potential as a corrosion inhibitor in various industries. Its unique properties, such as biocompatibility, biodegradability, and water solubility, make it an attractive alternative to conventional corrosion inhibitors. The molecular structure of HA, with its carboxyl and hydroxyl groups, enables it to form coordination complexes with metal ions, thereby inhibiting the corrosion process. The adsorption of HA onto metal surfaces forms a protective barrier that prevents corrosive species from attacking the metal substrate.

Experimental studies have shown that HA effectively reduces the corrosion rate and increases the polarization resistance of different metallic substrates. It follows the Langmuir adsorption isotherm, suggesting a monolayer coverage of the inhibitor. Moreover, the combination of HA with other compounds, such as cerium salts, or its incorporation into coatings and composite materials leads to synergistic effects and improved corrosion inhibition performance.

However, challenges remain that must be addressed to fully exploit the potential of HA as a corrosion inhibitor. Optimization of HA-based coatings and composites for long-term stability and durability under harsh environmental conditions is necessary. A comprehensive understanding of the corrosion mechanisms and the interactions between HA and different metal surfaces is vital for tailored corrosion protection solutions. Additionally, considerations regarding the scale-up and cost-effectiveness of HA production and its integration into corrosion protection systems are crucial for practical implementation.

Despite these challenges, the utilization of HA as a corrosion inhibitor offers a sustainable and environmentally friendly approach to corrosion protection. Its natural abundance, non-toxicity, and biodegradability make it an attractive option for industries seeking greener alternatives. Further research and development efforts, along with collaboration between academia, industry, and regulatory bodies, are essential to harnessing the full potential of HA in corrosion inhibition applications.

REFERENCES

[1] J. Li, Z. Xu, W. Lei, Z. Xie, and L. Zhang. "Corrosion inhibition effect of hyaluronic acid on mild steel in hydrochloric acid solution." *Progress in Organic Coatings*, vol. 114, 2018, pp. 178–185.
[2] F. Zhang, X. Cao, Y. Shi, Q. Zhang, and Y. Chen. "Corrosion inhibition of mild steel by hyaluronic acid in simulated concrete pore solution." *Corrosion Science*, vol. 134, 2018, pp. 55–63.
[3] H. Li, Y. Gu, G. Qian, C. Zhang, and L. Zhang. "Corrosion inhibition of mild steel by hyaluronic acid in hydrochloric acid solution." *Corrosion Science*, vol. 53, no. 3, 2011, pp. 1238–1245.

[4] M. Stern and J. W. Jedrzejas. "Hyaluronic acid: An integral player in the rheology of biopolymers." *Carbohydrate Polymers*, vol. 63, no. 2, 2006, pp. 210–215.

[5] Y. Cao, M. Tan, F. Zheng, C. Xie, and G. He. "Adsorption behavior of hyaluronic acid on hydroxyapatite." *Journal of Biomaterials Applications*, vol. 25, no. 7, 2011, pp. 703–714.

[6] A. Azad, V. Kucera, J. P. Celis, and M. Vereecken. "Corrosion testing techniques for characterization of metallic biomaterials: A review." *Materials Science and Engineering C*, vol. 33, no. 6, 2013, pp. 3652–3663.

[7] M. Al Qahtani, M. Saleh, N. Al Abdulsalam, et al. "Hyaluronic acid as an eco-friendly corrosion inhibitor for mild steel in hydrochloric acid solution." *Journal of Bio- and Tribo-Corrosion*, vol. 6, no. 3, 2020, article 49.

[8] M. Abou-Okeil, A. Abdel-Aziz, S. Abdel-Azim, and M. Mohamed. "Hyaluronic acid and chitosan as green corrosion inhibitors for carbon steel in acidic media." *Journal of Materials Science*, vol. 48, no. 22, 2013, pp. 7791–7801.

[9] M. Zulkifli, M. Kadhum, M. Lamin, and A. Saad. "Hyaluronic acid as a green inhibitor for mild steel in 1 M HCl solution: Electrochemical and surface studies." *Journal of Industrial and Engineering Chemistry*, vol. 32, 2015, pp. 1–11.

[10] Z. Cai, J. Zhang, L. Zhang, F. Li, and Q. Wang. "Hyaluronic acid as a green corrosion inhibitor for mild steel in 0.5 M HCl solution." *Journal of the Taiwan Institute of Chemical Engineers*, vol. 80, 2017, pp. 259–265.

[11] M. Abou-Okeil, A. Abdel-Aziz, S. Abdel-Azim, and M. Mohamed. "Hyaluronic acid and chitosan as green corrosion inhibitors for carbon steel in acidic media." *Journal of Materials Science*, vol. 48, no. 22, 2013, pp. 7791–7801.

[12] Z. Cai, J. Zhang, L. Zhang, F. Li, and Q. Wang. "Hyaluronic acid as a green corrosion inhibitor for mild steel in 0.5 M HCl solution." *Journal of the Taiwan Institute of Chemical Engineers*, vol. 80, 2017, pp. 259–265.

10 Chemically Modified Natural Alginate Polysaccharide as Green Corrosion Inhibitors

M. Rbaa, A. Rouifi, K. Alaoui, K. Dahmani, Z. Rouifi, M. Galai, Y. El Kacimi, and B. Lakhrissi
Ibn Tofail University

Elyor Berdimurodov
National University of Uzbekistan

10.1 INTRODUCTION

Pickling baths containing hydrochloric acid are commonly used to eliminate corrosion products from metal working. However, mild steels can be damaged by the corrosion process in these cases. To prevent over-pickling, steel specimens should be promptly removed from the pickling bath after removing the iron oxides. Above pickling can result in thoughtful destruction and compromise the physical assets of the metal [1].

To address this issue, corrosion inhibitors have been recognized as a highly effective and price-efficient method to safeguard metals alongside corrosion, particularly in acidic environments [2]. In recent years, many investigative studies have focused on exploring the use of biopolymers and their derivatives as environmentally friendly corrosion inhibitors [3–14].

As we are aware, this paper defines how the anti-corrosion effectiveness of alginate can be improved through chemical modification using a green condensation method with a substituted 8-hydroxy-quinoline.

The PL and PA, depicted in Figure 10.1, were characterized by means of NMR analysis. The corrosion inhibition performance of PL and PA was evaluated in an HCl medium by means of various methods [17,18].

DOI: 10.1201/9781003400059-10

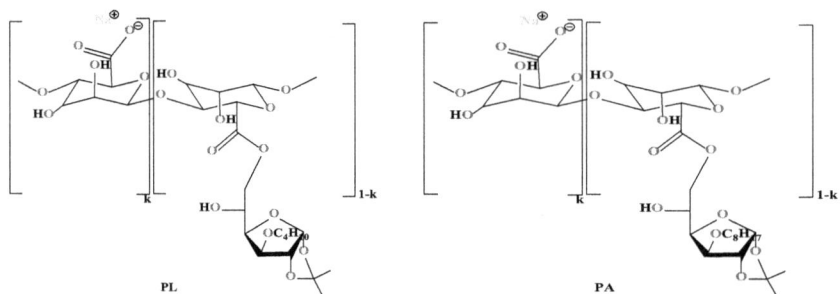

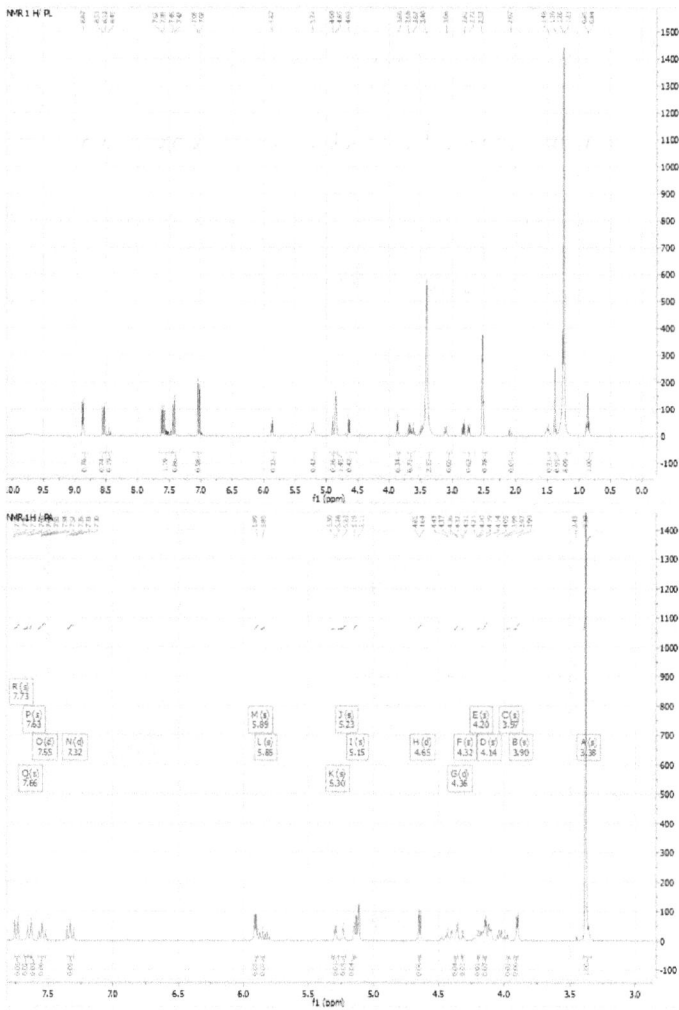

FIGURE 10.1   Chemical structures and NMR spectra of PL and PA.

10.2   CURRENT APPLICATION

Algae represent a significant reservoir of natural polysaccharides that possess distinctive physical properties. Additionally, algae contain a diverse array of nutrients, including vitamins, salts, iodine, and sterols. These organisms have been exploited for an extended period, primarily for food-related purposes, owing to their broad range of varieties. Alginates are also used in biomedical domains, bioplastic applications, packaging, textiles, paper, and wound dressing. Add-on, alginates are used also as corrosion inhibitors for material and its alloys in various mediums (Table 10.1).

TABLE 10.1
Alginate Application

Common Application	Properties	References
Foods applications (food additives in jams, jellies…)	The enhancement and stabilization of food structure primarily relied on their capacity to form gels.	[15]
	Additives capabilities of viscosifying, stabilizing, and emulsifying solutions.	[15]
Pharmaceutical industries	Purified alginates are employed as stabilizers for solutions and dispersions of solid materials.	[15]
Biomedical applications	Alginate possesses a variety of desirable properties, including bacteriostatic, antiviral, and fungistatic characteristics, as well as high absorbency, non-allergenicity, breathability, hemostasis, biocompatibility, and versatility for the inclusion of medications. Furthermore, it has excellent mechanical properties.	[16,19]
	Alginate is used for controlled drug release and cell encapsulation. It is also utilized as scaffolds for ligament and tendon tissue engineering and for the creation of molds in dentistry.	
Packaging applications	Active packaging solutions for microwave heating now include innovative products.	[20]
	Edible films with novel barrier properties and robust mechanical characteristics have been created using emulsifiable alginate. These films are capable of safeguarding encapsulated dynamic ingredients.	[21]
	Natural antibacterial agent delivers excellent edible films.	[22]
	Incorporating stable silver nanoparticles into alginate films has been shown to enhance the preservation of vegetables and fruits.	[23]
	Antibacterial films gained by integration of extruded white ginseng.	[24]
	Alginate films incorporating lactic acid bacteria were developed as antimicrobial packaging materials.	[25]
	Improvement of Antiradical properties using polycaprolactone and alginate-calcium film.	[26]
	A mixture of calcium-complexed alginate (which forms a loose gel) and starch has been suggested as a means of achieving high water retention in paper coatings in the paper industry.	[27]

(Continued)

TABLE 10.1 (*Continued*)
Alginate Application

Common Application	Properties	References
Alginates in textiles and medical textiles	Alginic acid and silver fibers is used to improve antimicrobial properties.	[28–30]
	Alginate-calcium fibers with strong tensile strength have been produced using wet spinning techniques and are suitable for the production of fabric materials.	[28–30]
	Nanoparticles of sodium alginate were loaded onto cotton fabrics.	[31]
Alginates in hemostatic material and wound dressing	Alginate is led in forward-thinking wound dressings with the aim to absorb odors, provide strong antibacterial properties, smooth pain and relieve irritation.	[30–33]
	Nanofibers of alginates used to improve processability.	
	Calcium alginate fibers take a new gel-forming ability.	
Alginate used as corrosion inhibitors	Sodium alginate as a polysaccharide biopolymer used as corrosion inhibitor for API X60 steel in neutral 3.5% NaCl.	[34]
	8-Hydroxyquinoline-grafted alginate is used as a biobased corrosion inhibitor of mild steel in 1.0 M HCl solution.	[35]

10.3 CASE STUDIES

The current environmental context is a lesson that clearly teaches us that we should design new generations of materials. Green sources for emerging corrosion inhibitors fascinate thoughtful interest in various corrosion treatments. Chemically modified natural alginate polysaccharide developed as biopolymer corrosion inhibitor was used to inhibit mild steel dissolution in hydrochloride acid medium, causing corrosion and drastic damage to material in various steel applications. These new approaches must minimize the environmental impact making possible.

The new alginate polysaccharide corrosion inhibitors (PL) and (PA).

10.3.1 POTENTIODYNAMIC POLARIZATION CURVES

By incorporating modified natural alginate polysaccharide, the polarization curves (displayed in Figure 10.2) exhibit an anodic breakdown potential, Eb. As the inhibitor concentration increases, Eb shifts toward the noble direction, indicating an elevated adsorption of inhibitors on the metallic surface. This change in Eb suggests the development of an anodic protective film on the electrode surface, which is observed at the tested concentrations of modified natural alginate polysaccharide [36–41]. However, for potentials exceeding −300 Mv/SCE relative to E_{corr}, the current/potential characteristics remain unaffected by the presence of modified natural alginate polysaccharide, and this point can be defined as the desorption potential.

The phenomenon observed can be attributed to the significant metal dissolution, which leads to the desorption of the inhibitor molecule from the electrode surface.

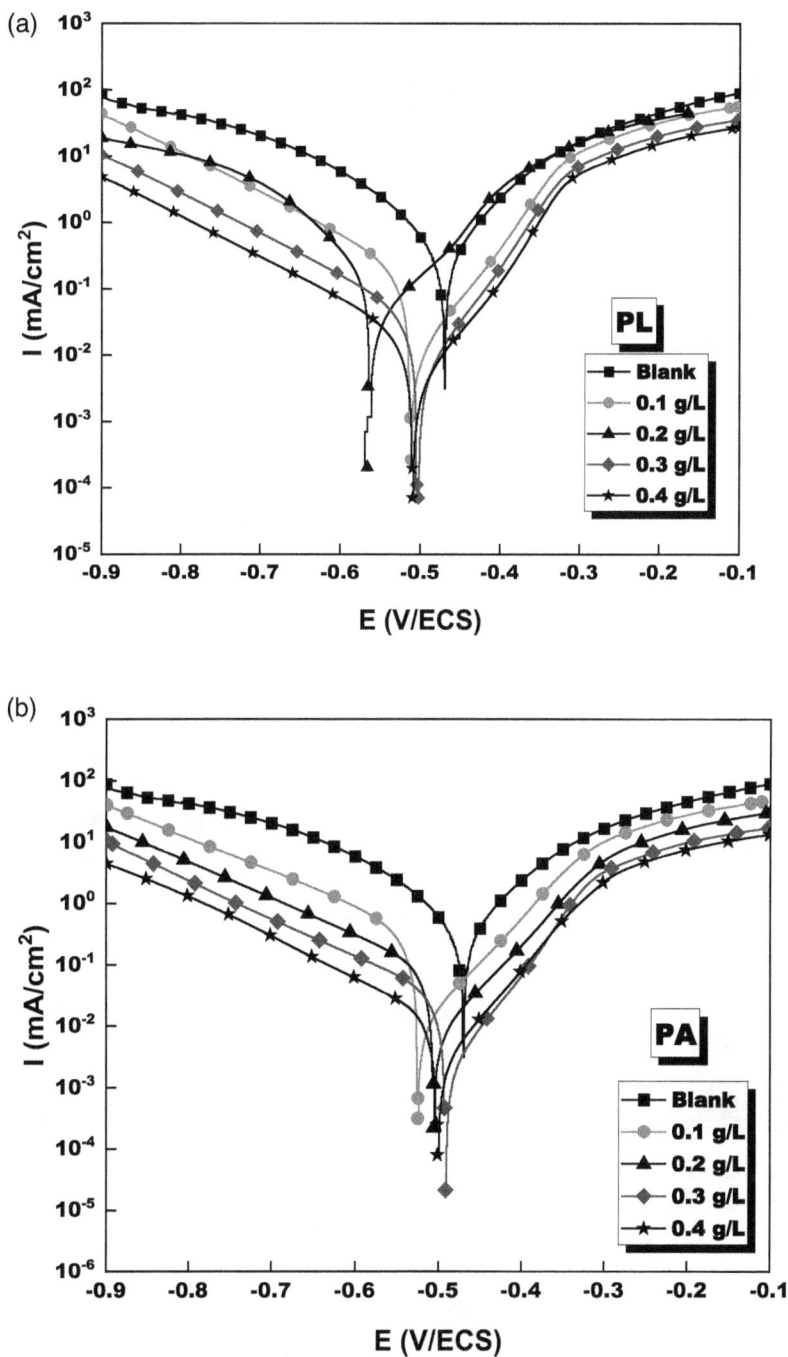

FIGURE 10.2 The potentiodynamic polarization curves were recorded for mild steel in 1.0 M HCl under two different conditions: (a) in the presence of PA and (b) in the presence of PL.

TABLE 10.2

Electrochemical Characteristics of Mild Steel in 1.0 M HCl under Different Inhibitor Concentrations

Conc. (g/L)	E_{corr} (mV vs. SCE)	i_{corr} (µA/cm²)	β_c (mV/dec)	β_a (mV/dec)	η_{pp} (%)	θ_{PP}
		Blank Solution (1.0 M HCl without Inhibitors)				
–	498	983	140	150	–	–
		PL/1.0 M HCl/Mild Steel				
0.1	513	67	139	140	93.1	0.931
0.2	563	63	135	137	93.5	0.935
0.3	505	54	138	142	94.5	0.945
0.4	509	50	132	136	94.9	0.949
		PA/1.0 M HCl/Mild Steel				
0.1	524	104	137	145	89.4	0.894
0.2	506	81	133	137	91.7	0.917
0.3	492	66	138	133	93.2	0.932
0.4	502	63	130	136	93.5	0.935

As a result, the desorption rate of inhibitors exceeds their adsorption rate causing a noticeable increase in corrosion current at higher potential [42] (Table 10.2).

The modified natural alginate polysaccharides under investigation, namely PL and PA, contain formyl and oxo groups that can react with HCl to produce protonated species. Additionally, as per Trethewey and Chamberlain [43], at high concentrations, anions in the solution can exhibit either inhibitive properties or block pores in a passive film.

In this scenario, electrostatic repulsion makes it difficult for the protonated molecules of the modified natural alginate polysaccharides to approach the positively charged surface of mild steel. However, the presence of Cl⁻ ions allows the anions to adsorb onto the mild steel surface, facilitating the approach of protonated molecules to the surface [44–46].

10.3.2 ELECTROCHEMICAL IMPEDANCE SPECTROSCOPY

Figure 10.3 shows the typical Nyquist plots for mild steel in 1.0 M HCl, both with and without various concentrations of PL and PA at the E_{OCP}. The impedance response exhibits a single capacitive loop, with a noticeable change in its diameter after the addition of all inhibitors. The maximum diameter is achieved at 0.4 g/l of modified natural alginate polysaccharide. This capacitive loop can be attributed to the charge transfer reaction (Figure 10.4).

Figure 10.5 displays the equivalent electrical circuit with a single time constant used in this study, comprising elements such as Rs, representing the solution resistance, Rct, denoting the charge transfer resistance, and CPEdl, the constant phase element of the double layer, as previously described.

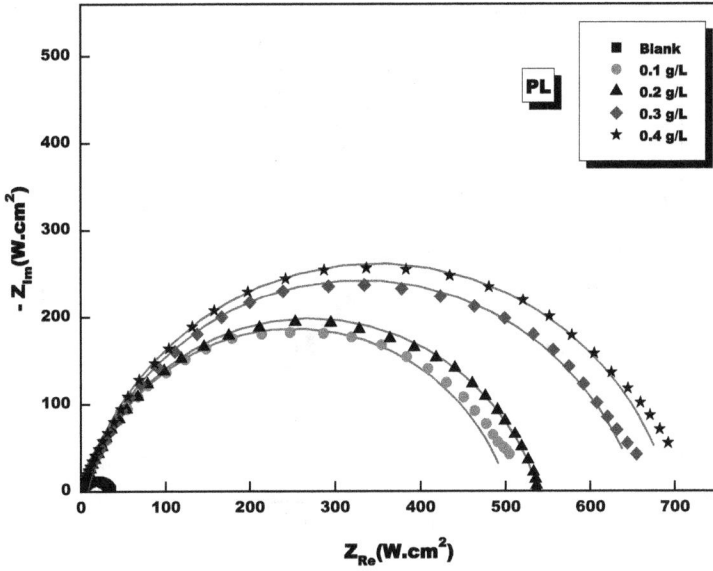

(a) PT

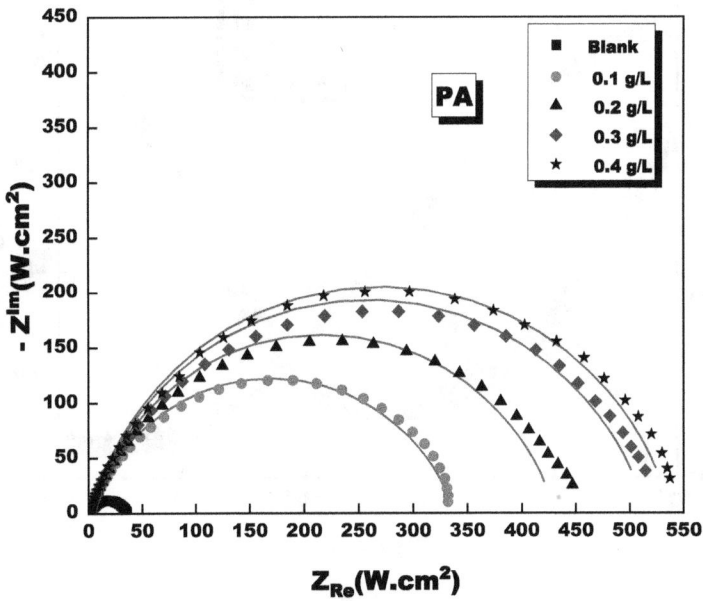

(b) PA

FIGURE 10.3 Nyquist plots were generated for mild steel in a 1.0 M HCl solution both with and without the presence of various concentrations of (a) PA and (b) PL at 298 ±2 K.

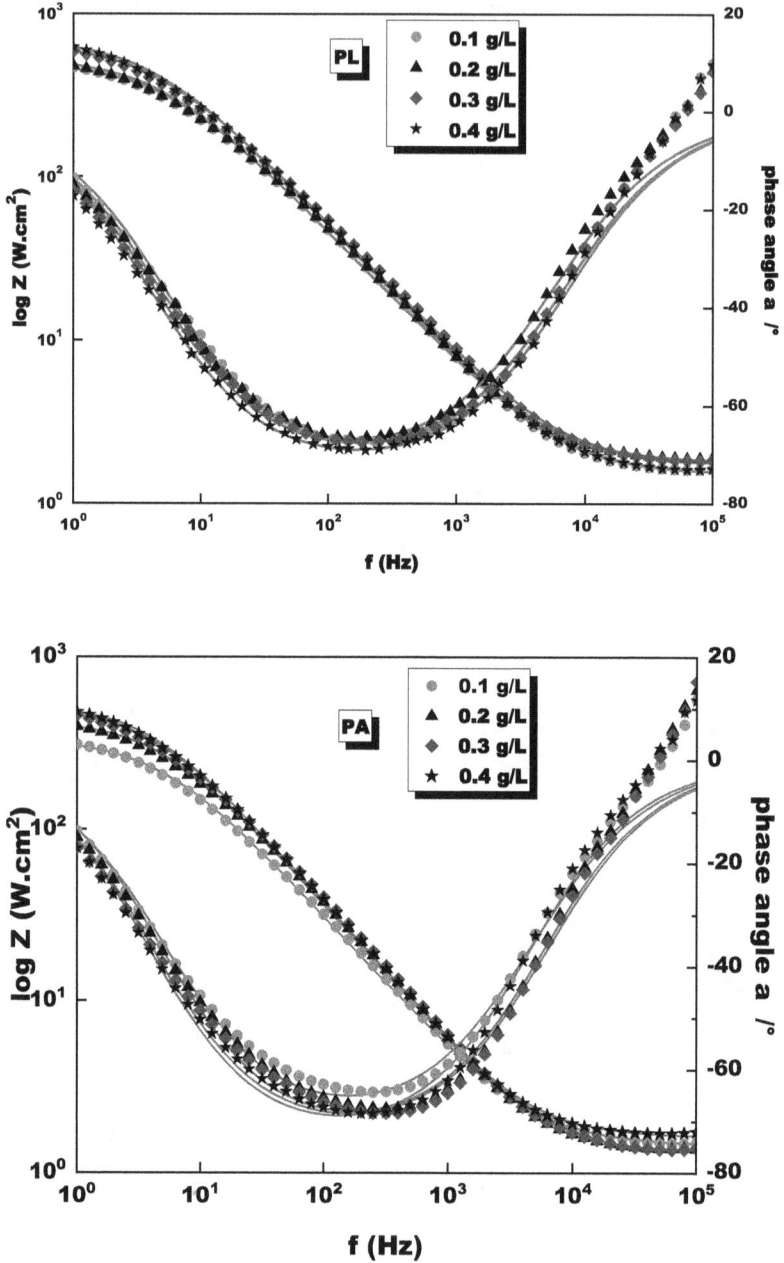

FIGURE 10.4 Bode plots for mild steel in 1.0 M HCl solution in the absence and presence of various concentrations of PA and PL at E_{corr} ($T = 298 \pm 2$ K): comparison of experimental (scatter) and fitting (red line) data.

$$CPE_{dl}$$

Rs
→>
R_{ct}

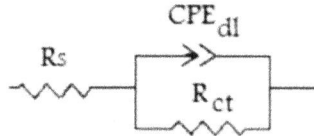

FIGURE 10.5 The metal/solution interface of mild steel can be modeled using electrical equivalent circuits.

TABLE 10.3
Electrochemical Characteristics of Mild Steel in a 1.0 M HCl Solution Were Analyzed Both with and without the Presence of Various Concentrations of Inhibitor Compounds at 298 ± 2 K

Concentration (M)	R_s (Ω cm²)	R_{ct} (Ω cm²)	C_{dl} (μF/cm²)	n_{dl}	η_{EIS} (%)
Blank Solution (1.0 M HCl/Mild Steel/without Inhibitors)					
00	1,107	34.8	420	0.772	–
PL/1.0 M HCl/Mild Steel					
0.1	1.517	501.3	112	0.815	93.0
0.2	1.7	536.8	116	0.8	93.5
0.3	1.722	654.0	101	0.812	94.6
0.4	1.521	692.7	97	0.823	94.9
PA/1.0 M HCl/Mild Steel					
0.1	1.439	336.5	184	0.800	89.6
0.2	1.300	429.6	136	0.822	91.8
0.3	1.292	514.2	131	0.821	93.2
0.4	1.604	537.2	126	0.831	93.5

The impedance parameters, such as R_s, R_{ct}, n, C_{dl} and inhibition efficiency η_{EIS} (%), obtained from fitting the *EIS* data using the equivalent circuit of Figure 10.3, are calculated and listed in Table 10.3.

Upon examining the data presented in Table 10.3, it can be observed that the R_{ct} value exhibits a positive correlation with the concentration of the inhibitors. Furthermore, the C_{dl} value demonstrates a consistent pattern of variation in response to changes in the concentration of modified natural alginate polysaccharides. These fluctuations in the measured parameters can be attributed to the gradual displacement of water molecules on the metallic surface by PL and PA molecules, resulting in a reduction in the number of active sites that are essential for the corrosion reaction. [47–54].

The rise in R_{ct} value is attributed to the development of a protective film at the interface between the metal and the solution. The constant value of 'n' within the range of '0.8–0.831' indicates that the dissolution mechanism of mild steel is controlled by charge transfer, both in the presence and absence of PL and PA inhibitors [48].

The impedance plots were analyzed using an electrical equivalent circuit, as illustrated in Figure 10.5.

Influence of temperature: Changes in temperature can lead to modifications of the steel electrode and the interface between PA and PL compounds. Figure 10.6 displays the polarization curves that depict the behavior of mild steel in 1.0 M HCl

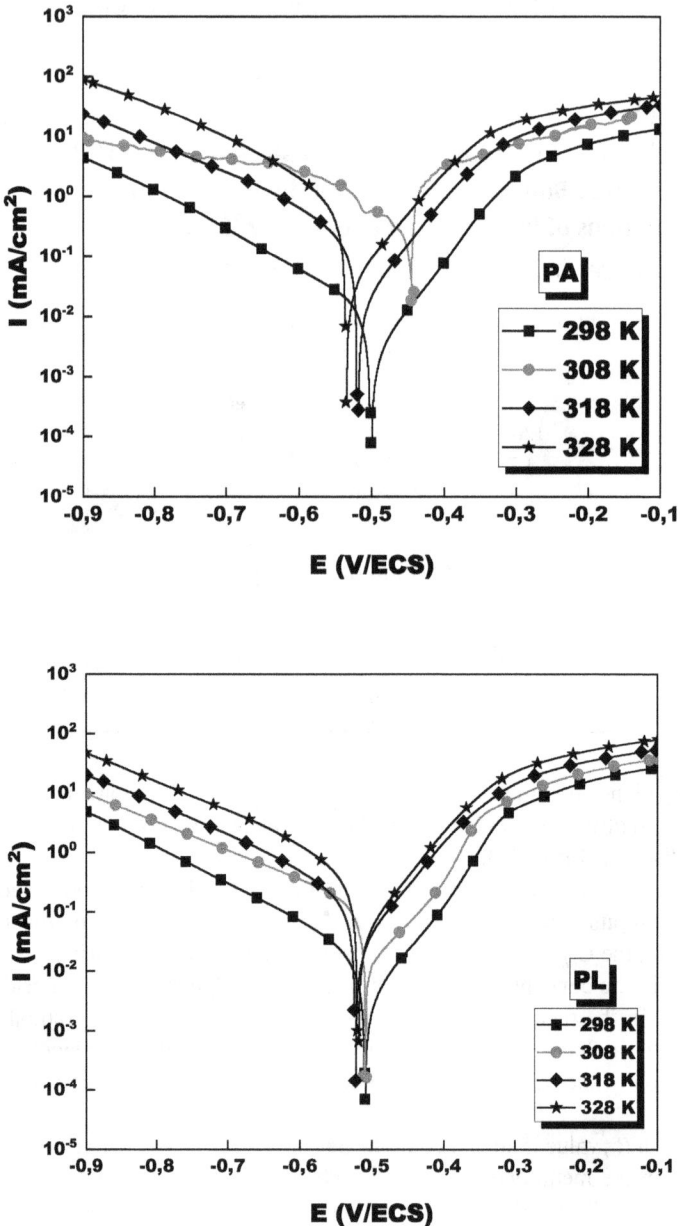

FIGURE 10.6 Investigation of the effect of temperature on the polarization curves of mild steel in 1.0 M HCl with the presence of PA and PL.

with PA and PL molecules, and their corresponding data are presented in Table 10.4. The polarization curves show slight variations in the cathodic and anodic branches. Table 10.4 reveals an increase in the i_{corr} of MS with increasing temperature for both inhibited and uninhibited solutions. Additionally, the corrosion current density of MS increases more rapidly with temperature in the absence of PL and PA compounds. These results confirm that the PL and PA compounds are effective inhibitors for all temperature ranges studied [49].

Table 10.5 presents the values of activation energy (E_a), enthalpy (ΔH_a), and entropy (ΔS_a), which were evaluated based on the polarization curves data obtained in the absence and with the optimal concentration of PA and PL.

Figure 10.7 illustrates a linear relationship between the natural logarithm of the corrosion current density and 1,000/T. The slope of the line was used to determine the activation energy values for molar hydrochloride acid with and without PA and PL compounds. Similarly, Figure 10.8 displays a linear relationship between the natural logarithm of i_{corr}/T and 1,000/T.

TABLE 10.4

The Impact of Temperature on the Electrochemical Characteristics of MS in 1.0 M HCl with 10^{-3} M Concentrations of PA and PL Compounds

Medium	Temp (K)	$-E_{corr}$ (mV/SCE)	i_{corr} (μA/cm²)	$-\beta_c$ (mV/dec)	β_a (mV/dec)	η_{PDP} (%)
HCl	298±2	497	984	93	105	–
	308±2	492	1,202	193	122	–
	318±2	474	1,381	72	123	–
	328±2	470	2,202	164	127	–
PL	298±2	509	50	132	136	94.9
	308±2	508	90	165	107	92.5
	318±2	524	145	163	129	90.0
	328±2	520	266	149	128	87.9
PA	298±2	502	63	130	136	93.5
	308±2	441	105	179	109	91.2
	318±2	520	149	156	139	89.7
	328±2	536	299	154	125	86.4

TABLE 10.5

Activation Parameters of the Dissolution of MS in 1.0 M HCl

Medium	E_a (kJ/mol)	ΔH_a (kJ/mol)	ΔS_a (J/mol/K)
Blank	21.0	18.5	−126
PA	44.5	41.9	−71.6
PL	40.6	38.0	−83.0

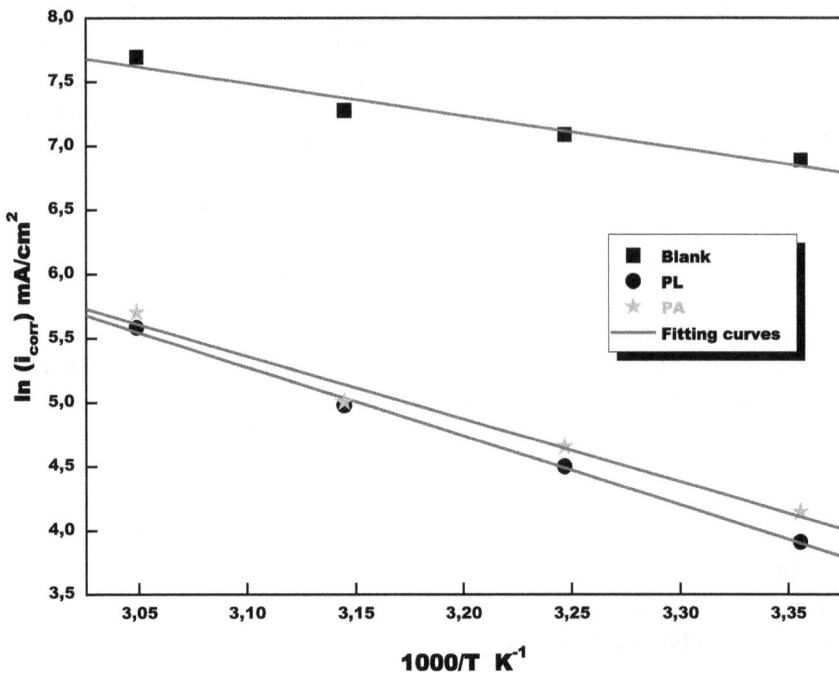

FIGURE 10.7   Arrhenius plots for MS in 1.0 M HCl.

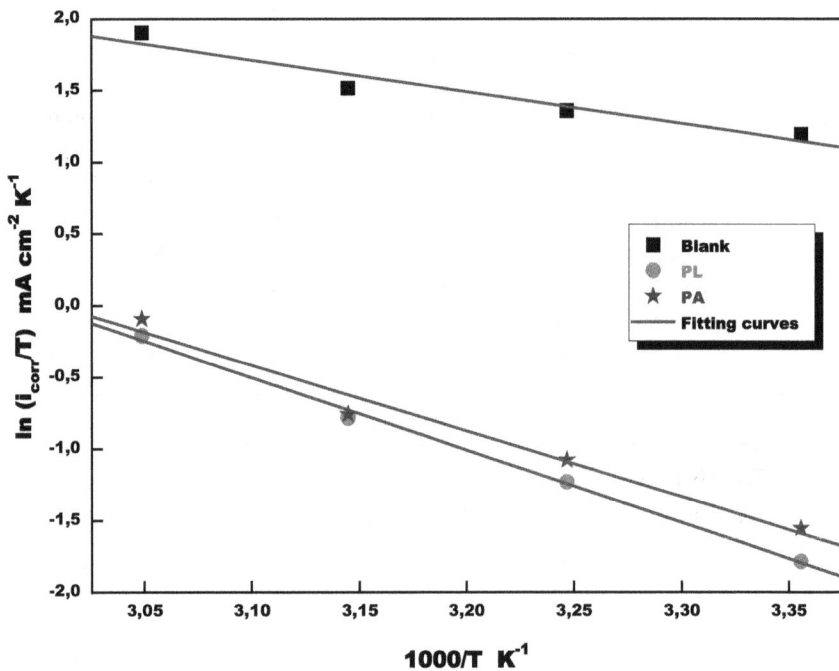

FIGURE 10.8   Transition state plot MS in 1.0 M HCl.

TABLE 10.6

Adsorption Parameters for Substituted Quinolines Compounds on Mild Steel in 1.0 M HCl at 298 ± 1 K Using the Langmuir Isotherm

Inhibitors	K_{ads} (L/mol)	R^2	ΔG°_{ads} (kJ/mol)
PL	268.0	0.99998	−23.7
PA	150.6	0.99999	22.3

The inhibition mechanism depends on the adsorption of inhibitors on the metal surface, which is a critical step. The level of corrosion inhibition is affected by the extent of inhibitor adsorption on the mild steel surface. In order to understand the adsorption of PA and PL on the mild steel surface, different adsorption isotherms were analyzed. The Langmuir isotherm provided the best fit with regression coefficient (R^2) values near unity for both PA and PL compounds [49,54].

Computed values for each examined inhibitor are presented in Table 10.6. The significantly high Kads values for PA and PL compounds suggest that their adsorption onto the MS surface, which involves the displacement of water molecules, is a favorable process. This indicates that the organic compounds exhibit strong adsorption onto the metal surface. Additionally, the trend in the Kads and ΔG values suggest that PA > PL, which corresponds to their relative inhibitory efficacy.

10.4 FUTURE PREDICTIONS

Exploring the corrosion performance of modified natural alginate polysaccharides of different metals in an alternative acid medium: H_2SO_4, H_3PO_4, etc. would be of significant interest. This investigation would provide further insight into the potential applications and efficacy of the modified natural alginate polysaccharide.

10.5 CONCLUSION

All of the investigated modified natural alginate polysaccharides were identified as genuine corrosion inhibitors for mild steel. Potentiodynamic polarization curves demonstrated that these compounds act as mixed-type inhibitors, with inhibition efficiency increasing with concentration up to a maximum of 0.4 g/L. Additionally, the presence of inhibitors alters the mechanisms of hydrogen reduction and iron dissolution reactions. Inhibition occurs by adsorption on the metallic surface, with the inhibitor's effectiveness being influenced by the presence of formyl and oxo groups in their structures, as well as the size difference between PL and PA molecules. PL molecules were found to be the most effective inhibitor among the studied compounds. The inhibition efficiencies of the compounds investigated were consistent across all measurement techniques.

REFERENCES

[1] Y. El Kacimi, M. Galai, K. Alaoui, R. Touir, M. Ebn Touhami, Surface morphology studies and kinetic-thermodynamic characterization of steels treated in 5.0 M HCl medium: hot-dip galvanizing application. *Anti-Corrosion Methods and Materials*, 65:176–189 (2018). doi:1108/ACMM-11-2016-1735.

[2] M. Rbaa, M. Galai, M. Ouakki, R. Hsissou. A. Berisha, S. Kaya, E. Berdimurodov, B. Lakhrissi, A. Zarrouk, Synthesis of new halogenated compounds based on 8-hydroxy-quinoline derivatives for the inhibition of acid corrosion: Theoretical and experimental investigations. *Materials Today Communications*, 12:104654 (2022). doi:1016/j.mtcomm.2022.104654.

[3] E. Berdimurodov, A. Kholikov, K. Akbarov, L. Guo, S. Kaya, K. P. Katin, D. K. Verma, M. Rbaa, O. Dagdag, Novel cucurbit[6]uril-based [3]rotaxane supramolecular ionic liquid as a green and excellent corrosion inhibitor for the chemical industry. *Colloids and Surfaces A: Physicochemical and Engineering*, 2:127837 (2021). doi:1016/j.colsurfa.2021.127837

[4] A. O. James, N. C. Oforka, K. Abiola, Inhibition of acid corrosion of mild steel by pyridoxal and pyridoxol hydrochlorides. *International Journal of Electrochemical Science*, 2:278–284 (2007).

[5] M. Rbaa, F. Benhiba, R. Hssisou, Y. Lakhrissi, B. Lakhrissi, M. E. Touhami, A. Zarrouk, Green synthesis of novel carbohydrate polymer chitosan oligosaccharide grafted on d-glucose derivative as bio-based corrosion inhibitor. *Journal of Molecular Liquids*, 114549 (2021). doi:1016/j.molliq.2020.114549.

[6] Y. el Kacimi, M. Achnin, A. Younas, M. Ebn Touhami, A. Alami, R. Touir, M. Sfaira, D. Chebabe, A. Elachqar, B. Hammouti, Inhibition of mild steel corrosion by some phenyltetrazole substituted compounds in hydrochloric acid. *Portugaliae Electrochimica Acta*, 30(1):53–65 (2012). doi:10.4152/pea.201201053

[7] Y. El Kacimi, M. A. Azaroual, R. Touir, M. Galai, K. Alaoui, M. Sfaira, M. Ebn Touhami, S. Kaya, Corrosion inhibition studies for mild steel in 5.0 M HCl by substituted phenyltetrazole. *Euro-Mediterranean Journal for Environmental Integration*, 2017:2(1):1–11. doi:10.1007/s41207-016-0011-8

[8] M. Rbaa, M. Fardioui, C. Verma, A. S. Abousalem, M. Galai, E. E. Ebenso, A. Zarrouk, 8-Hydroxyquinoline based chitosan derived carbohydrate polymer as biodegradable and sustainable acid corrosion inhibitor for mild steel: Experimental and computational analyses. *International Journal of Biological Macromolecules*, 155:645–655 (2020). doi:1016/j.ijbiomac.2020.03.200.

[9] M. Rbaa, F. Benhiba, A. S. Abousalem, M. Galai, Z. Rouifi, H. Oudda, A. Zarrouk, Sample synthesis, characterization, experimental and theoretical study of the inhibitory power of new 8-hydroxyquinoline derivatives for mild steel in 1.0 M HCl. *Journal of Molecular Structure*, 1213:128155 (2020). doi:1016/j.molstruc.2020.128155.

[10] M. Rbaa, P. Dohare, A. Berisha, O. Dagdag, L. Lakhrissi, M. Galai, A. Zarrouk, New Epoxy sugar based glucose derivatives as ecofriendly corrosion inhibitors for the carbon steel in 1.0 M HCl: Experimental and theoretical investigations. *Journal of Alloys and Compounds*, 833:154949 (2020). doi:1016/j.jallcom.2020.154949.

[11] M. Rbaa, A. S. Abousalem, M. E. Touhami, I. Warad, F. Bentiss, B. Lakhrissi, A. Zarrouk, Novel Cu (II) and Zn (II) complexes of 8-hydroxyquinoline derivatives as effective corrosion inhibitors for mild steel in 1.0 M HCl solution: Computer modeling supported experimental studies. *Journal of Molecular Liquids*, 290:111243 (2019). doi:1016/j.molliq.2019.111243.

[12] N. Errahmany, M. Rbaa, A. Tazouti, E. H. EL Kafssaoui, B. Lakhrissi, R. Touir, M. Ebn Touhami, Electrochemical and theoretical studies of novel quinoxaline derivatives as corrosion inhibitor for mild steel in HCl solution. *Analytical and Bioanalytical Electrochemistry*, 11(8):1032–1056 (2019).

[13] M. Rbaa, B. Lakhrissi, Novel oxazole and imidazole based on 8-hydroxyquinoline as a corrosion inhibition of mild steel in hcl solution: Insights from experimental and computational studies. *Surfaces and Interfaces*, 15:43–59 (2019). doi:1016/j.surfin.2019.01.010.

[14] M. E. Azhar, B. Mernari, H. Elattari, M. Traisnel, F. Bentiss, M. J. Lagrenee, Inhibiting effects of 3,5-bis(n-pyridyl)-4-amino-1,2,4-triazoles on the corrosion for mild steel in 1 M HCl medium. *Corrosion Science*, 40:391 (1998).

[15] W. Sabra, W.-D. Deckwer, Alginate: A polysaccharide of industrial interest and diverse biological functions. In: S. Dumitriu (Ed.), *Polysacharides: Structural Diversity and Functional Versatility*, 2d edn., Marcel Dekker, New York, pp. 515–533, 2005.

[16] Fundueanu, G. et al. Prepration and characterization of Ca-alginate microspheres by a new emulsification method. *International Journal of Pharmaceutics* 170:11–21 (1998).

[17] Y. Yi, R. J. Neufeld, D. Poncelet, Immobilization of cells in polysaccharide gels. In: S. Dumitriu (Ed.), *Polysacharides: Structural Diversity and Functional Versatility*, 2d edn., Marcel Dekker, New York, pp. 867–891, 2005.

[18] K. Y. Lee, D. J. Mooney, Alginate: Properties and biomedical applications. *Progress in Polymer Science*, 7, 106–126 (2012)

[19] K. I. Draget, C. Taylor, Chemical, physical and biological properies of alginates and their biomedical implications. *Food Hydrocolloids*, 25, 251–256 (2011).

[20] H. Kong, D. Mooney, Polysaccharide-based hydrogels in tissue engineering. In: S. Dumitriu (Ed.), *Polysacharides: Structural Diversity and Functional Versatility*, 2d edn., Marcel Dekker, New York, pp. 817–837, 2005.

[21] A. Albert, A. Salvador, S. M. Fiszman, A film of alginate plus salt as an edible susceptor in microwaveable food. *Food Hydrocolloids*, 27:421–426 (2012).

[22] A. Hambleton, F. Debeaufort, A. Bonnote, A. Voilley, Influence of alginate emulsion-based films structure on its barrier properties and on the protection of microencapsulated aroma compound. *Food Hydrocolloids*, 23:2116–2124 (2009).

[23] Y. Pranoto, V. M. Salokhe, SK. Rakshit, Physical and antibacterial properties of alginate-based film incorporated with garlic oil. *Food Research International*, 38:267–272 (2005).

[24] A. Fayaz, K. Balaji, M. Girilal, P. T. Kalaichelvan, R. Venkatesan, Mycobased synthesis of silver nanoparticles and their incorporation into sodium alginate films for vegetable and fruit preservation. *Journal of Agricultural and Food Chemistry*, 57:6246–52 (2009).

[25] K. Norajit, G. H. Ryu, Preparation and properties of antibacterial alginate films incorporating extruded white ginseng extract. *Journal of Food Processing and Preservation*, 35:387–393 (2011).

[26] A. Concha-Meyer, R. Schoebitz, C. Brito, R. Fuentes, Lactic acid bacteria in an alginate film inhibit Listeria monocytogenes growth on smoked salmon. *Food Control*, 22:485–489 (2011).

[27] S. Salmieri, M. Lacroix, Physicochemical properties of alginate/polycaprolactone-based films containing essential oils. *Journal of Agricultural and Food Chemistry*, 54:10205–10214 (2006).

[28] M. Joyce, S. A. Gilbert, Effect of Ca^{+2} on the water retention of alginate in paper coatings. *Journal of Pulp and Paper Science*, 22:J126–J130 (1996).

[29] Y. Qin, The characterization of alginate wound dressing with different fiber and textile structures. *Journal of Applied Polymer Science*, 100:2516–2520 (2006).

[30] Y. Qin, Alginate fibres: An overview of the production processes and applications in wound management. *Polymer International*, 57:171–108 (2008).

[31] Y. Qin, The gel swelling properties of alginate fibers and their applications in wound management. *Polymer for Advanced Technologies*, 19:6–14 (2008).

[32] K. Alaoui, N. Dkhireche, M. Ebn Touhami, Y. El Kacimi, Review of application of imidazole and imidazole derivatives as corrosion inhibitors of metals, *New Challenges and Industrial Applications for Corrosion Prevention and Control*, 120–133 (2020). doi:10.4018/978-1-7998-2775-7.ch005

[33] S. Petrulyte, Advanced textile materials and biopolymers in wound management. *Danish Medical Bulletin*, 55(1):72–77 (2008).

[34] I. B. Obota, Ikenna B. Onyeachub, A. Madhan Kumara, Sodium alginate: A promising biopolymer for corrosion protection of API X60 high strength carbon steel in saline medium. *Carbohydrate Polymers*, 178:200–208 (2017).

[35] M. Fardioui, M. Rbaa, F. Benhiba, M. Galai, T. Guedira, B. Lakhrissi, I. Warad, A. Zarrouk, Bio-active corrosion inhibitor based on 8-hydroxyquinoline-grafted-alginate: Experimental and computational approaches, *Journal of Molecular Liquids*, 323:114615. (2021).

[36] M. Stern, A. L. Geary, Electrochemical Polarization I: A theoretical analysis of the shape of polarization curves. *Journal of the Electrochemical Society*, 104:56–63 (1957).

[37] I. B. Obot, N. O. Obi-Egbedi, Acenaphtho [1,2–b] quinoxaline as a novel corrosion inhibitor for mild steel in 0.5 M H_2SO_4. *Corrosion Science*, 52:923–926 (2010).

[38] K. Alaoui, A. Abousalem, B. Tüzün, Y. El Kacimi, Triazepines compounds as novel synthesized corrosion inhibitors: State of play and perspectives. *New Challenges and Industrial Applications for Corrosion Prevention and Control*, 120–133 (2020). doi:10.4018/978-1-7998-2775-7.ch007

[39] Y. Abboud, A. Abourriche, T. Saffaj, M. Berrada, M. Charrouf, A. Bennamara, N. Al Himidi, H. Hannache, Corrosion inhibition of carbon steel in acidic media by bifurcaria bifurcata extract. *Materials Chemistry and Physics*, 105:1–5 (2007).

[40] F. Bentiss, C. Jama, B. Mernari, H. El Attari, L. El Kadi, M. Lebrini, M. Traisnel, M. Lagrenée, Corrosion control of mild steel using 3,5-bis(4-methoxyphenyl)-4-amino-1,2,4-triazole in normal hydrochloric acid medium. *Corrosion Science*, 51:1628–1635 (2009).

[41] H. H. Hassan, E. Abdelghani, M. A. Amin, Inhibition of mild steel corrosion in hydrochloric acid solution by triazole derivatives: Part I: Polarization and EIS studies. *Electrochimca Acta*, 52:6359–6366 (2007).

[42] A. A. Aksüt, W. J. Lorenz, F. Mansfeld, The determination of corrosion rates by electrochemical d.c. and a.c. methods-II. Systems with discontinuous steady state polarization behavior. *Corrosion Science*, 22:611–619 (1982).

[43] K. R. Trethewey, J. Chamberlain, *Corrosion for Science and Engineering*, 2nd edn., Edinburgh Gate Harlow, Essex CM20 2JE, England, 1995.

[44] M. Behpour, S. M. Ghoreishi, N. Soltani, M. Salavati-Niasari, M. Hamadanian, A. Gandomi, Electrochemical and theoretical investigation on the corrosion inhibition of mild steel by thiosalicylaldehyde derivatives in hydrochloric acid solution. *Corrosion Science*, 50:2172–2181 (2008).

[45] R. Solmaz, Investigation of the inhibition effect of 5-((E)-4-phenylbuta-1,3-dienylideneamino)-1,3,4-thiadiazole-2-thiol Schiff base on mild steel corrosion in hydrochloric acid. *Corrosion Science*, 52:3321–3330 (2010).

[46] A. S. EL-Gaber, A. S. Fouda, A. Desoky, Synergistic inhibition of Zn corrosion by some anions in aqueous media. *Ciência & Tecnologia dos Materiais*, 20:71–77 (2008).

[47] Y. El Kacimi, R. Touir, M. Galai et al., Effect of silicon and phosphorus contents in steel on its corrosion inhibition in 5 M HCl solution in the presence of Cetyltrimethylammonium/KI. *Journal of Materials and Environmental Science*, 7:371–381 (2016).

[48] D. A. Lopez, S. N. Simison, S. R. Sanchez, The influence of steel microstructure on CO_2 corrosion. EIS studies on the inhibition efficiency of benzimidazole. *Electrochimica Acta*, 48:845–854 (2003).

[49] M. Ouakki, M. Rbaa, M. Galai, B. Lakhrissi, E. H. Rifi, M. Cherkaoui, experimental and quantum chemical investigation of imidazole derivatives as corrosion inhibitors on mild steel in 1.0 M hydrochloric acid. *Journal of Bio-and Tribo-Corrosion*, 4(3):35 (2018). doi:1007/s40735-018-0151-2).

[50] A. Y. Musa, A. A. H. Kadhum, A. B. Mohamad, M. S. Takriff, Experimental and theoretical study on the inhibition performance of triazole compounds for mild steel corrosion. *Corrosion Science*, 52:3331–3340 (2010).

[51] N. Labjar, M. Lebrini, F. Bentiss, N. Chihib, S. ElHajjaji, C. Jama, Corrosion inhibition of carbon steel and antibacterial properties of aminotris-(methylenephosphonic) acid. *Materials Chemistry and Physics*, 119:330–336 (2010).

[52] X. Zheng, S. Zhang, W. Li, L. Yin, J. He, J. Wua, Investigation of 1-butyl-3-methyl-1H-benzimidazolium iodide as inhibitor for mild steel in sulfuric acid solution. *Corrosion Science*, 80:383–392 (2014).

[53] K, Alaoui, Y. El Kacimi, M. Galai, K. Dahmani, R. Touir, A. El Harfi, M. Ebn Touhami, Poly (1-phenylethene): As a novel corrosion inhibitor for carbon steel/hydrochloric acid interface. *Analytical and Bioanalytical Electrochemistry*, 8:830–847 (2016).

[54] K. Alaoui, N. Dkhireche, M. E. Touhami, Y. E. Kacimi, Review of application of imidazole and imidazole derivatives as corrosion inhibitors of metals (2020). doi:10.4018/978-1-7998-2775-7.ch005.

11 Lignin as Corrosion Inhibitors

Taiwo W. Quadri and Eno E. Ebenso
University of South Africa

11.1 INTRODUCTION

Metallic corrosion poses a significant and persistent challenge in various industries, leading to substantial economic losses and serious environmental concerns. The corrosion phenomenon is a chemical process that flourishes in the presence of liquid chemicals, gases, acids, bases, salts, water, and aggressive metal polishes. Common industrial metals and alloys such as iron, mild steel (MS), carbon steel (CS), zinc, aluminum, brass, tin, and copper are predisposed to corrosive electrolytes in the environment. Consequently, these metals and alloys experience depreciation in their useful properties [1,2]. One of the most viable methods for curbing the progress of industrial metallic corrosion is the introduction of chemical additives widely known as corrosion inhibitors. Several decades of intensive research have witnessed the development of thousands of chemical inhibitors with varying inhibition efficiencies (%IE) for different tested metals in different conditions. The classification of chemical inhibitors into inorganic or organic origin is widely accepted. Inorganic molecules such as chromates, phosphates, and nitrites are widely employed as industrial inhibitors because of their selective advantages [3]. However, the recent clamor for environmental sustainability has caused a shift from the promotion of inorganic inhibitors to organic inhibitors. Generally, organic molecules impede the process of corrosion by adhering to the metal sample surface and forming a chemical bond with it. Organic inhibitors are identified as compounds possessing heteroatoms, π-electrons, multiple bonds, and cyclic rings, which can adhere to the sample surface to impede corrosion reaction at low doses. Several reports have documented the explorations of heterocyclic organic compounds as suppressors of metallic corrosion in different media [4–7]. Whereas studies have shown the superior inhibition performance of organic inhibitors as compared to inorganic ones, however, many of these synthetic organic molecules still pose some threat to the environment [8,9]. This valid concern has promoted the continued search for safe and effective inhibitor compounds.

Nowadays, much attention is on the design and development of "green" corrosion inhibitors (GCI) sequel to the birth and advancement of green chemistry. The science of green chemistry is governed by the philosophy of sustainability, minimal adverse environmental impacts, and preservation of natural resources for upcoming generations [9]. The greenness of chemical inhibitors is often in relation to certain properties such as sustainability, low cost, ready availability, non-toxicity, and environmental friendliness. GCIs can be disposed with no fear of environmental harm

148   DOI: 10.1201/9781003400059-11

as they contain no harmful substances. They are rich in polar atoms and electrons necessary for strong adsorption to metal surfaces. GCIs could be in organic or inorganic forms with each having its unique functionality and applicability. Some popular organic GCIs include dyes, surfactants, ionic liquids, plant extracts, polymers, and drugs [10–12].

An important aspect of corrosion management strategy is corrosion monitoring, which entails data gathering using different methods. Popular corrosion monitoring methods include chemical, electrochemical, surface analytical, and computational approaches. In many cases, a combination of these methods is required to gain proper insight into the corrosion mechanism of metals [13]. The cheapest and conventional way of analyzing corrosion processes and studying inhibition behavior of chemical additives is the chemical method, otherwise known as the weight loss measurement (WLM). In furtherance to WLM, advanced electrochemical instruments have been developed to measure corrosion rates and unravel the kinetics of corrosion reactions. As a result, techniques such as electrochemical impedance spectroscopy (EIS), which supplies information on the charge transfer resistance (R_{ct}); potentiodynamic polarization scans (PDP), which provides information on the corrosion current density (i_{corr}); and linear polarization resistance (LPR), which gives information on the polarization resistance (R_p) have become popularized. Generally, the open circuit potential (OCP) or corrosion potential (E_{corr}) is often measured before electrochemical tests to ascertain the stability of the system. Despite the theoretical diversity of the various electrochemical approaches, the results obtained from them often show remarkable similarity in terms of the %IE indicating that the results are reasonable and reliable. Surface studies are analytical approaches that provide information on the surface characteristics of the metal substrates in the presence and absence of GCIs. In the presence of plain electrolytes only, the outcome of surface studies shows a rough and degraded metal surface. However, upon introducing GCIs, the surface tends to be smoother in comparison to the degraded surface immersed in only plain electrolyte. This observation is often attributed to the adherence of the GCI on the surface. Analytical studies are often conducted using techniques such as scanning electron microscopy (SEM), atomic force microscopy (AFM), Fourier transform infrared spectroscopy (FT-IR), and energy dispersive X-ray spectroscopy (EDS). Apart from the application of these experimental approaches, the use of computational chemistry tools has become desirable in investigating inhibition potentials of GCIs in different electrolytes. As a result of relative low cost and deeper insights associated with computational chemistry, these techniques have gained special interest in the inhibition study of GCIs. Electronic and structural parameters that are useful in explaining inhibition processes are often obtained from computational modeling approaches such as density functional theory (DFT) and molecular dynamic simulation (MDS) [9,10,14–16].

11.2 BRIEF HISTORY, STRUCTURE, AND PROPERTIES OF LIGNIN

Lignin, which means wood, is derived from the Latin word "lignum" [17]. In the process of treating wood with alkaline solutions and nitric acid in 1838, Anselme Payen discovered lignin [18]. Soluble and insoluble fractions were obtained from the wood processing. The former and the later fractions were referred to as incrusting material and cellulose, respectively. The former was later renamed as lignin by Franz

Schulze in 1857 [19]. Since then, several progressive studies have been conducted on lignin. Lignin is a well-known natural biopolymer that is generally present in plants. It forms about 15–40 wt.% of woody parts of plants. Some of the multifarious functions of lignin in wood include acting as a structural material by providing elasticity and strength to the plant tissue, transporting water, determining the chemical and physical characteristics of the wood, protecting the wood from harmful microbes, and providing thermal stability to the plant [17,20].

Lignin is still rated as the second most abundant biopolymer on Earth after cellulose. Despite this fact, lignin has not been accorded widespread investigations as much as other natural macromolecules due to its extreme complexity and liability [15,21]. Based on available reports, lignin is associated with a high annual output level of almost 100 million tons on a global scale, but its value-added utilization is less than 5% as it is often treated as a waste material [22,23]. Lignin is a highly oxygenated complex amorphous biopolymer possessing aromatic and aliphatic units. It has β-O-4 as the most substantial linkage. Lignin is a product of several biochemical reactions involving radical crosslinking polymerization of three monolignols, namely, p-coumaryl, coniferyl, and sinapyl alcohols. Figures 11.1 and 11.2 represent the chemical representation of lignin and its monolignols, respectively.

Lignin is widely sourced from wood, agricultural residues, bagasse, cotton, pulp, paper, and so on. Based on the adopted isolation techniques, lignin has a molecular size of 500–50,000 g/mol, a glass transition temperature of 90°C–170°C, is optically inert, and is insoluble in water. Lignin is known for its mechanical, electrical, and thermal properties. Several functional groups such as aliphatic and phenolic alcohols, carboxylic acids, aromatic units, and carbonyls make lignin a choice compound for numerous applications. To gain an understanding of the chemical and structural compositions of lignin-based compounds, researchers have employed various characterization instruments including FT-IR, ultraviolet visible spectroscopy (UV-vis), nuclear magnetic resonance (NMR), gel permeation chromatography (GPC), etc.

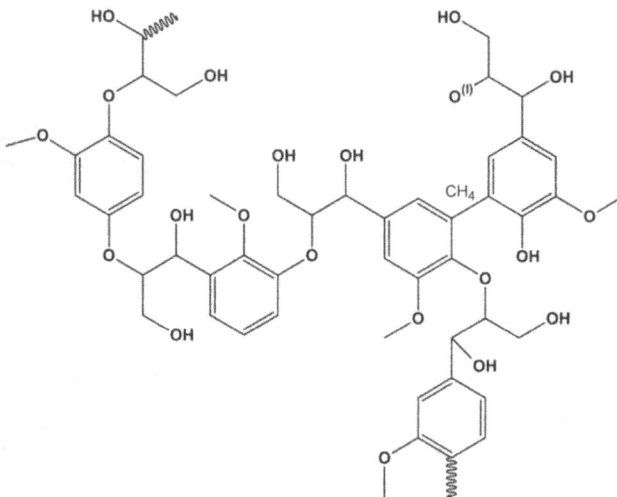

FIGURE 11.1 Chemical structure of lignin.

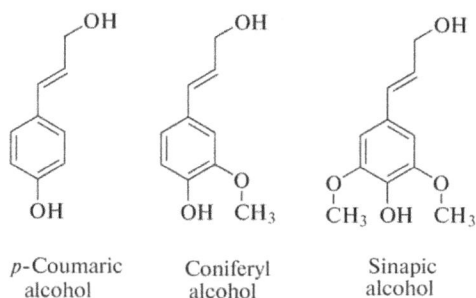

FIGURE 11.2 Structural representations of monolignols.

11.3 DELIGNIFICATION, MODIFICATION, AND APPLICATIONS OF LIGNIN

Lignocellulosic biomass, a naturally abundant organic base for sustainable production of high-value products comprises three basic components, namely, cellulose, hemicellulose, and lignin. To utilize lignin, it is imperative to isolate it from its other counterparts. Lignin extraction or delignification or pulping is a process often carried out to separate lignin from lignocellulosic biomass. This could either be through a biological, physical, or chemical method. The extracted lignin is often referred to as technical lignin as opposed to the native lignin, which is biosynthesized in nature. The selected pulping process has a significant impact on the lignin yield, structure, properties, and purity. Moreover, the lignin yield is dependent on the extraction method, reaction medium, processing time, and temperature conditions [17,24]. Chemical pulping is the most widely utilized delignification process as it dissolves lignin in the black liquor without altering the carbohydrate component. The popular chemical methods include kraft pulping, sulfite pulping, soda pulping, and organosolv pulping [25]. In the kraft process, lignin is diffused in sodium sulfide and sodium hydroxide at high temperature (170°C) and pH (13–14). Usually, this process involves acid-mediated precipitation. Sulfite pulping is carried out at low pH levels (1–2) and at a temperature range of 125°C–170°C with the use of acid bisulfite or sulfur dioxide. In soda process, biomass digestion occurs at a temperature range of 150°C–170°C with the use of aqueous sodium hydroxide as catalyst. Lastly, in organosolv process, organic solvents such as methanol, ethanol, and acetic acid are utilized as delignifying agents at a high temperature range of 170°C–200°C [17,25,26]. In addition to the briefly discussed extraction methods, studies have shown the emergence of other methods such as sugar platform and novel processes. A summary of the extraction procedure is presented in Figure 11.3.

As a complex structure, lignin can undergo chemical modification to produce new compounds of interest. Without modification, lignin is known to demonstrate industrial applications, which is, however, limited due to its weak mechanical properties and thermal stability. To enhance the usefulness of lignin, it is often modified via lignin depolymerization, functionalization of hydroxyl moieties, formation of new active sites, formation of lignin nanoparticles, and lignin graft copolymerization [17,27]. Further information on the chemistry, properties, and extraction of lignin can be found elsewhere [24].

Biomass separation process

Sulfur pulping | Sulfur-free pulping | Biorefinery separation

Sulfite pulping | Kraft pulping | Solvent pulping | Soda pulping | Sugar platform | Novel processes

pH = 1-2
140°C, H_2SO_3

pH = 13
170°C, NaOH + Na_2S

100 - 200°C
Acetic acid / formic acid / water / ethanol etc.

pH = 11 - 13
150 - 170°C, NaOH

Acrid hydrolysis, enzymatic treatment, steam explosion, etc.

Ionic liquids extraction, supercritical solvents, etc.

Lignosulfonate | Kraft lignin | Organosolv lignin | Soda lignin | Hydrolysis lignin | Ionosolv lignin, aquasolv lignin, etc.

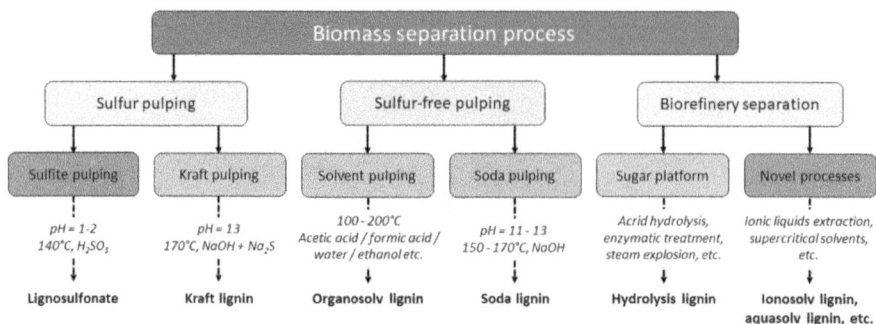

FIGURE 11.3 Extraction of lignin and their products [25].

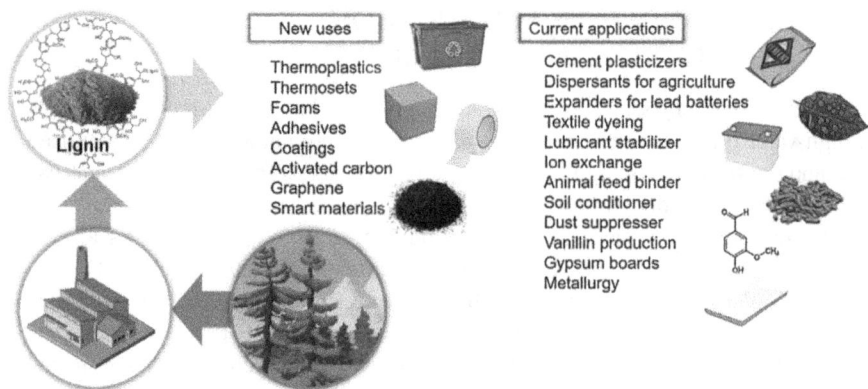

Lignin

New uses

Thermoplastics
Thermosets
Foams
Adhesives
Coatings
Activated carbon
Graphene
Smart materials

Current applications

Cement plasticizers
Dispersants for agriculture
Expanders for lead batteries
Textile dyeing
Lubricant stabilizer
Ion exchange
Animal feed binder
Soil conditioner
Dust suppresser
Vanillin production
Gypsum boards
Metallurgy

FIGURE 11.4 Present and potential applications of lignin-based compounds [29].

Lignin is known to have wide-ranging applications as depicted in Fig 9.4. Its industrial application spans the rubber industry, waste treatment, biofuel, cement industry, cosmetic industry, and food industry. Furthermore, it has found wide biomedical applications in the field of wound dressing, antimicrobial, drug delivery, and tissue engineering. Its applications are dependent on its source, physiochemical characteristics, and chemical modifications. It is interesting to note that it is only in recent years that the potential application of lignin as anticorrosive materials began to gain wide attention among corrosion experts [15,20,28]. This recent foray into the utilization of lignin as anticorrosive materials is largely connected to its admirable properties such as film-forming ability, hydrophobicity, and water impermeability, in addition to its ready availability, low cost, and eco-friendly nature [22]. By these demonstrated properties of lignin, it is suitably qualified to be regarded as a GCI. Nevertheless, lignin is known to be insoluble, which poses a challenge in its usage as a solution-phase inhibitor. To circumvent this challenge, researchers have introduced lignin modification to improve its solubility, thereby enhancing its inhibition performance toward metallic corrosion. At present, several modification methods are in use to upscale the inhibitive effect of lignin-based compounds (Figure 11.4).

11.4 LITERATURE REVIEW ON THE INHIBITIVE POTENTIAL OF LIGNIN-BASED CORROSION INHIBITORS

One of the earliest studies on the protection performance of lignin and its derivatives was conducted by Srivastav and Srivastava [30] in 1981. Using WLM, the duo examined the anticorrosive potentials of different plant materials in $2M$ H_2SO_4, $2M$ HCl, and 3% NaCl toward MS dissolution and computed the protection efficiencies of lignin as 76.4%, 81.5%, and 61.4%, respectively. In $2M$ HCl, lignin retarded Al dissolution to the tune of 67.2%. These results present lignin as an effective corrosion control agent in different media. Another WLM study emerged in 1983 by Forostyan and Prosper [31] where it was shown that the number of carboxyl moieties has a positive influence on the inhibition characteristics of lignin in the dissolution of steel in H_2SO_4. Comparatively, ammoniated hydrolysis lignin was reported to yield the highest inhibition effect. The capability of aqueous lignin to mitigate corrosion in low CS has also been demonstrated using WLM and several surface morphological instruments [32].

Othman et al. showed in their comparative work that the corrosion protection of CS offered by rice straw extract (RSE) in 3.5% NaCl is greater than the protection performance afforded by commercial lignin. The experimental outcome from electrochemical instruments showed that RSE inhibited steel deterioration to the tune of 92% after 42 days of exposure to NaCl. Conducted surface studies using SEM–EDS and optical camera delineated that a ferric-RSE protective film was formed to keep the steel substrate away from corrosive ions. Furthermore, X-ray diffraction (XRD) and cyclic polarization studies were utilized to demonstrate the protection of the steel from pitting corrosion [33]. In another study, the research team studied the protection properties of RSE, ethylene glycol, and lignin toward the deterioration of steel in NaCl in a multiple flow loop condition [34]. The outcome showed that the suppressive abilities of the compounds are a function of the chemical structure, composition, constituents, and functional groups present in the compounds. The protection performance of the tested samples was found to follow the order: Ethylene glycol < Lignin < RSE.

Oil palm (*Elaeis guineensis*) belongs to a species of palm trees generally found in tropical and subtropical regions of Africa, Asia, and South America. It is one of the richest sources of biofuels and can be exploited for its food and household products. However, it generates a large amount of waste with serious environmental concern annually [35,36]. Oil palm fronds (OPF) are abundant lignocellulosic materials from oil palm trees that have been identified as one of the major contributors of biomass waste around the world; others include oil palm empty fruit bunch (EFB) and the trunk. Recently, numerous researchers have advocated and investigated the profitable utilization of biomass waste. In alignment with this objective, Hussin et al. in 2014 carried out lignin extraction via soda, kraft, and organosolv processes from OPF and characterized the extracted components using ^{31}P NMR and GPC. The alkaline process was found to speed up the separation of lignin molecule from OPF biomass as it gave the highest yield of 15.86%. The anticorrosion and antioxidant abilities of the lignin extracts were performed and correlated [37]. The impedance study revealed the potency of the extracted lignin to retard corrosion in the following order:

soda lignin (85.81%) > organosolv lignin (81.84%) > kraft lignin (67.85%) at 500 ppm. In another study conducted in 2017, the team established the outstanding corrosion protection behavior of depolymerized OPF for MS in 1 M HCl using popular techniques [38]. Here, the protection performance decreased in the presented order: soda lignin (97.70%) > organosolv lignin (94.10%) > kraft lignin (93.90%) at 400 ppm. Langmuir adsorption model was used in fitting the electrochemical data with obtained excellent fits, which suggest a strong interaction between the steel samples and the depolymerized lignin inhibitors.

In another study, Akbarzadeh et al. extracted soda and kraft lignin from EFB toward the dissolution of MS in 3.5% NaCl at pH of 6 and 8 [39]. WLM demonstrated that with 800 ppm, a maximum %IE of 91.02% and 89.38% can be obtained for soda and kraft lignin, respectively. Polarization tests showed an attenuation of the i_{corr} from 28.91 µA/cm (3.5% NaCl) to 2.13 (soda lignin) and 3.16 µA/cm (kraft lignin) with 800 ppm at pH 6 and an i_{corr} reduction from 17.75 µA/cm (3.5% NaCl) to 2.47 (soda lignin) and 1.93 µA/cm (kraft lignin) with 800 ppm at pH 8. The presentation of Tafel parameters for the three selected monomers of lignin indicate that p-coumaric acid and ferulic acid with carboxyl moieties offered a more effective corrosion protection on account of their electron donating tendency to iron than 4-hydroxybenzaldehyde. The values of ΔG_{ads} depicted that the adsorption of the lignin samples was via physical and chemical adsorption. SEM and FTIR confirmed the emergence of a deposit on the sample substrate, thereby forestalling further corrosion, while EDS revealed the drop in oxygen content as the concentration of lignin increased. The team has also investigated the inhibition potential of three lignin types derived from OPF via ultrafiltration method for the corrosion control of MS in HCl [40]. The study showed that with as little as 500 ppm, it was possible to obtain excellent %IE for the three lignin inhibitors following the order: soda lignin (87.76%) > organosolv lignin (83.94%) > kraft lignin (81.11%) at 500 ppm. The team also reported the effectiveness of soda and organosolv lignin extracted from OPF biomass to repress cannonball deterioration. The successful extraction of the two lignin types was analyzed using ^{13}C NMR, FTIR, and TGA. An extensive investigation of the effect of doses, reaction time, and pH was carried out using FT-IR. Using the represented electrochemical instrument (Figure 11.5), authors conducted PDP scans of the rust sample, which attained an %IE of 86.9% with 7 wt.% [41]. The percentage of rust conversion revealed that 7 wt.% soda lignin yielded a higher rust transformation of 92.73% than organsolv lignin with 90.91%. The 7 wt.% soda lignin was further investigated using SEM–EDS and XRD and showed that the treated rust powder was almost completely transformed to amorphous form.

Using soda pulping extraction technique, Zultiniar et al. [42] reported the protection performance of lignin isolated from coconut fiber in relation to the dissolution of ASTM36 steel in HCl. UV-vis and FT-IR confirmed the successful extraction of lignin. Immersion tests and surface examinations were employed to investigate the anticorrosive properties of the extracted lignin. With 3 g/L, a peak %IE of 78.11% was obtained. An investigation into the anticorrosive potential of lignin isolated from termite frass using Klason's technique in NaOH solution was reported in a recent study [43]. Study outcomes after morphological and experimental assessments showed that lignin obtained from a sustainable source (termite frass) could serve as an effective chemical inhibitor with an %IE of 70% at 600 ppm.

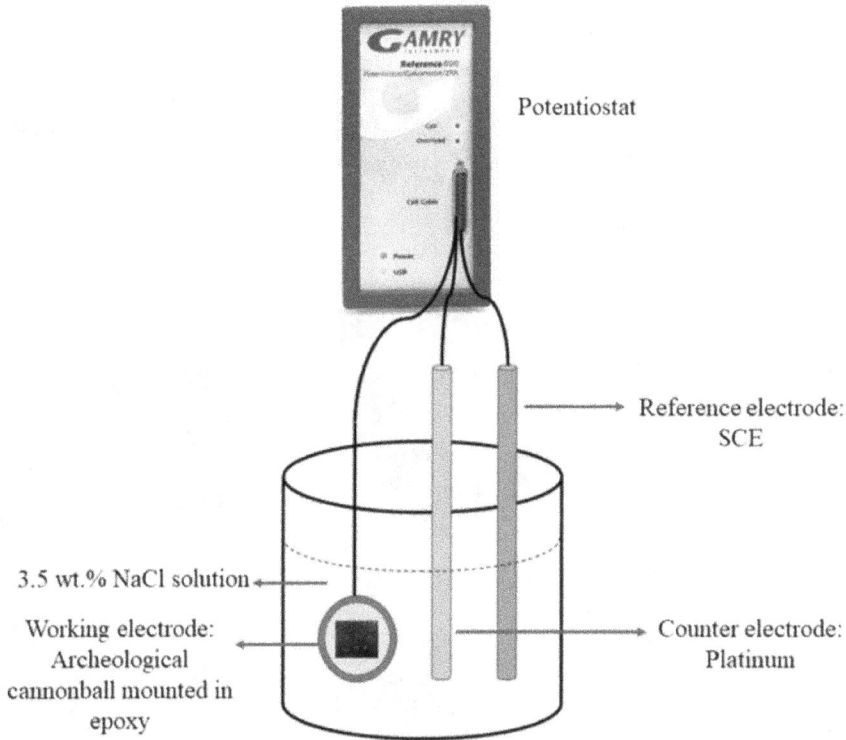

FIGURE 11.5 Electrochemical workstation for PDP scan [41].

Gao et al. [44] researched the corrosion protection of steel samples using a newly synthesized lignin-based quaternary ammonium as inhibitor in the presence of acid solution. The chemical analysis of the synthesized materials was probed and established using modern analytical techniques. The corrosion examination was conducted using WLM, OCP, EIS, and PDP. WLM experiments revealed that the optimum protection capacity of 87.54% was obtainable at 75 ppm, beyond which the protection efficiency depreciates. The OCP prior to electrochemical measurement was done for 1 hour, and the graphical representation revealed the stability of the system. In the case of 1 M HCl, stability was reached at 1,000 seconds, while for 75 ppm, it was attained after 2,000 seconds. The polarization curves of the inhibited concentrations were found to be in parallel with the non-inhibited solution and did not exhibit a major change in the anode and cathodic branches. This observation suggests that the introduction of lignin–DMC did not alter the mechanism of anodic and cathodic reactions, but rather retarded the active point of the reactions by establishing a protective layer on the steel. Since the displacement in the E_{corr} was lower than 85 mV coupled with the reduction in the i_{corr} at both branches, lignin–DMC was categorized as a mixed-type chemical inhibitor. In addition, the inhibitor assessment was strengthened with EIS, SEM, and AFM. In another study, the team examined the corrosion-suppression properties of another lignin-based compound on steel in HCl.

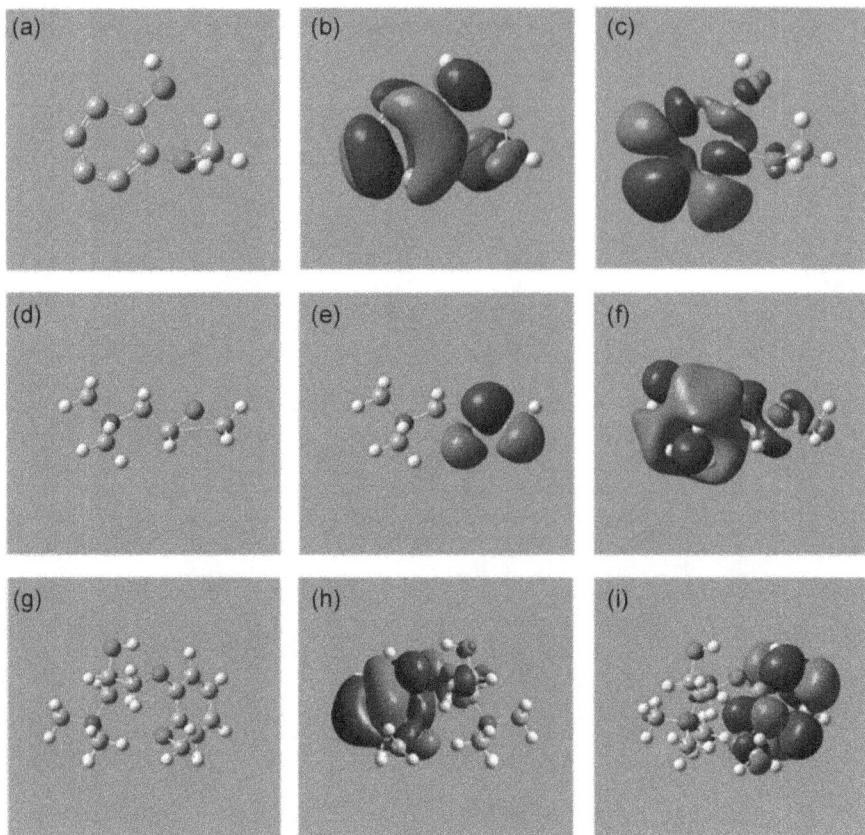

FIGURE 11.6 Images showing (a) the optimized structure of lignin, (b) HOMO of lignin, (c) LUMO of lignin, (d) the optimized structure of EPTAC, (e) HOMO of EPTAC, (f) LUMO of EPTAC, (g) the optimized structure of lignin–EPTAC, (h) HOMO of lignin–EPTAC, and (i) LUMO of lignin–EPTAC [45].

Here, the team grafted (2,3-epoxypropyl)trimethyl ammonium chloride (EPTAC) into kraft lignin, analyzed the product, and carried out corrosion-suppression studies using electrochemical and computational tools. In this study, DFT and MDS were utilized to describe the mechanism of action of EPTAC, lignin, and lignin–EPTAC. Figure 11.6 represents the frontier molecular orbital representations for the optimized molecules with their highest occupied molecular orbital (HOMO) and lowest unoccupied molecular orbital (LUMO) [45]. Based on the energy gap (ΔE) values obtained for the three compounds, lignin–EPTAC would act as the highest-performing anti-corrosive agent. Furthermore, the stable configuration of the lignin–EPTAC on the steel sample is displayed in Figure 11.7. Other authors have also demonstrated the use of computational chemistry tools in explaining the protection mechanism of lignin extracted from agricultural waste with similar outcomes [46,47].

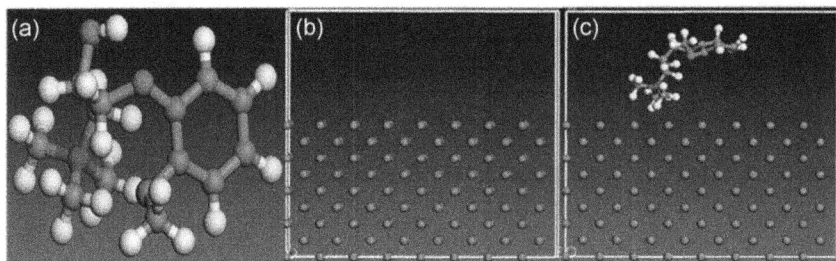

FIGURE 11.7   MDS diagram for lignin–EPTAC configuration on Fe (001) [45].

The inhibitive behavior of kraft lignin for the dissolution of ASTM 61 steel in water distribution systems using EIS, PDP, and LPR has also been evaluated [48]. The PDP and EIS outcomes portended that the sulfonated lignin afforded superior protection to the steel substrate than phosphate. Based on the concentration study, an increase in the dosage from 0.1 to 0.4 mg/L leads to a corresponding rise in the R_{ct} values and in the %IE from 47.0% to 86.4%. However, an excess dosage of the lignin-based inhibitor was found to promote steel deterioration. Authors declared the mechanism of inhibition undefined, which requires further investigation to uncover the protective mechanism. El-Deeb et al. [49] carried out the extraction of lignin from wheat straw using soda pulping. This was followed by lignin modification via acid hydrolysis and carboxylation and characterization using thermal and spectroscopic techniques. The modified extracted lignin was subjected to electrochemical and surface analytical tests to determine their anticorrosive properties for Al dissolution in 1 M NaOH. The lignin acted as an excellent chemical inhibitor. Lignin possessing carboxyl groups with higher adsorption centers was found to exhibit superior barrier to corrosion than the lignin compound with the hydroxyl group. The obtained adsorption data was best fitted with the Langmuir adsorption isotherm. Visualized images from the SEM revealed that the Al surface identified with numerous pits and cavities became smoother and had reduced roughness after the addition of the lignin inhibitor.

Lignin generated from sulfite pulping is designated as lignosulfonates and are often formed by using sulfite acid/sulfite salt at varying low pH levels and high temperatures. These lignin-based compounds account for approximately 90% of the entire market of commercial lignin. Generally, lignosulfonates are water-soluble due to fragmentation and the introduction of sulfate moieties into native lignin [50,51]. Their solubility in water promotes their usage as solution-phase anticorrosive agents for metal samples. In a study conducted in 2006, Ouyang explored the scale and corrosion-suppression properties of sodium lignosulphonate [52]. In another study, Vagin employed electropolymerized lignosulfonate to retard the progress of CS in acidic electrolyte [53] and achieved a significant reduction in the corrosion rate of the steel sample that was comparably effective than a commercial inhibitor. Other studies have demonstrated the potencies of calcium lignosulfonate and sodium lignosulphonate to reinforce the corrosion resistance of Q235 steel CS in simulated core pore solutions [54,55]. Another study explored the corrosion control abilities

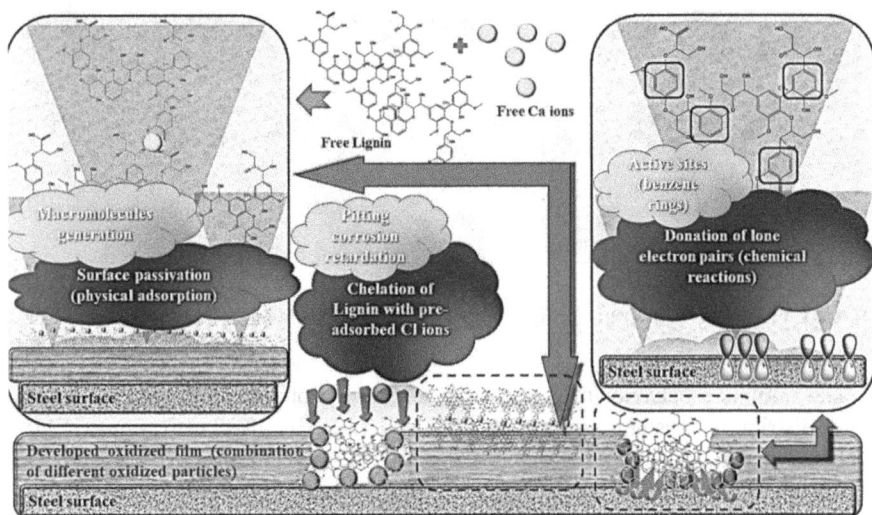

FIGURE 11.8 Mechanism of corrosion protection of sodium sulfonate [56].

of sodium lignosulfonate for steel-rebar in 3.5% NaCl [56]. According to this study, the mechanism of action follows this order: the negatively charged sodium lignosulfonate compounds competes with chloride ions for adherence on the steel substrate; after dissolution of sodium lignosulfonate in the solution, an electrostatic force is formed between the steel substrate and the negatively charged lignosulfonate; lastly, electron sharing occurs between vacant iron orbitals and the lone pairs of electrons of sodium lignosulfonate. As a result of these three processes, a stable layer of sodium lignosulfonate emerges on the steel surface. Also, an interaction occurs between the free Ca^{2+} in the solution and the sodium lignosulfonate molecules, which results in lignin-Ca^{2+} macromolecules forming a physical layer that passivates the steel surface. A representation of the protection mechanism is displayed in Figure 11.8.

The inhibitive potential of lignin copolymers [57] and lignin terpolymer [58] can also be found in the literature. The contribution of three lignin monomers as effective corrosion control agents was studied for MS in 3.5% NaCl using experimental and design of experiment method [59]. The factorial experimental design was employed to assess the main and interaction impacts of the monomers. The study concluded that the %IE increased in the order: hydroxybenzaldehyde (48.01%) < ferulic acid (87.53%) < p-coumaric (88.51%) at 400 ppm. P-coumaric acid and ferulic acid displayed the greatest adsorption and interaction effect. The study was validated with SEM analyses. Besides its use as excellent anticorrosive agent, lignin has been reported as an effective antioxidant and tribological agent in some recent studies [60,61]. The authors prepared a green lignin–ionic liquid composite and explored its multifunctional properties by conducting solubility, thermal stability, rheological, tribological, oxidation resistance, and corrosion tests. The outcomes revealed that the tested lignin composite displayed excellent anticorrosion, anti-wear, and anti-friction properties. The π-electrons of the benzene rings and the delocalized

TABLE 11.1

Lignin-Based Compounds as Aqueous Corrosion Suppressors

S/N	Lignin and Its Source	Metal/Medium	Max Conc	Max %IE	References
1.	Hydrolyzed kraft lignin from EFB	MS/1 M HCl	600 ppm	91.35%	[62]
2.	Lignin extract of sunflower	CS/1 M H_2SO_4	15 g/L	78.8%	[63]
3.	Lignin extract from pulp and paper	MS/0.5 M H_2SO_4	1,000 ppm	95.00%	[64]
4.	Lignin from bagasse	MS/1 M HCl	10 g/L	80.79%	[65]
5.	Alkali lignin	CS/1 M HCl	1,500 ppm	58.58%	[66]
6.	Lignin sulfonate	MS/2 M HCl	1,000 ppm	87.88%	[67]
7.	Lignin-phenylhydrazone	CS/ 3.5% NaCl 0.1 M HCl	600 ppm	96.45%	[68]
8.	Sodium lignosulfonate	Zn/0.01 M HCl	5 mM	88.70%	[69]
9.	Extracted alkali lignin	Al/ 1 M HCl	1,500 mg/L	86.7%	[70]
10.	Lignin from Siam weed	MS/ 1 M HCl	4,000 ppm	92.39%	[71]
11.	Lignin	CS/ 1 M HCl	2,500 ppm	85.00%	[72]
12.	Lignin	CS/ 1 M H_2SO_4	2,500 ppm	70.00%	[72]
13.	Lignin	CS/ 1 M HCl	2,500 ppm	80.00%	[73]

imidazole electrons were found to coordinate with the unfilled d-orbital of iron to form an adherence film on the steel surface. Table 11.1 presents a list of some lignin-based compounds utilized as solution-phase inhibitors in different media.

11.5 CONCLUSIONS AND OUTLOOK

The industrial world has witnessed an upsurge in the utilization of metallic materials in the last few decades, and with the upsurge comes the need to develop organic corrosion inhibitors that are safe, non-toxic, and eco-friendly. As presented in this chapter, lignin-based inhibitors have been identified as powerful alternatives to the widespread hazardous inorganic inhibitors owing to their effectiveness in controlling corrosion as well as their eco-friendly nature. Several studies in literature have shown via experimental and computational routes that lignin, a by-product of biomass, is capable of interacting with metal surfaces to form protective covers from corrosive species. The numerous cyclic rings, π electron densities, heteroatoms, and functional moieties present in lignin-based compounds increase their ability to act as effective inhibitors in different metal/electrolyte systems. This chapter has clearly shown that lignin-based compounds are poised as worthy replacements for toxic, environmentally harmful, and expensive chemical inhibitors.

Despite the numerous advantages and successes recorded with the utilization of lignin in controlling metallic deterioration, much work is still required to uncover the mechanistic insights into the corrosion protection of lignin. Furthermore, most researchers have limited the studies of lignin-based inhibitors to corrosion control of steel in HCl or H_2SO_4 with scanty studies reported on other metal/electrolyte systems.

Also, the corrosion-impeding ability of lignin in high-temperature range has not been fully explored in literature, which is necessary to provide rich thermodynamic insights into lignin-based compounds as industrially relevant corrosion inhibitors.

REFERENCES

[1] Quadri TW, Akpan ED, Olasunkanmi LO, Fayemi OE, Ebenso EE. Fundamentals of corrosion chemistry. In: C M Hussain, C Verma and J Aslam (Eds.), *Environmentally Sustainable Corrosion Inhibitors*. Elsevier, 2022, pp. 25–45.

[2] Quadri TW, Olasunkanmi LO, Fayemi OE, Ebenso EE. Chapter 13 Industrial corrosion inhibitors: Nanostructured carbon allotropes as ideal substitutes. In: Jeenat Aslam, Chandrabhan Verma, Dakeshwar Kumar Verma and Ruby Aslam (Eds.), *Carbon Allotropes: Nanostructured Anti-Corrosive Materials*. Berlin, Boston: De Gruyter, 2022, pp. 327–354. doi: 10.1515/9783110782820-013.

[3] Panchal J, Shah D, Patel R, Shah S, Prajapati M, Shah M. Comprehensive review and critical data analysis on corrosion and emphasizing on green eco-friendly corrosion inhibitors for oil and gas industries. *Journal of Bio-and Tribo-Corrosion*. 2021;7(3):1–29.

[4] Chen L, Lu D, Zhang Y. Organic compounds as corrosion inhibitors for carbon steel in HCl solution: A comprehensive review. *Materials*. 2022;15(6):2023.

[5] Goyal M, Kumar S, Bahadur I, Verma C, Ebenso EE. Organic corrosion inhibitors for industrial cleaning of ferrous and non-ferrous metals in acidic solutions: A review. *Journal of Molecular Liquids*. 2018;256:565–573.

[6] Kamaruzzaman WMIWM, Nasir NAM, Hamidi NASM, Yusof N, Shaifudin MS, Suhaimi AMAAM, et al. Frontiers in organic corrosion inhibitors for chloride and acidic media: A review. *Journal of Bio-and Tribo-Corrosion*. 2022;8(2):1–18.

[7] Quraishi MA, Chauhan DS, Saji VS. *Heterocyclic Organic Corrosion Inhibitors: Principles and Applications*. Elsevier, 2020.

[8] Quadri TW, Olasunkanmi LO, Fayemi OE, Ebenso EE. *Nanomaterials and Nanocomposites as Corrosion Inhibitors. Sustainable Corrosion Inhibitors II: Synthesis, Design, and Practical Applications*. ACS Publications, 2021, pp. 187–217.

[9] Thakur A, Sharma S, Ganjoo R, Assad H, Kumar A, editors. Anti-corrosive potential of the sustainable corrosion inhibitors based on biomass waste: A review on preceding and perspective research. *Journal of Physics: Conference Series*. 2022;2267:012079.

[10] Popoola LT. Organic green corrosion inhibitors (OGCIs): A critical review. *Corrosion Reviews*. 2019;37(2):71–102.

[11] Popoola LT. Progress on pharmaceutical drugs, plant extracts and ionic liquids as corrosion inhibitors. *Heliyon*. 2019;5(2):e01143.

[12] Sastri VS. *Green Corrosion Inhibitors: Theory and Practice*. John Wiley & Sons, 2012.

[13] Quadri TW, Fayemi OE, Olasunkanmi LO, Ebenso EE. Survey of different electrochemical and analytical techniques for corrosion measurements. *Electrochemical and Analytical Techniques for Sustainable Corrosion Monitoring*. Elsevier, 2023, pp. 293–323.

[14] Aslam J, Verma C, Aslam R. *Grafted Biopolymers as Corrosion Inhibitors: Safety, Sustainability, and Efficiency*. John Wiley & Sons, 2023.

[15] Shahini M, Ramezanzadeh B, Mohammadloo HE. Recent advances in biopolymers/carbohydrate polymers as effective corrosion inhibitive macro-molecules: A review study from experimental and theoretical views. *Journal of Molecular Liquids*. 2021;325:115110.

[16] Quadri TW, Olasunkanmi LO, Fayemi OE, Akpan ED, Verma C, Sherif E-SM, et al. Quantitative structure activity relationship and artificial neural network as vital tools in predicting coordination capabilities of organic compounds with metal surface: A review. *Coordination Chemistry Reviews*. 2021;446:214101.

[17] Sana SS, Nguyen THC, Van Le Q, Haldhar R, Abhishek N, Chakravorty A, et al. Bio-based lignin and its applications. In: V K Gupta (Ed.), *Valorization of Biomass to Bioproducts*. Elsevier, 2023, pp. 441–474.

[18] Payen A. Memoir on the composition of the tissue of plants and of woody material. *CR Biology* 1838;7:1052–1056.

[19] Schulze HJF. Neuenburg. Eine geschichtlich-staatsrechtliche Skizze nebst einer Beleuchtung der neuesten schweizerischen Denkschrift, vom 7. December. 18561857.

[20] Dehghani A, Ramezanzadeh B, Mahdavian M. Current applications of fatty acids, lignin, and lipids as green corrosion inhibitors. In: C Verma and D KVerma (Eds.), *Handbook of Biomolecules*. 2023, pp. 535–550.

[21] Pikovskoi I, Kosyakov D, Shavrina I, Ul'Yanovskii N. Study of nettle (*Urtica dióica*) lignin by atmospheric pressure photoionization orbitrap mass spectrometry. *Journal of Analytical Chemistry*. 2019;74:1412–1420.

[22] Dastpak A, Yliniemi K, de Oliveira Monteiro MC, Höhn S, Virtanen S, Lundström M, et al. From waste to valuable resource: Lignin as a sustainable anti-corrosion coating. *Coatings*. 2018;8(12):454.

[23] Stewart D. Lignin as a base material for materials applications: Chemistry, application and economics. *Industrial Crops and Products*. 2008;27(2):202–207.

[24] Liao JJ, Abd Latif NH, Trache D, Brosse N, Hussin MH. Current advancement on the isolation, characterization and application of lignin. *International Journal of Biological Macromolecules*. 2020;162:985–1024.

[25] Ruwoldt J, Blindheim FH, Chinga-Carrasco G. Functional surfaces, films, and coatings with lignin: A critical review. *RSC Advances*. 2023;13(18):12529–12553.

[26] Yu O, Kim KH. Lignin to materials: A focused review on recent novel lignin applications. *Applied Sciences*. 2020;10(13):4626.

[27] Mondal S, Jatrana A, Maan S, Sharma P. Lignin modification and valorization in medicine, cosmetics, environmental remediation and agriculture: A review. *Environmental Chemistry Letters*. 2023;21:1–27.

[28] Jędrzejczak P, Collins MN, Jesionowski T, Klapiszewski Ł. The role of lignin and lignin-based materials in sustainable construction-A comprehensive review. *International Journal of Biological Macromolecules*. 2021;187:624–650.

[29] Wang Z, Ganewatta MS, Tang C. Sustainable polymers from biomass: Bridging chemistry with materials and processing. *Progress in Polymer Science*. 2020;101:101197.

[30] Srivastav K, Srivastava P. Studies on plant materials as corrosion inhibitors. *British Corrosion Journal*. 1981;16(4):221–223.

[31] Forostyan YN, Prosper M. Inhibiting properties of lignin. *Chemistry of Natural Compounds*. 1983;19:358–359.

[32] Yahya S, Othman N, Daud A, Jalar A. Surface morphology studies of low carbon steel treated in aqueous lignin. *Sains Malays*. 2013;42:1793–1798.

[33] Othman NK, Yahya S, Ismail MC. Corrosion inhibition of steel in 3.5% NaCl by rice straw extract. *Journal of Industrial and Engineering Chemistry*. 2019;70:299–310.

[34] Yahya S, Othman N, Ismail M. Corrosion inhibition of steel in multiple flow loop under 3.5% NaCl in the presence of rice straw extracts, lignin and ethylene glycol. *Engineering Failure Analysis*. 2019;100:365–380.

[35] Murphy DJ, Goggin K, Paterson RRM. Oil palm in the 2020s and beyond: Challenges and solutions. *CABI Agriculture and Bioscience*. 2021;2(1):1–22.

[36] Hussin MH, Rahim AA, Ibrahim MNM, Brosse N. Physicochemical characterization of alkaline and ethanol organosolv lignins from oil palm (Elaeis guineensis) fronds as phenol substitutes for green material applications. *Industrial Crops and Products*. 2013;49:23–32.

[37] Hussin MH, Shah AM, Rahim AA, Ibrahim MNM, Perrin D, Brosse N. Antioxidant and anticorrosive properties of oil palm frond lignins extracted with different techniques. *Annals of Forest Science*. 2015;72:17–26.

[38] Shah AM, Rahim AA, Ibrahim MNM, Hussin MH. Depolymerized oil palm frond (OPF) lignin products as corrosion inhibitors for mild steel in 1 M HCl. *International Journal of Electrochemical Science.* 2017;12:9017–9039.

[39] Akbarzadeh E, Ibrahim MM, Rahim AA. Corrosion inhibition of mild steel in near neutral solution by kraft and soda lignins extracted from oil palm empty fruit bunch. *International Journal of Electrochemical Science* 2011;6(11):5396–5416.

[40] Hussin MH, Rahim AA, Ibrahim MNM, Brosse N. The capability of ultrafiltrated alkaline and organosolv oil palm (Elaeis guineensis) fronds lignin as green corrosion inhibitor for mild steel in 0.5 M HCl solution. *Measurement.* 2016;78:90–103.

[41] Osman LS, Hamidon TS, Abd Latif NH, Elias NHH, Saidin M, Shahidan S, et al. Rust conversion of archeological cannonball from Fort Cornwallis using oil palm frond lignin. *Industrial Crops and Products.* 2023;192:116107.

[42] Sandy MK, Wibowo E, Heltina D, editors. Coconut fiber extraction using soda pulping method as green corrosion inhibitor for ASTM A36 steel. *Journal of Physics: Conference Series.* 2021;2049:012094.

[43] Kulkarni P, Ponnappa CB, Doshi P, Rao P, Balaji S. Lignin from termite frass: A sustainable source for anticorrosive applications. *Journal of Applied Electrochemistry.* 2021;51:1491–1500.

[44] Gao C, Wang S, Dong X, Liu K, Zhao X, Kong F. Construction of a novel lignin-based quaternary ammonium material with excellent corrosion resistant behavior and its application for corrosion protection. *Materials.* 2019;12(11):1776.

[45] Gao C, Zhao X, Liu K, Dong X, Wang S, Kong F. Construction of eco-friendly corrosion inhibitor lignin derivative with excellent corrosion-resistant behavior in hydrochloric acid solution. *Materials and Corrosion.* 2020;71(11):1903–1912.

[46] Shahmoradi A, Talebibahmanbigloo N, Javidparvar A, Bahlakeh G, Ramezanzadeh B. Studying the adsorption/inhibition impact of the cellulose and lignin compounds extracted from agricultural waste on the mild steel corrosion in HCl solution. *Journal of Molecular Liquids.* 2020;304:112751.

[47] Srivastava M, Tiwari P, Srivastava S, Kumar A, Ji G, Prakash R. Low cost aqueous extract of Pisum sativum peels for inhibition of mild steel corrosion. *Journal of Molecular Liquids.* 2018;254:357–368.

[48] Abu-Dalo MA, Al-Rawashdeh NA, Ababneh A. Evaluating the performance of sulfonated Kraft lignin agent as corrosion inhibitor for iron-based materials in water distribution systems. *Desalination.* 2013;313:105–114.

[49] El-Deeb MM, Ads EN, Humaidi JR. Evaluation of the modified extracted lignin from wheat straw as corrosion inhibitors for aluminum in alkaline solution. *International Journal of Electrochemical Science* 2018;13:4123–4138.

[50] Hosseinpour A, Rezaei Abadchi M, Mirzaee M, Ahmadi Tabar F, Ramezanzadeh B. Recent advances and future perspectives for carbon nanostructures reinforced organic coating for anti-corrosion application. *Surfaces and Interfaces.* 2021;23:100994.

[51] Aro T, Fatehi P. Production and application of lignosulfonates and sulfonated lignin. *ChemSusChem.* 2017;10(9):1861–1877.

[52] Ouyang X, Qiu X, Lou H, Yang D. Corrosion and scale inhibition properties of sodium lignosulfonate and its potential application in recirculating cooling water system. *Industrial & Engineering Chemistry Research.* 2006;45(16):5716–5721.

[53] Vagin MY, Trashin SA, Karyakin AA. Corrosion protection of steel by electropolymerized lignins. *Electrochemistry Communications.* 2006;8(1):60–64.

[54] Wang Y, Zuo Y, Zhao X, Zha S. The adsorption and inhibition effect of calcium lignosulfonate on Q235 carbon steel in simulated concrete pore solution. *Applied Surface Science.* 2016;379:98–110.

[55] Li J, Liu W, Xie W. Influence of sodium lignosulfonate on the corrosion-inhibition behavior of Q235 steel in simulated concrete pore solutions. *International Journal of Electrochemical Science.* 2020;15:7136–7151.

[56] Mohammadkhah S, Dehghani A, Ramezanzadeh B. Sodium lignosulfonate biomolecule application as an effective steel-rebar corrosion inhibitor in a chloride induced-simulated concrete pore solution. *Biomass Conversion and Biorefinery.* 2023:1–13.

[57] Gao C, Zhao X, Fatehi P, Dong X, Liu K, Chen S, et al. Lignin copolymers as corrosion inhibitor for carbon steel. *Industrial Crops and Products.* 2021;168:113585.

[58] Ren Y, Luo Y, Zhang K, Zhu G, Tan X. Lignin terpolymer for corrosion inhibition of mild steel in 10% hydrochloric acid medium. *Corrosion Science.* 2008;50(11):3147–3153.

[59] Akbarzadeh E, Ibrahim M, Rahim A. Monomers of lignin as corrosion inhibitors for mild steel: Study of their behaviour by factorial experimental design. *Corrosion Engineering, Science and Technology.* 2012;47(4):302–311.

[60] Zhang C, Yang Z, Huang Q, Wang X, Yang W, Zhou C, et al. Lignin composite ionic liquid lubricants with excellent anti-corrosion, anti-oxidation, and tribological properties. *Friction.* 2023;11(7):1239–1252.

[61] Yu Q, Yang Z, Huang Q, Lv H, Zhou K, Yan X, et al. Lignin composite ionic liquid lubricating material as a water-based lubricating fluid additive with excellent lubricating, anti-wear and anti-corrosion properties. *Tribology International.* 2022;174:107742.

[62] Akbarzadeh E. Corrosion inhibition of thermal hydrolyzed lignin on mild steel in hydrochloric acid. *The Journal of Corrosion Science and Engineering.* 2014;16.

[63] Alaneme KK, Olusegun SJ. Corrosion inhibition performance of lignin extract of sun flower (*Tithonia diversifolia*) on medium carbon low alloy steel immersed in H2SO4 solution. *Leonardo Journal of Sciences.* 2012;20(11):59–70.

[64] Shivakumar M, Dharmaprakash M, Manjappa S, Nagashree K. Corrosion inhibition performance of lignin extracted from black liquor on mild steel in 0.5 m H2SO4 acidic media. *Portugaliae Electrochimica Acta.* 2017;35(6):351–358.

[65] Rahayu P, Sundari C, Farida I, editors. Corrosion inhibition using lignin of sugarcane bagasse. *IOP Conference Series: Materials Science and Engineering.*2018; 434:012087.

[66] Zulkafli R, Othman NK, Rahman IA, Jalar A. Effect of rice straw extract and alkali lignin on the corrosion inhibition of carbon steel. *Malaysian Journal of Analytical Sciences.* 2014;18(1):204–211.

[67] Abu-Dalo MA, Al-Rawashdeh NA, Mutlaq AA. Green approach to corrosion inhibition of mild steel by lignin sulfonate in acidic media. *Journal of Iron and Steel Research International.* 2016;23(7):722–732.

[68] Aissiou N, Bounoughaz M, Djeddi A. Lignin-phenylhydrazone as a corrosion inhibitor of API X52 carbon steel in 3.5% NaCl and 0.1 mol/L HCl medium. *Chemical Research in Chinese Universities.* 2021;37:718–728.

[69] Altwaiq A, Abdel-Rahem R, AlShamaileh E, Al-luaibi S, Khouri Si. Sodium lignosulfonate as a friendly-environment corrosion inhibitor for zinc metal in acidic media. *Eurasian Journal of Analytical Chemistry.* 2015;10(1):10–18.

[70] mnim Altwaiq A, Sa'ib JK, Al-luaibi S, Lehmann R, Drücker H, Vogt C. The role of extracted alkali lignin as corrosion inhibitor. *Journal of Materials and Environmental Science.* 2011;2(3):259–270.

[71] Nwosu F, Muzakir M. Thermodynamic and adsorption studies of corrosion inhibition of mild steel using lignin from siam weed (*Chromolaena odorata*) in acid medium. *Journal of Materials and Environmental Science.* 2016;7(5):1663–1673.

[72] Yahya S, Othman N, Daud A, Jalar A, editors. The inhibition of carbon steel corrosion by lignin in HCl and H2SO4. *AIP Conference Proceedings.* 2013; 1571:132–135.

[73] Yahya S, Othman NK, Daud AR, Jalar A, Ismail R. The influence of temperature on the inhibition of carbon steel corrosion in acidic lignin. *Anti-Corrosion Methods and Materials.* 2015;62(5):301–306.

12 Cellulose as Corrosion Inhibitors

Shveta Sharma
Himachal Pradesh University

Richika Ganjoo
Lovely Professional University

Ashish Kumar
Government of Bihar

12.1 INTRODUCTION

Corrosion is one of the most urgent problems affecting all over the globe since it has a negative influence on the economy of both developing nations and those that have already achieved economic success [1,2]. Corrosion was the cause of enormous financial losses as well as significant harm. Even continuous research is going on in the field of corrosion mitigation, even then it remains the major issue. Even using minerals in acid pickling and surface treatments might lead to corrosion [3–7]. For this reason, aggressive solutions in various engineering and industrial processes need the use of corrosion inhibitors [8]. Organic compounds are often employed as corrosion inhibitors owing to their ease of synthesis and application, as well as their cheap cost and strong inhibition effect. Polar functional groups present in these molecules, such as nitrogen and oxygen, may be found to adsorb on the surfaces of metals [9,10]. Corrosion inhibitors by using their electrons, may physically or chemically adsorb on the metal, can easily protect them by forming a thin protective layer on the surface of metals under study [11,12]. But toxicity issue is always there, as these conventional inhibitors are synthesized by using the chemicals, which are not environment friendly and are hazardous to nature. In recent years, a significant amount of effort has been put into the quest for corrosion inhibitors that are not only cheaper and less hazardous to the environment, but also non-toxic [13]. Therefore, cellulose and the composites made up of cellulose are becoming very popular to be used as corrosion inhibitors [14,15]. At present, keeping in mind the high demand for high-performance, effective, and eco-friendly corrosion inhibitors for various applications, researchers have used biopolymers, such as along with other nanoparticles, to form novel benign high-performance corrosion inhibitors. Cellulose and its derivatives have lately been employed in the field of corrosion control owing to the economic and industrial availability of these substances. Because of their natural and biological origin, they are regarded to be among the most environmentally friendly

DOI: 10.1201/9781003400059-12

chemicals, from which various other compounds can be obtained, that are utilized in the process of corrosion control [16]. Additionally, the inclusion of a variety of functional groups having polarity, such as –OH, boosts the efficiency of corrosion prevention by expanding the size of the molecules involved and elevating their solubility in polar electrolytes [17]. Therefore, in this chapter, cellulose and its derivatives taken into consideration and their role in corrosion mitigation for various metals have been briefly explained.

12.2   CELLULOSE AS CORROSION INHIBITOR

Toghan et al. [18] prepared nano crystalline cellulose (NCC), and the raw material for this was macrocrystalline cellulose (CEL), and hydrolysis was performed with sulfuric acid at a specific temperature for 1 hour. Further, characterization of newly formed NCC and CEL was done with the help of Fourier transform infrared (FT-IR), field emission scanning electron microscopy (FE–SEM), transmission electron microscopy (TEM), and Raman spectroscopy. FT-IR was performed to get an idea about the progress in the reaction and finally the properties of the newly formed NCC. As given in the Figure 12.1, the band for OH was smaller in case of NCC as compare to CEL, which depicted the increased strength of hydrogen bonds and increased crystallinity. Typical bands of cellulose were also there in both such as bands at 1,430, 1,321, 1,062, and 897 cm^{-1}, and the values at 1,062 and 897 cm^{-1} also confirmed the high cellulosic amount in NCC as compare to CEL [19]. The hydrolysis caused by sulfuric acid (H_2SO_4) in NCC forms the esters and new S=O bonds were formed, which gave a band value at 1,205 cm^{-1} [20]. Further, Raman spectroscopy also predicted the increase in crystallinity when NCC were formed from CEL. SEM and TEM analysis verified that on hydrolysis, there is a sharp decrease in the molecular size, and now chips-like structure got converted into rod-like structure,

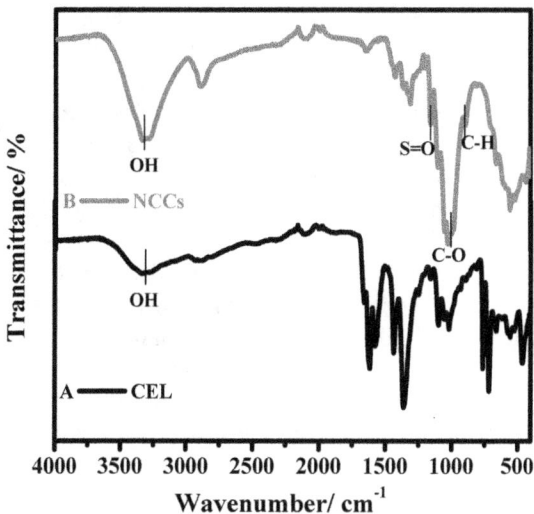

FIGURE 12.1   FT-IR spectra for NCC and CEL [18].

and amorphous part has been evacuated. Corrosion mitigation capacity of NCC and CEL was also investigated in hydrochloric acid for steel samples. Weight loss and electrochemical investigations were performed according to ASTM standard method [21]. In weight loss, the optimal protective capability was found to be above 94.3% for CEL and 97.9% for NCC when a maximum concentration of 200 mg/L was used.

Potentiodynamic polarization (PDP) study validated the corrosion inhibition tendency of both the cellulose types. The PDP curves for both CEL and NCC, and both the cathodic and anodic sides saw a change to lower current values when the CEL and NCC inhibitors were added to test solution. This caused the corrosion to be slowed down significantly. At a concentration of 200 ppm, the NCC gave higher efficiency (96.3%) as compared to CEL (93.1%) to stop the reaction. Electrochemical Impedance Spectroscopy (EIS) study was carried out in order to acquire more data about the mechanism of corrosion processes and to validate the prior results that were obtained through weight loss and PDP measurements. The fact that Nyquist plots drawn were having only a single semicircle and that the diameter of the loops increased as the inhibitor concentration rises demonstrates that the corrosion pathway is primarily determined by a charge transfer technique of measurement [22]. According to the results of the DFT calculation, both the inhibitor molecules CEL and NCC form bond with the sample surface via the mechanism of donor–acceptor attractions. In addition, the results of the MC simulations investigation demonstrated that CEL and NCC adsorb rapidly through the sites which are electron-rich in their structures.

Gouda et al. [23] investigated CeO_2-nanoparticle-loaded carboxymethyl cellulose (CMC) for its corrosion inhibition property for mild steel (MS) in acidic solution by employing various electrochemical and surface morphological techniques. The first nanoparticle-loaded cellulose was synthesized incorporating CeO_2 nanoparticles into an aqueous CMC solution and was coated on the specimen under investigation. SEM, TEM, and selected area electron analysis (SAED) were performed for characterizing the sample prepared: SEM images are mentioned in Figure 12.2a and b, TEM images are illustrated in Figure 12.2c–e, and at last, the SAED image is illustrated in Figure 12.2f. According to the findings of the SEM study, CMC has a homogenous quality with no discernible isolated particles. After the inclusion of CeO_2, the surface becomes heterogeneous, with particles of varying sizes; this is an indication that CeO_2 was successfully incorporated into the top layer of the CMC to construct a CMC-CeO_2 framework. In TEM analysis, it is possible to observe the three-dimensional structure of the polymer that was used (CMC), which validates the shape of the CMC-CeO_2 compound. Light was thrown on the fact that which part is there (amorphous and crystalline parts) in the produced material, and the crystalline phase of the prepared composite are both shown in the SAED picture, which verifies the existence of crystalline oxide even after it has been incorporated into the organic (CMC) framework which is amorphous in nature.

The thermal gravimetric analysis (TGA) was also performed for the CMC-CeO_2 framework and its constituent parts, CMC and CeO_2. In the case of CeO_2, the amount of weight lost is insignificant. Two exothermic peaks were present in case of CMC and CMC-CeO_2, revealing water evaporation and organic compound decomposition at low temperature and high-temperature peaks, respectively. The difference in weight loss and exothermic peak locations between CMC and CMC-CeO_2 shows

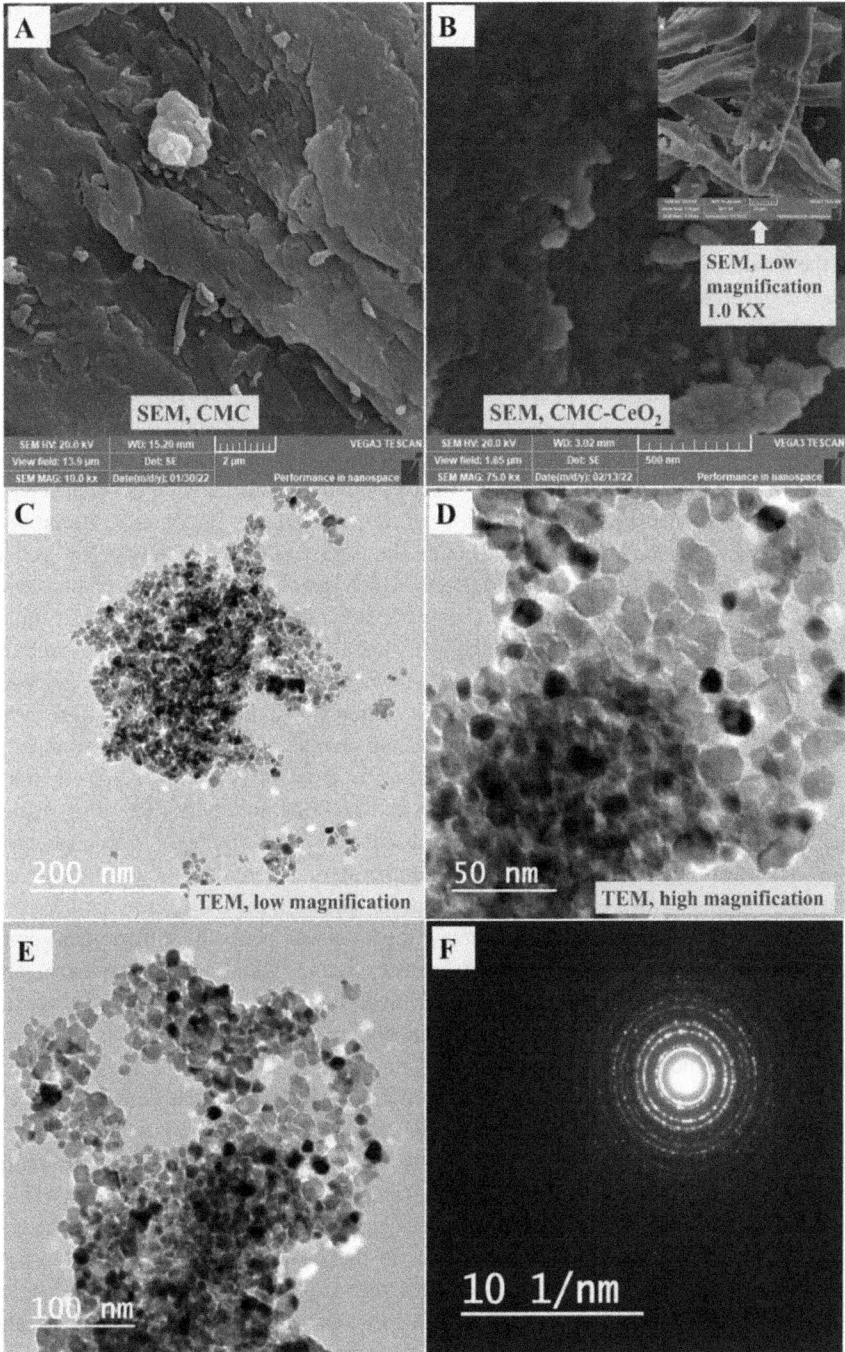

FIGURE 12.2 SEM, TEM, and SAED images of the samples [23].

that there was the bond formation between the polymeric component (CMC) and the oxide portion (CeO_2). Because of this postulated bond formation between these two, the polymer contents may have improved thermal stability.

After 50 minutes of immersion in HCl at a specific temperature, the PDP graphs of the uncoated sample and coated samples with varied concentrations of CeO_2-CMC were also studied, and changes were discovered in both the reactions(occurring both at anode and cathode) as the percentage of CeO_2 grew in the coating applied on the metallic sample; these changes mean that now coated layers inhibit the corrosion; reaction goes on the anode of the metallic sample as well as reduces the hydrogen generation on the cathode of the sample [24,25]. The highest inhibition efficiency of the sample came out to be 96.3% and was documented with 3.0% CeO_2–CMC–coated layer. In the instance of pure MS, the EIS research reveals a singular capacitive loop, which is represented by a single semicircle. This loop is attributed to the corrosion process of charge transfer, while when coated with polymer oxide matrix, there is formation of two capacitive loops [26,27], which is indicative of frequency dispersion effect caused by roughness and inhomogeneity of metallic surface. When compared to the diameter of Nyquist plots on an untreated MS surface, metal with varying percentages of coated material was much greater. This leads one to believe that the films coated with CeO_2–CMC generate a protective layer on the steel interface, which resulted in an increased corrosion resistance. The results of the theoretical analysis (DFT calculations and the MC simulations) indicated that applied film of the inhibitor were consistently adsorbed over the sample. In addition, the theoretical parameters including E_{HOMO}, E_{LUMO}, and ΔE were in accordance with the empirical data, which demonstrated that thin applied films were successfully adsorbing onto the metal surface and saving it from corrosion. Nwanonenyi et al. [28] studied hydroxypropyl cellulose (HPC) for its capacity to mitigate Al corrosion in acidic medium by employing weight loss and PDP measurements. Results revealed that HPC behaved as an excellent corrosion inhibitor with the increase in its concentration, but a dip was there with the rise in temperature. Therefore, it can be predicted that adsorption was favored at lower temperature, and with the increase in temperature, desorption was there. Further, corrosion mitigation was better in sulfuric acid compared to hydrochloric acid. Adsorption was in accordance with Langmuir isotherm, and inhibitor successfully controlled the cathodic and anodic reactions. Quantum chemical technique was employed to verify the results obtained from the different experiments performed and the results of E_{HOMO}, E_{LUMO}, and ΔE_{gap} that were obtained had a high degree of concordance with the inhibition efficiency (IE%) that was determined from the experimental findings. The Fukui function was used to do an analysis of the chemical reactivity of the HPC molecule by pointing out the locations of potential nucleophilic and electrophilic attack sites on the molecule. The Highest Occupied Molecular Orbitals (HOMO) orbital provides the locations for the electrophilic attack, which reflect the portions of the metal surface and inhibitor molecule that have the greatest potential to connect with one another. The nucleophilic attack sites are located in the Lowest Occupied Molecular Orbitals (LUMO) orbital, and in this location, the inhibitor molecule and the metal surface display anti-bonding orbital to produce a back bonding. Kiey et al. [29] produced and evaluated cellulose tetrazole (CTZ) for corrosion control property in acidic medium for carbon steel by utilizing various techniques of characterization along with electrochemical techniques. The findings that were published demonstrate that these molecules were adsorbed onto the surface of the selected metal, resulting in the formation of an adsorption film. Experiments using EIS showed that CTZ is effective as a high-performance and environmentally friendly carbon steel inhibitor in a media containing 1 M HCl. Its effectiveness reached its maximum of 94.2% when it was present at 100 ppm. Adsorption was also confirmed by the various experiments and was in accord to Langmuir adsorption. Inhibitor molecule's extent of reactivity and stability was further studied by theoretical investigations. According to the computational calculations, the value of E for the CTZ molecule was the lowest possible value, coming in at 7.20062 eV. This might have made its adsorption on the metal surface simpler, which would have increased its inhibitory effectiveness. It has been discovered that a rise in temperature results in an increase in the rate of corrosion. As a consequence of this, the inhibitory effectiveness of CTZ drops, and the corrosive medium becomes more severe at increased temperatures [30]. As the temperature increases, there is a disruption in the equilibrium that exists among the adsorption and desorption of inhibitor molecules on the surface of the metal, which results in a shift toward desorption until a new equilibrium can be formed [31]. Additionally, the standard activation energy calculated when CTZ was added as inhibitor molecule was larger than the value measured without inhibitor, which validates the presence of the inhibitor, which was reacting on the surface of the carbon steel [32]. Entropy value was also calculated, and the negative value confirmed the decreased disorder after the complex formation [33]. Calculated Gibb's free energy was −42.31, which was high, and negative depicted adsorption is chemical in nature through the formation of coordination bond [34].

In one of the studies, Arukalam et al. [35] worked on MS for mitigating its corrosion in acidic medium ethyl hydroxyethyl cellulose (EHEC); weight loss investigations were performed on 200 mL of test solution, and time of immersion was taken from 24 hours to 5 days. According to these findings, the rate of corrosion of metallic sample in acid solution reduces with time. At the concentration of 0.5 g/L EHEC that was investigated, it was discovered that EHEC was able to retard the rate of corrosion. This impact becomes more noticeable as the quantity of the inhibitor increases, indicating that the inhibition mechanism is likely to be quite responsive to the concentration of the additive that is now present. Isotherms of adsorption, which are dependent on the amount of surface covering, may offer information on the modes of adsorption that organic inhibitors have on the surface of the metal that is corroding. Results revealed the high degree of surface coverage. The adsorption approach is considered to be a substitution process between the inhibitor molecules present in the aqueous solution and the water that is adsorbed on the surface of the metallic material from a theoretical point of view [36,37]. A plot was drawn between C/θ versus C, and it gave a straight line, which was confirming the Langmuir adsorption [38]. According to the results of an EIS investigation, the introduction of EHEC into an acidic corrodent results in an increase in R_{ct} and a drop in Q_{dl}, which indicates that the corrosion process is slowed down. In the PDP study, it was noticed that the introduction of EHEC and EHEC + KI into the acid solution caused a modest shift in the corrosion potentials of both inhibited systems in the direction of a little negative shift. Additionally, in both instances, the introduction of EHEC and EHEC + KI inhibited both the anodic as well as the cathodic reaction. Literature suggests that if the value of E_{corr} is >85 mV, the inhibitor can be regarded as a cathodic or anodic, and if the E_{corr} is <85 mV, the inhibitor behave as mixed type [39]. Arukalam et al. [40] studied effectiveness of hydroxyethyl cellulose (HEC) for controlling the copper corrosion in hydrochloric and sulfuric acid medium. Efficiency calculated on the basis of weight loss technique was 95% in hydrochloric acid and was 96% in sulfuric acid. Electrochemical analysis also proved the corrosion control by the inhibitor in different test solutions. Further the experimental results were also cross-verified by the theoretical analysis. HEC was also studied for its anti-corrosion property by Mobin and Rizvi [41] for CS in acidic medium by employing various experimental and theoretical techniques. Investigations were performed when only HEC was added and when surfactant was added in to the HEC. Results validated the corrosion mitigation of the sample after the addition of HEC, and inhibition efficacy was increased with the increased amount of HEC, but temperature had opposite effect on the inhibition efficacy. After the addition of surfactants, the synergic effect was there, and the efficiency was higher as compared to the efficiency with HEC. The enthalpy value also increased after the addition of surfactant, indicating toward the physical adsorption, and positive value of entropy validated the increased solvent entropy and higher desorption of water. In EIS study, after adding the HEC or HEC and surfactant, the size of the semicircular loops increased. SEM investigation was carried out to get an idea about the surface of the metal. Samples were kept in the test solution for 6 hours. When sample was kept in the pure acid solution, the level of roughness or damage was very high. When HEC was dissolved in the test solution, surface was comparatively smoother, but few pits were there on the top of the sample; finally, after the addition of both HEC and surfactant in the test solution, the surface was fully protected with least pits and high level of smoothness. Atomic force microscopy (AFM) was also taken into consideration, and the results were the same as that of SEM. The roughness value was least when HEC and surfactants both were added. The height profile curve becomes smoother when the corrosive assault of 1 M HCl was stopped by HEC and HEC + surfactant, respectively, in contrast to the specimen when it was immersed in 1 M HCl [42]. Sangeetha et al. [43] worked on aminated hydroxyethyl cellulose (AHEC): first it was synthesized and FT-IR was employed to characterize it, and IR spectra of HEC gave bands for hydroxy, $-CH_2$ group at 3,396 and 2,929 cm^{-1}, respectively. Further peaks for C–O–C stretching were also present [44]. Spectra of AHEC showed the shift in the hydroxy peak: now it was shown at 3,417 cm^{-1}, and other peaks were there confirming the presence of amino group, CH, CH_2, and CH_3 groups of AHEC [45]. After that, newly formed compound was investigated for its corrosion-reducing property on steel sample in acidic test solution. Weight loss revealed that corrosion-reducing tendency of AHEC increased with increased amount of it, and maximum efficacy was 91.8%. Transition state equation was used for getting the information about the enthalpy and entropy [46]. Positive value of enthalpy assured the endothermic reaction along with restriction to corrosion reaction [47]; also, the enthalpy values are less than the calculated activation energy value, validating the fact of presence of gaseous reactions [48]. PDP confirmed the decreased corrosion current after the addition of inhibitor, and mixed type of nature of AHEC was also confirmed. EIS confirmed the bonding of AHEC on the metallic surface, resulting in increased charge transfer resistance and decreased dielectric constant, which resulted in reduced double-layer capacitance, ultimately validating the inhibitor's adsorption. Finally, SEM and AFM results revealed the presence of a thin protective layer of AHEC, which goes well with the fact that AHEC protected the metallic surface. Few of the studies on cellulose and its derivatives are summarized in Table 12.1 [18,23,28,29,35,40,41,43].

12.3 CONCLUSION

Cellulose and its derivatives are investigated by many researchers for their corrosion inhibition tendencies in different media for different metals, and many studies are summarized in this chapter. Numerous approaches, including weight loss, electrochemical analysis, surface study using SEM, TEM, and AFM, are employed to investigate the corrosion mitigation propensity.. For characterizing newly synthesized cellulose derivative, FT-IR, UV-vis, Raman spectra, etc. were employed. Almost all the cellulose derivative behaved as excellent inhibitors. Researchers have worked on increasing the efficiency of natural polymers including cellulose either by small modifications or by graft copolymerization, but the groups which are grafted did not have enough functional groups to increase the efficiency of parent molecule; secondly, solubility issues are there. Therefore, few researchers are preferring chemical modification to overcome the previous drawbacks. But still more work can be done in future in order to enhance the inhibition efficiency of cellulose and its derivative with special attention to its efficacy at more aggressive media and at high temperatures.

TABLE 12.1
Cellulose and Its Derivatives as Corrosion Inhibitors for Different Metals

Name of the Inhibitor	Metal	Test Solution	Techniques Employed	Efficiency
NCC and CEL	Steel	HCl	Weight loss, electrochemical, DFT SEM, TEM, Raman spectroscopy	96.3% 93.1%
CeO$_2$-CMC	MS	HCl	PDP, EIS, SEM, TEM, TGA, FT-IR	96.3%
Hydroxypropyl cellulose (HPC)	Al	HCl/ H$_2$SO$_4$	Weight loss, PDP, quantum chemical analysis	
Cellulose tetrazole (CTZ)	CS	HCl	Weight loss, electrochemical, DFT SEM	94.2%
Ethyl hydroxyethyl cellulose (EHEC) + KI	MS	H$_2$SO$_4$	Weight loss, electrochemical, DFT SEM	60.19%
Hydroxyethyl cellulose (HEC)	Cu	HCl/ H$_2$SO$_4$	Weight loss, electrochemical, quantum chemical analysis	95% and 96%
Hydroxyethyl cellulose (HEC)	CS	HCl	Weight loss, electrochemical, quantum chemical analysis	85.8%
Aminated hydroxyethyl cellulose (AHEC)	MS	HCl	Weight loss, electrochemical, quantum chemical analysis	91.8%

REFERENCES

[1] Verma, C., Ebenso, E. E., & Quraishi, M. A. (2017). Ionic liquids as green and sustainable corrosion inhibitors for metals and alloys: An overview. *Journal of Molecular Liquids*, *233*, 403–414.

[2] Ahamad, I., & Quraishi, M. A. (2009). Bis (benzimidazol-2-yl) disulphide: An efficient water soluble inhibitor for corrosion of mild steel in acid media. *Corrosion Science*, *51*(9), 2006–2013.

[3] Sharma, S., Saha, S. K., Kang, N., Ganjoo, R., Thakur, A., Assad, H., & Kumar, A. (2022). Multidimensional analysis for corrosion inhibition by Isoxsuprine on mild steel in acidic environment: Experimental and computational approach. *Journal of Molecular Liquids*, *357*, 119129.

[4] Nwankwo, H. U., Akpan, E. D., Olasunkanmi, L. O., Verma, C., Al-Mohaimeed, A. M., Al Farraj, D. A., & Ebenso, E. E. (2021). N-substituted carbazoles as corrosion inhibitors in microbiologically influenced and acidic corrosion of mild steel: Gravimetric, electrochemical, surface and computational studies. *Journal of Molecular Structure*, *1223*, 129328.

[5] Hasanov, R., Sadıkoğlu, M., & Bilgiç, S. (2007). Electrochemical and quantum chemical studies of some Schiff bases on the corrosion of steel in H2SO4 solution. *Applied Surface Science*, *253*(8), 3913–3921.

[6] Olasunkanmi, L. O., Kabanda, M. M., & Ebenso, E. E. (2016). Quinoxaline derivatives as corrosion inhibitors for mild steel in hydrochloric acid medium: Electrochemical and quantum chemical studies. *Physica E: Low-dimensional Systems and Nanostructures*, *76*, 109–126.

[7] Saranya, J., Sounthari, P., Parameswari, K., & Chitra, S. (2016). Acenaphtho [1, 2-b] quinoxaline and acenaphtho [1,2-b] pyrazine as corrosion inhibitors for mild steel in acid medium. *Measurement*, *77*, 175–186.

[8] Jiang, L., Qiang, Y., Lei, Z., Wang, J., Qin, Z., & Xiang, B. (2018). Excellent corrosion inhibition performance of novel quinoline derivatives on mild steel in HCl media: Experimental and computational investigations. *Journal of Molecular Liquids*, *255*, 53–63.

[9] Dagdag, O., Safi, Z., Erramli, H., Cherkaoui, O., Wazzan, N., Guo, L., ... & El Harfi, A. (2019). Adsorption and anticorrosive behavior of aromatic epoxy monomers on carbon steel corrosion in acidic solution: Computational studies and sustained experimental studies. *RSC Advances*, *9*(26), 14782–14796.

[10] Dagdag, O., El Harfi, A., Cherkaoui, O., Safi, Z., Wazzan, N., Guo, L., ... & Jalgham, R. T. (2019). Rheological, electrochemical, surface, DFT and molecular dynamics simulation studies on the anticorrosive properties of new epoxy monomer compound for steel in 1 M HCl solution. *RSC Advances*, *9*(8), 4454–4462.

[11] Cherrak, K., Benhiba, F., Sebbar, N. K., Essassi, E. M., Taleb, M., Zarrouk, A., & Dafali, A. (2019). Corrosion inhibition of mild steel by new Benzothiazine derivative in a hydrochloric acid solution: Experimental evaluation and theoretical calculations. *Chemical Data Collections*, *22*, 100252.

[12] Ramezanzadeh, B., Karimi, B., Ramezanzadeh, M., & Rostami, M. (2019). Synthesis and characterization of polyaniline tailored graphene oxide quantum dot as an advance and highly crystalline carbon-based luminescent nanomaterial for fabrication of an effective anti-corrosion epoxy system on mild steel. *Journal of the Taiwan Institute of Chemical Engineers*, *95*, 369–382.

[13] Sherif, E. S. M., Erasmus, R. M., & Comins, J. D. (2007). Corrosion of copper in aerated synthetic sea water solutions and its inhibition by 3-amino-1,2,4-triazole. *Journal of Colloid and Interface Science*, *309*(2), 470–477.

[14] Abdelraof, M., Ibrahim, S., Selim, M., & Hasanin, M. (2020). Immobilization of L-methionine γ-lyase on different cellulosic materials and its potential application in green-selective synthesis of volatile sulfur compounds. *Journal of Environmental Chemical Engineering*, *8*(4), 103870.

[15] Abu-Elghait, M., Hasanin, M., Hashem, A. H., & Salem, S. S. (2021). Ecofriendly novel synthesis of tertiary composite based on cellulose and myco-synthesized selenium nanoparticles: Characterization, antibiofilm and biocompatibility. *International Journal of Biological Macromolecules, 175,* 294–303.

[16] Akl, E. M., Dacrory, S., Abdel-Aziz, M. S., Kamel, S., & Fahim, A. M. (2021). Preparation and characterization of novel antibacterial blended films based on modified carboxymethyl cellulose/phenolic compounds. *Polymer Bulletin, 78,* 1061–1085.

[17] Dacrory, S., Abou-Yousef, H., Kamel, S., & Turky, G. (2019). Development of biodegradable semiconducting foam based on micro-fibrillated cellulose/Cu-NPs. *International Journal of Biological Macromolecules, 132,* 351–359.

[18] Toghan, A., Gouda, M., Shalabi, K., & El-Lateef, H. M. A. (2021). Preparation, characterization, and evaluation of macrocrystalline and nanocrystalline cellulose as potential corrosion inhibitors for ss316 alloy during acid pickling process: Experimental and computational methods. *Polymers, 13*(14), 2275.

[19] Neto, W. P. F., Silvério, H. A., Dantas, N. O., & Pasquini, D. (2013). Extraction and characterization of cellulose nanocrystals from agro-industrial residue-Soy hulls. *Industrial Crops and Products, 42,* 480–488.

[20] Lu, P., & Hsieh, Y. L. (2010). Preparation and properties of cellulose nanocrystals: Rods, spheres, and network. *Carbohydrate Polymers, 82*(2), 329–336.

[21] ASTM G3-89 A. *Standard Practice for Conventions Applicable to Electrochemical Measurements in Corrosion Testing.* ASTM International: West Conshohocken, PA, 2010; pp. 1–9.

[22] Singh, A., Ansari, K. R., Quraishi, M. A., Lgaz, H., & Lin, Y. (2018). Synthesis and investigation of pyran derivatives as acidizing corrosion inhibitors for N80 steel in hydrochloric acid: Theoretical and experimental approaches. *Journal of Alloys and Compounds, 762,* 347–362.

[23] Gouda, M., Khalaf, M. M., Al-Shuaibi, M. A., Mohamed, I. M., Shalabi, K., El-Shishtawy, R. M., & El-Lateef, H. M. A. (2022). Facile synthesis and characterization of CeO2-nanoparticle-loaded carboxymethyl cellulose as efficient protective films for mild steel: A comparative study of experiential and computational findings. *Polymers, 14*(15), 3078.

[24] Abd El-Lateef, H. M. (2016). Synergistic effect of polyethylene glycols and rare earth Ce4+ on the corrosion inhibition of carbon steel in sulfuric acid solution: Electrochemical, computational, and surface morphology studies. *Research on Chemical Intermediates, 42,* 3219–3240.

[25] Abd El-Lateef, H. M., Ismael, M., & Mohamed, I. M. (2015). Novel Schiff base amino acid as corrosion inhibitors for carbon steel in CO2-saturated 3.5% NaCl solution: Experimental and computational study. *Corrosion Reviews, 33*(1–2), 77–97.

[26] Hamadi, L., Kareche, A., Mansouri, S., & Benbouta, S. (2020). Corrosion inhibition of Fe-19Cr stainless steel by glutamic acid in 1M HCl. *Chemical Data Collections, 28,* 100455.

[27] Rammelt, U., Duc, L. M., & Plieth, W. (2005). Improvement of protection performance of polypyrrole by dopant anions. *Journal of Applied Electrochemistry, 35*(12), 1225–1230.

[28] Nwanonenyi, S. C., Obasi, H. C., & Eze, I. O. (2019). Hydroxypropyl cellulose as an efficient corrosion inhibitor for aluminium in acidic environments: Experimental and theoretical approach. *Chemistry Africa, 2,* 471–482.

[29] Al Kiey, S. A., Hasanin, M. S., & Dacrory, S. (2021). Potential anticorrosive performance of green and sustainable inhibitor based on cellulose derivatives for carbon steel. *Journal of Molecular Liquids, 338,* 116604.

[30] Popova, A., Sokolova, E., Raicheva, S., & Christov, M. (2003). AC and DC study of the temperature effect on mild steel corrosion in acid media in the presence of benzimidazole derivatives. *Corrosion Science, 45*(1), 33–58.

[31] Zou, C., Yan, X., Qin, Y., Wang, M., & Liu, Y. (2014). Inhibiting evaluation of β-cyclodextrin-modified acrylamide polymer on alloy steel in sulfuric solution. *Corrosion Science*, *85*, 445–454.

[32] Kaur, H., Sharma, J., Jindal, D., Arya, R. K., Ahuja, S. K., & Arya, S. B. (2018). Crosslinked polymer doped binary coatings for corrosion protection. *Progress in Organic Coatings*, *125*, 32–39.

[33] Kumar, S., Vashisht, H., Olasunkanmi, L. O., Bahadur, I., Verma, H., Singh, G., ... & Ebenso, E. E. (2016). Experimental and theoretical studies on inhibition of mild steel corrosion by some synthesized polyurethane tri-block co-polymers. *Scientific Reports*, *6*(1), 30937.

[34] Rehim, S. S. A., Hazzazi, O. A., Amin, M. A., & Khaled, K. F. (2008). On the corrosion inhibition of low carbon steel in concentrated sulphuric acid solutions. Part I: Chemical and electrochemical (AC and DC) studies. *Corrosion Science*, *50*(8), 2258–2271.

[35] Arukalam, I. O., Madu, I. O., Ijomah, N. T., Ewulonu, C. M., & Onyeagoro, G. N. (2014). Acid corrosion inhibition and adsorption behaviour of ethyl hydroxyethyl cellulose on mild steel corrosion. *Journal of Materials*, *2014*, 1–11.

[36] Ameer, M. A., Khamis, E., & Al-Senani, G. (2002). Effect of temperature on stability of adsorbed inhibitors on steel in phosphoric acid solution. *Journal of Applied Electrochemistry*, *32*, 149–156.

[37] Obot, I. B. (2009). Synergistic effect of nizoral and iodide ions on the corrosion inhibition of mild steel in sulphuric acid solution. *Portugaliae Electrochimica Acta*, *27*(5), 539–553.

[38] Eddy, N. O., Odoemelam, S. A., & Odiongenyi, A. O. (2009). Ethanol extract of Musa species peels as a green corrosion inhibitor for mild steel: Kinetics, adsorption and thermodynamic considerations. *Electronic Journal of Environmental, Agricultural and Food Chemistry*, *8*(4), 243–255.

[39] Satapathy, A. K., Gunasekaran, G., Sahoo, S. C., Amit, K., & Rodrigues, P. V. (2009). Corrosion inhibition by *Justicia gendarussa* plant extract in hydrochloric acid solution. *Corrosion Science*, *51*(12), 2848–2856.

[40] Arukalam, I. O., Madufor, I. C., Ogbobe, O., & Oguzie, E. E. (2014). Acidic corrosion inhibition of copper by hydroxyethyl cellulose. *British Journal of Applied Science & Technology*, *4*(9), 1445.

[41] Mobin, M., & Rizvi, M. (2017). Adsorption and corrosion inhibition behavior of hydroxyethyl cellulose and synergistic surfactants additives for carbon steel in 1 M HCl. *Carbohydrate Polymers*, *156*, 202–214.

[42] Zhang, B., He, C., Chen, X., Tian, Z., & Li, F. (2015). The synergistic effect of polyamidoamine dendrimers and sodium silicate on the corrosion of carbon steel in soft water. *Corrosion Science*, *90*, 585–596.

[43] Sangeetha, Y., Meenakshi, S., & Sundaram, C. S. (2016). Corrosion inhibition of aminated hydroxyl ethyl cellulose on mild steel in acidic condition. *Carbohydrate Polymers*, *150*, 13–20.

[44] Selim, I. Z., Basta, A. H., Mansour, O. Y., & Atwa, A. I. (1994). Hydroxyethyl cellulose. II. IR spectra and their relation with the dielectric properties of hydroxyethyl celluloses. *Polymer-Plastics Technology and Engineering*, *33*(2), 161–174.

[45] Liesiene, J., & Kazlauske, J. (2013). Functionalization of cellulose: Synthesis of water-soluble cationic cellulose derivatives. *Cellulose Chemistry and Technology*, *47*, 515–525.

[46] Sangeetha, Y., Meenakshi, S., & Sundaram, C. S. (2016). Interactions at the mild steel acid solution interface in the presence of O-fumaryl-chitosan: Electrochemical and surface studies. *Carbohydrate Polymers*, *136*, 38–45.

[47] Zerga, B., Attayibat, A., Sfaira, M., Taleb, M., Hammouti, B., Ebn Touhami, M., ... & Rais, Z. (2010). Effect of some tripodal bipyrazolic compounds on C38 steel corrosion in hydrochloric acid solution. *Journal of Applied Electrochemistry*, *40*, 1575–1582.

[48] Ekanem, U. F., Umoren, S. A., Udousoro, I. I., & Udoh, A. P. (2010). Inhibition of mild steel corrosion in HCl using pineapple leaves (Ananas comosus L.) extract. *Journal of Materials Science*, *45*, 5558–5566.

13 Pectin as Corrosion Inhibitors

Saman Zehra, Mohammad Mobin, and Babar Khan
Aligarh Muslim University

13.1 INTRODUCTION

Corrosion is considered a never-ending, expensive issue that is hard to solve entirely. It is a significant issue that affects engineering applications in the chemical, motorized, metallurgical, natural, and medical fields as well as, more specifically, in the design of a much more comprehensive range of mechanical parts that also differ in size, functionality, and valuable lifespan [1–3]. Metal and alloy buildings subject to corrosion may sustain life-threatening damage, necessitating costly renovation and replacement work, product losses, safety risks, and environmental contamination [4]. Metal degradation may result in severe production loss because corroded machinery breaks down, and aqueous corrosion products contaminate major industrial products (such as chemical compounds). Product leakage causes a reduction in efficiency. The industry might experience equipment failure if the corrosion-contaminated materials are not handled [5]. Corrosion is an undesirable observable occurrence that must be avoided because of these negative impacts [6,7]. The goal of prevention rather than total eradication would be more practical and achievable. The usage of the inhibitor against corrosion is the best-known and effective way to prevent corrosion among the many ways to stop the deterioration or disintegration of metal surfaces.

When mixed in small amounts to a corroding medium, a compound known as a corrosion inhibitor slows down or completely prevents corrosion [8]. These compounds have the ability to chemically and physically adsorb at the metal-solution interface, preventing the metal from making contact with a corrosive environment [9,10]. Through physisorption or chemisorption processes, an effective corrosion inhibitor should be able to bind quickly to the metal surface [11].

Recent years have seen a significant shift toward the utility of natural polymeric structures procured through the extract of leaves or seeds as eco-friendly corrosion inhibitors. These chemicals are renewable sources of materials that are secure, affordable, and cost-effective [9,10,12–14]. The primary cell walls of higher plants include pectin, a naturally occurring polysaccharide [15]. The chemical profile of pectin includes a substantial quantity of connected D-galacturonic acid remnants. D-galacturonic acid is a cyclical monosaccharide having four hydroxyl side groups and one carboxylic acid. The cyclic monosaccharide's remaining carboxyl groups are partially esterified as side chains and other deprotonation results in anions. Pectin's nontoxicity, cheap manufacturing costs, and several other qualities have also made it suitable for usage as a drug delivery system controller [12,13] and as an excipient

DOI: 10.1201/9781003400059-13

in the food sector [14,15] due to its biodegradability and favorable environmental nature. Pectin may also be extracted from natural products using a highly concentrated solution of HCl since it is soluble in acid [16,17]. These distinctive qualities have made pectin suitable for use as a corrosion inhibitor by creating a protective thin film of pectin molecules on the metal or alloy surface which need to be protected, which then prevents metallic dissolution due to the corrosive action. A protective layer might be formed by swapping out the molecules of water adsorbed with chains of pectin adsorbed.

So, this chapter encompasses the applicability of pectin as a corrosion inhibitor and aims to cover precisely the available literature in the field.

13.2 PECTIN: AN INTRODUCTION

Pectin belongs to a group of polysaccharides that are complex and exist in the cell walls of higher plants, where it acts as a hydrating agent and a binder for the cellulose network [18]. These substances generated commonly during the early stages of primary cell wall synthesis, make up about one-third of the cell walls in the dry sections of dicotyledonous and certain monocotyledonous plants [19–21]. The primary exceptions are the cell walls of the Gramineae family, which may include very trace quantities of pectin with a typical structure [22]. Other monocotyledonous families are less well known, although at least some of them are known to contain conventional or non-conventional pectin in typical amounts [23]. Pectin may be found in most plants in the intercellular space between neighboring cells' main cell walls. With a progressive decline through the primary cell wall toward the direction of plasma membrane, the middle lamella of the cell wall has the largest concentration of pectin. Soft plant tissues with fast development and increased moisture content have comparatively significant levels of pectin.

Pectin is the primary, middle lamella component involved in intercellular adhesion and also contributes to the hardness and structure of plant tissue, same as that of the intercellular material of animal origin (such as collagen) (Figure 13.1) [24]. The orientation, mechanical characteristics, and connections between pectic chemicals and cellulose fiber determine how strong a plant cell wall is.

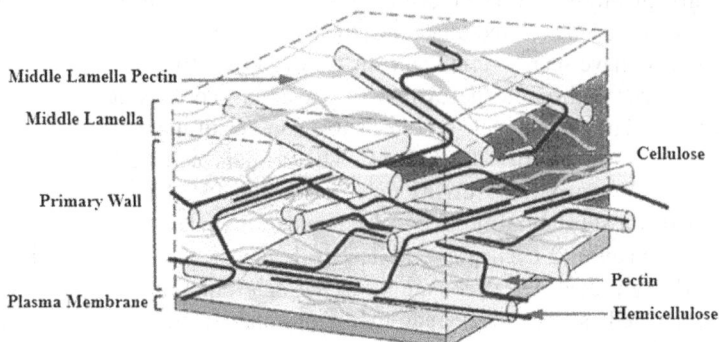

FIGURE 13.1   Plant cell wall.

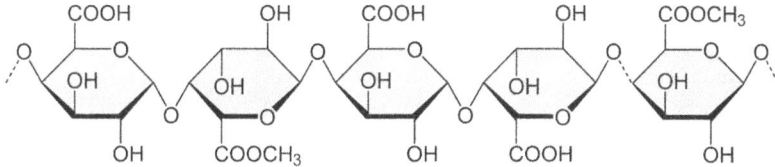

FIGURE 13.2 Structure of citrus pectin.

13.3 STRUCTURE OF PECTIN

One of the essential elements impacted by the many interplay features among cell wall components is pectin, a complex heteropolysaccharide [25]. In certain monocotyledons and dicotyledons, it makes up one-third of the cell wall's dry material. Pectin makes up around 5% of the main cells in non-graminaceous monocots and dicots, 35% of the cells in grass, and 2%–10% of the cells in woody tissues [26].

Understanding the structure of pectin is crucial to comprehending its application in various fields. Pectins are diverse in terms of both chemical structure and molecular weight because, similar to other polysaccharides, pectins are polymolecular and polydisperse as well. The source, the terms of extraction, the location, and other environmental elements all affect how they are made up [27]. Pectin compounds at the main cell wall contain significantly more oligosaccharide chains present on their backbone and considerably longer side chains than the pectin present on the middle lamella [28].

As shown in Figure 13.2, pectin is an acidic hetero-polysaccharide obtained from citrus fruits and is composed of esterified D-galacturonic acid that is present in an alpha-(1–4) chain. It is used as a source of dietary fiber, in food, cosmetics, and medications [29]. Depending on the extraction process and the source, Pectin's molecular weight ranges from 60,000 to 130,000 g/mol.

13.4 PECTIN AND THEIR DERIVATIVES ACT AS CORROSION INHIBITORS

Pectin is a biodegradable, benign promising candidate for green corrosion inhibitors. Pectin's chemistry, like that of other polysaccharides, contributes to its capacity to lessen metal corrosion. Pectin has functional groups on its carbohydrate backbone that are carboxylic (–COOH) and carboxymethyl (–COOCH₃), making it a flair chemical to reduce corrosion and scaling in various environments [30].

In a neutral NaCl solution, pectin and ascorbic acid have been studied for their ability to suppress tin corrosion [31]. At room temperature and 200 ppm ascorbic acid + pectin, the binary system showed a greater anticorrosive effect. The protective complex that formed at the tin/solution interaction is what caused the inhibitory effect. The binary pectin or ascorbic acid system was acting as a cathodic type inhibitor, according to the results of the potentiodynamic polarization method, which demonstrated that the system's inhibitive activity solely affected the cathodic process. Pectin's functional chemistry may be carefully adjusted for a variety of purposes, just like other carbohydrate biopolymers. By adding one or more of the adsorption sites that are capable of adhering to the surface of the metal and displacing the water molecule from the metal/solution interface that acts as a corrosive molecule, structural changes to some anticorrosive polymers enhance the overall material performances.

For mild steel inhibition in 3.5 wt.% NaCl solution, pectin has been grafted to polyacrylamide and polyacrylic acid. The resulting materials protected steel to a greater than 85% degree. The graft polymerization process is only slightly different from that described by Mishra et al. [32], that produces final hydrogel products which is pH-sensitive using pectin and acrylic and acrylamide acids as precursors. After polymer production and modification, corrosion was tested using Electrochemical Impedance Spectroscopy (EIS) and Tafel polarization, and the protective coating that had developed on the steel surface was subsequently evaluated analytically with Fourier transform infrared (FTIR) Spectroscopy and Scanning Electron Microscopy (SEM). Products that modify pectin have also reportedly been employed as water treatment anti-scalants [32]. This technique was only intended to develop a copolymer graft with an anti-scalant potential for carbonates, sulfates, and phosphates by lowering the molecular weight of pectin by acid hydrolysis prior to grafting to acrylamide. The hydrolyzed pectin-based grafted material functioned brilliantly in the scale remediation experiment against the carbonates' precipitation; however, this reaction was neither temperature- nor pH-dependent.

Pectin (from citrus peel) has been used for Al metal reduction in acidic (0.5–2 M HCl) environments, according to Fares et al [33]. At a concentration of 8.0 g/L pectin, a $\% h$ greater than 91% was achieved, and it was noted that this percentage gradually decreased with temperature. With a rise in pectin content, $\% h$ values rose, which was mirrored in comparable increases in activation energy, enthalpy, and entropy. Pectin's ability to prevent corrosion was thought to be due to its ability to bind to metal surfaces, and SEM studies supported this theory. In this acidic medium, it was adsorbed using the Langmuir isotherm. According to Figure 13.3, the physical adsorption involved the electrostatic interaction of the pectin macromolecule's

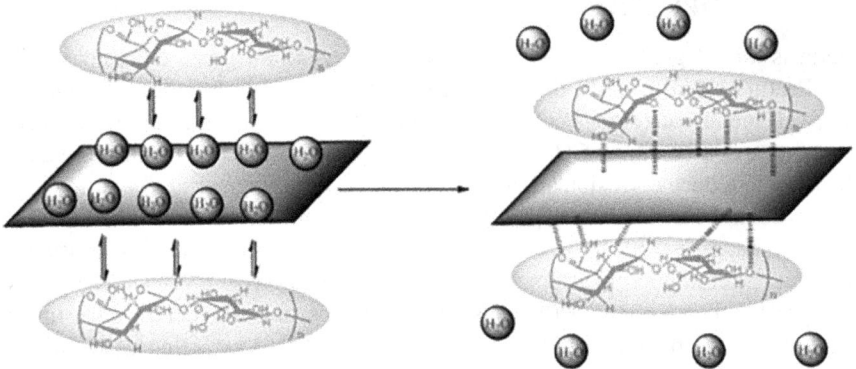

FIGURE 13.3 Schematic representation for the replacement of water molecules with pectin macromolecules and protective layer formation.

partially negatively charged oxygen atoms with partially positively charged aluminum ions, which released adsorbed water molecules and resulted in the barrier formation against the highly corrosive HCl molecules.

Figure 13.4 shows pictures taken with a scanning electron microscope that show the topological alterations of an aluminum surface that were examined with a 2.0 M HCl and an inhibitor. As demonstrated in Figure 13.4a and b, the Al-surface has been badly corroded by the HCl solution. The development of longitudinal parallel grooves and the abundance of large, irregular deep gaps spread throughout the sheet supported the idea that HCl plays a part in causing three-dimensional corrosion. In contrast, Figure 13.4c displays the pectin solution protective layers that had developed on the surface. The dark nodes on the metal surface might be pectin macromolecules gathered around a few uneven humps. Aluminum surface along with pectin inhibitor and HCl solution which is corrosive in nature is shown in Figure 13.4d.

In a different study, Fiori-Bimbi et al. [34] examined pectin for its capacity to inhibit the corrosive assault on mild steel during acid-cleaning operations. They extracted pectin in many steps using acid from the peel of fresh lemon. The pectin extract was tested using chemical and electrochemical methods for its ability to prevent mild steel from corroding in 1 M HCl. Pectin was shown to minimize mild steel corrosion, and this effect persisted as the temperature soared. The Tafel plot results showed that the pectin in this investigation was a mixed-type inhibitor, and UV spectroscopic examination proved that the geometric blocking effect

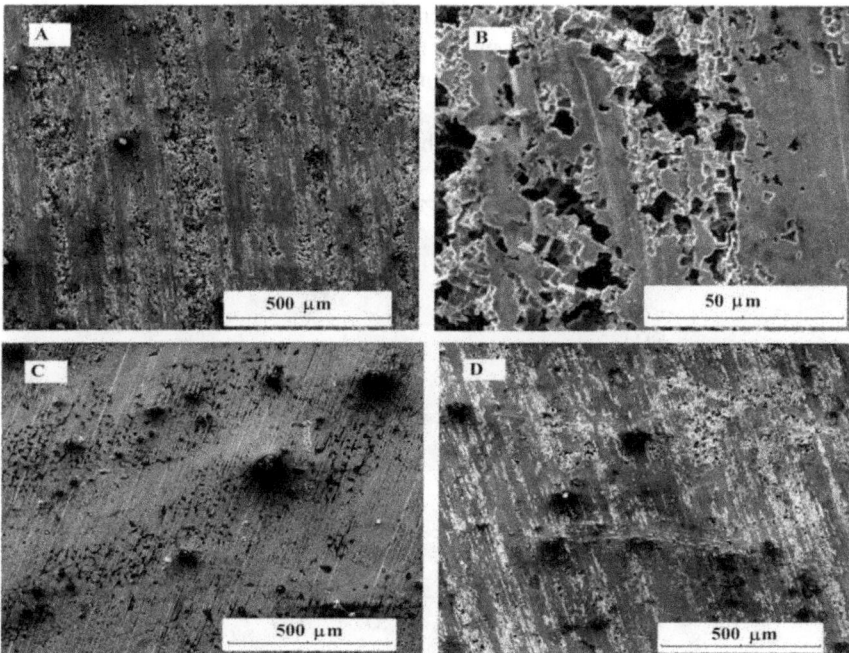

FIGURE 13.4 Scanning electron micrographs of; (a) and (b) aluminum surface in 2.0 M corrosive HCl solution, (c) aluminum surface in 8.0 g/L pectin solution, and (d) and (e) aluminum surface in 2.0 M HCl corrosive solution and in presence of 8.0 g/L pectin solution.

brought on by complexes or species of the pectin-Fe^{2+} type that had chemisorbed at the metal/solution interface was what was causing pectin inhibition. Using the electrochemical data, the thermodynamics and kinetics of adsorption were immediately calculated, and the trend in values further revealed information about the adsorption process.

Even at extremely low concentrations, pectin was employed without the addition of any other chemicals and worked as an effective mild steel corrosion inhibitor. The results show that temperature rises have a positive impact on the effectiveness of corrosion inhibition (Figure 13.5).

The mild steel surface's chemisorption of pectin results in corrosion inhibition, which acts as a mixed-type inhibitor that affects both cathodic and anodic processes. The geometric blocking effect of the deposited inhibitive species at the metal surface is also a possible cause of this behavior, which is compatible with an inhibition mode. According to the UV spectroscopic analysis, a complex between pectin and the Fe^{2+} ions produced during the corrosion event is expected to form (Figure 13.6).

X60 pipeline steel in HCl medium was used as the test material in a research study performed by Umoren et al. [35] where pectin (commercial pectin from apples) was observed using chemical and electrochemical methods. Temperature and pectin concentration were discovered to affect the corrosion inhibition efficiency (%h); greater magnitudes of %h were achieved at higher temperatures and pectin concentrations, with 79% being the highest %h ever recorded for 1,000 ppm pectin. According to the outcomes of potentiodynamic polarization, the anodic and cathodic responses are influenced by pectin inhibition but most of the time cathodic inhibition is observed. SEM and measurements of the water contact angle corroborated the theory that the suppression of corrosion of the X60 steel pipeline was caused by the formation of the protective thin layer through the adsorption phenomenon onto the metal

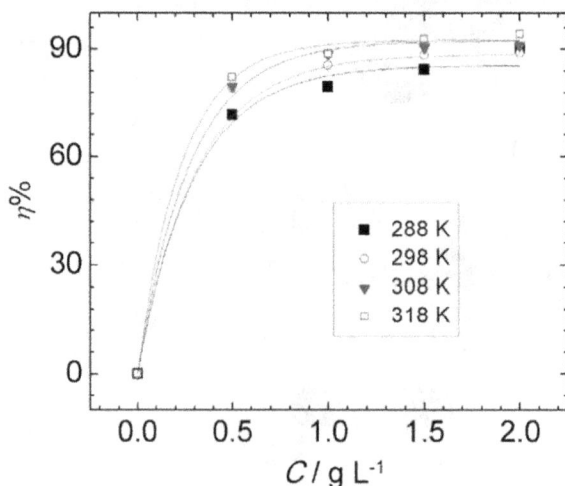

FIGURE 13.5 Corrosion inhibition of steel expressed as inhibition efficiency in percentage $\eta(\%)$ as a function of pectin concentration in 1 M HCl solution at different temperatures.

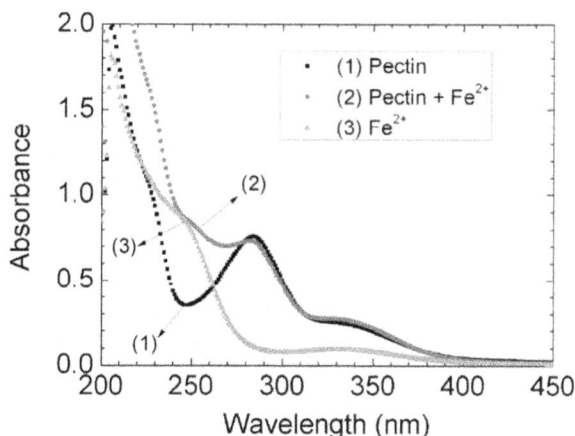

FIGURE 13.6 UV-visible spectra of (squares) 1 M HCl with 1.0 g L^{-1} pectin added; (circles) 1 M HCl with 1.0 g L^{-1} pectin added after mild steel immersion for 6 days at 298 K; (triangles) 1 M HCl after mild steel immersion for 6 days at 298 K.

surface. Pectin adsorption was tracked using the Langmuir adsorption isotherm, and DFT quantum chemical calculations were utilized to explain how pectin inhibited adsorption.

Pectin extract from Opuntia cladodes has also been utilized to lessen mild steel corrosion in 1 M HCl using weight loss, electrochemical impedance spectroscopy, and potentiodynamic polarization methods [36]. Charge-transfer resistance and reduced double-layer capacitance values showed that corrosion inhibition increased with pectin content. The maximum %h (96%) was observed when the concentration of pectin was 1 g/L at 35°C, indicating that pectin from this source functioned as a mixed type of inhibitor. On the metallic substrate, pectin adsorption occurred after the Langmuir adsorption isotherm.

Propyl phosphonic acid (PA) and Zn^{2+} ions in neutral conditions effectively prevent pectin for carbon steel, as shown by Prabakaran et al. [37] using chemical and electrochemical methods. Experiments on corrosion that include pectin inhibition are not limited to media that have been exposed to acid. Pectin was reported to increase the functionality of the secondary components (PA and Zn^{2+} additives) to block. The optimal pectin concentration was found using a weight loss approach, and it was also demonstrated that the addition of PA and Zn^{2+} enhanced pectin's ability to defend against corrosion. The findings of potentiodynamic polarization showed that pectin inhibition affected both cathodic and anodic processes and that its adsorption was clearly associated with a reduction in corrosion current density values. We used surface analytics (SEM, AFM, and XPS) and spectroscopic (FTIR) methods to investigate the development of complexes/films on the surface of carbon steel due to the pectin type of complexes.

In a study done by Grassino et al. [38], scientists used waste tomato (Lycopersicum esculentum) to extract pectin for the first time. Following its extraction from this source, pectin was further characterized using Fourier Transform Infrared (FTIR) and nuclear magnetic resonance spectroscopy (NMR) investigations, in addition to rheological testing and color analysis. The results were contrasted with an industry-standard pectin reference. As determined by the extent of esterification, the isolated pectin was of the methoxy pectin type (about 82%). The pectin extract was used to evaluate the inhibition of tin corrosion in a solution of 2% NaCl, 1% acetic acid, and 0.5% citric acid. Values of percent IE up to 73% were reported at extremely low concentrations (4 g/L) using EIS and potentiodynamic polarization method. The polarization curves for tin in test solution (2% NaCl, 1% acetic acid, and 0.5% citric acid) at 25 (C in the absence and presence of various concentrations of pectin (CAP and A-I) are presented in Figure 13.7. Tafel results indicated that pectin in this study is a mixed-type inhibitor (Figure 13.7a). The plots demonstrate the relationship between current density and pectin concentration, i.e., that current density declines as pectin concentration rises. The cathodic polarization curves are more affected by the addition of both pectins (CAP and A-I) than the anodic polarization curves. Tin's anodic curves in Figure 13.7b show that the addition of this pectin has no effect on the dissolution of anodic metals because their form is similar to that of the curve produced in a blank solution.

The effectiveness of the soluble polymers, pectin and poloxamer, in preventing the corrosion of steel composed of carbon in 1.0 M HCl solution was examined by Abdallah et al. [39] by the usage of four different methods. Lowering the temperature and boosting the tested polymer's concentration both improve inhibitory potency. The corrosion parameter result from all used measurements highlights the potency of these two chemicals as an inhibitor. An adsorbing-coated layer that shields the steel surface from hostile solutions was thought to be the cause of the inhibition. Pectin adheres to the steel surface according to the Freundlich isotherm. Physical and chemical adsorption are both used in the adsorption process. Additionally, by shifting the pitting potential in a positive direction, poloxamer and pectin compounds prevent the corrosion of C-steel in Cl- ions and solutions. Due to its larger molar mass and thus better surface coverage on the surface of C-steel, pectin exhibits more inhibitory efficiency than poloxamer.

13.5 CONCLUSION AND FUTURE PERSPECTIVE

The ability of pectin, a polymer derived from nature, to prevent a corrosive assault on different metals and corrosive conditions has been studied. The following conclusions may be taken from the results: Even at extremely low concentrations, pectin used alone, without the addition of other chemicals, effectively inhibits corrosion for various metals, Pectin is categorized as a mixed-type inhibitor since it affects both cathodic and anodic processes. The geometrical blocking effect of the accumulated hindering species at the metal surface also explains this behavior, which is compatible with an inhibition mode. Despite being present in a wide variety of plant species, there are extremely few commercial sources of pectin. To get pectin with the needed quality qualities, it is necessary to investigate additional sources of pectin or alter the present sources. It is possible to alter pectins in vivo using contemporary scientific technologies like genetic engineering.

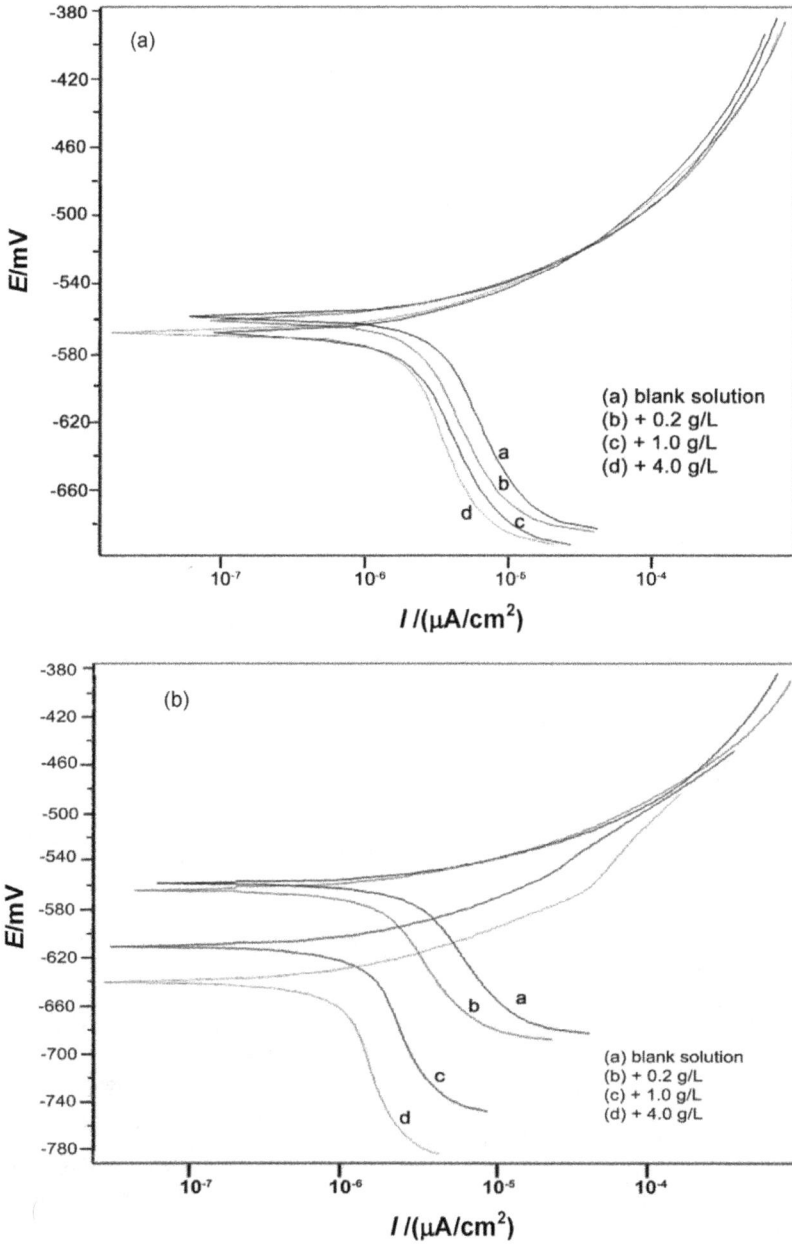

FIGURE 13.7 Potentiodynamic polarization curves obtained for tin at 25°C, in 2% NaCl, 1% acetic acid and 0.5% citric acid solution in the absence and presence of various concentrations (0.2 e4.0 g/L) of (a) commercial apple pectin (CAP) and (b) pectin extracted from tomato peel (A-I).

REFERENCES

[1] W.D. Callister, *Materials Science and Engineering: An Introduction*. 7th edn., Wiley, 2007.

[2] R.W. Revie, H.H. Uhlig, *Uhlig's Corrosion Handbook*. Wiley, 2011.

[3] S. Zehra, M. Mobin, R. Aslam, An overview of the corrosion chemistry. Environmentally sustainable corrosion inhibitors. In: C. M. Hussain, C. Verma, and J. Aslam (Eds.), *Fundamentals and Industrial Applications*. Elsevier, 2021, pp. 3–23.

[4] K.E. Heusler, D. Landolt, S. Trasatti, Electrochemical corrosion nomenclature (Recommendations 1988). *Electrochim. Acta* 35 (1990) 295.

[5] H.H. Uhlig, The cost of corrosion to the U.S. *Chem. Eng. News Arch.* 27(39) (1949) 2764.

[6] L.H. Bennet, J. Kruger, R.I. Parker, E. Passiglia, C. Reimann, A.W. Ruff, H. Yakowitz, E.B. Berman, *Economic Effects of Metallic Corrosion in the United States*. Vol. 511, National Bureau of Standards Special Publication, 1978.

[7] G.H. Koch, M.P.H. Brongers, N.G. Thompson, Y.P. Virmani, J.H. Payer, Corrosion costs and preventive strategies in the United States. *Mater. Perf.* 42(Supplement) (2002). https://rosap.ntl.bts.gov/view/dot/39217

[8] S. Zehra, M. Mobin, J. Aslam, Corrosion inhibitors: An introduction, book title: Environmentally sustainable corrosion inhibitors. In: C. M. Hussain, C. Verma, and J. Aslam (Eds.), *Fundamentals and Industrial Applications*. Elsevier, 2022, pp. 47–67. doi:10.1016/B978-0-323-85405-4.00022-7.

[9] M. Mobin, S. Zehra, R. Aslam, l-Phenylalanine methyl ester hydrochloride as a green corrosion inhibitor for mild steel in hydrochloric acid solution and the effect of surfactant additive. *RSC Adv.*, 6 (2016) 5890–5902.

[10] M. Mobin, S. Zehra, M. Parveen, L-Cysteine as corrosion inhibitor for mild steel in 1 M HCl and synergistic effect of anionic, cationic and non-ionic surfactants. *J. Mol. Liq.* 216 (2016) 598–607.

[11] R. Aslam, M. Mobin, J. Aslam, H. Lgaz, I.-M. Chung, S. Zehra, Synergistic inhibition behavior between rhodamine blue and cationic gemini surfactant on mild steel corrosion in 1 M HCl medium. *J. Mol. Struct.* 1228 (2021) 129751.

[12] M.M. Fares, S.M. Assaf, Y.M. Abul-Haija, Pectin grafted poly(N-vinylpyrrolidone): optimization and in vitro controllable theophylline drug release. *J. Appl. Polym. Sci.* 117 (2010) 1945–1954.

[13] S.M. Assaf, Y.M. Abul-Haija, M.M. Fares, Versatile pectin grafted poly (N-isopropylacrylamide); modulated targeted drug release. *J. Macromol. Sci. A Pure Appl. Chem.* 48(6) (2011) 493–502.

[14] L. Liu, M.L. Fishman, J. Kost, K.B. Hicks, Pectin-based systems for colon-specific drug delivery via oral route. *Biomaterials* 24 (2003) 3333–3343.

[15] M.M. Fares, Y.R. Tahboub, S.T. Khatatbeh, Y.M. Abul-Haija, Eco-friendly, vascular shape and interpenetrating poly (acrylic acid) grafted pectin hydrogels; biosorption and desorption investigations. *J. Polym. Environ.* 19 (2011) 431–439.

[16] F. Kar, N. Arslan, Effect of temperature and concentration on viscosity of orange peel pectin solutions and intrinsic viscosity–molecular weight relationship. *Carbohydr. Polym.* 40(4) (1999) 277–284.

[17] U. Kalapathy, A. Proctor, Effect of acid extraction and alcohol precipitation conditions on the yield and purity of soy hull pectin. *Food Chem.* 73(4) (2001) 393–396.

[18] G. Muralikrishna, R.N. Taranathan, Characterization of pectin polysaccharides from pulse husks. *Food Chem.* 50 (1994) 87.

[19] D.H. Northcote, Chemistry of the plant cell wall. *Annu. Rev. Plant Physiol.* 23 (1972) 113.

[20] M.C. Jarvis, W. Forsyth, H.J. Duncan, A survey of the pectin content of nonlignified monocot cell walls. *Plant Physiol.* 88 (1988) 309.

[21] J.E. Hoff, M.D. Castro, Chemical composition of potato cell wall. *Agric. Food Chem.* 17 (1969) 1328.

[22] N. Shibuya, T. Iwasaki, Polysaccharides and glycoproteins in the rice endosperm cell wall. *Agric. Biol. Chem.* 42 (1978) 2259.

[23] R.G. Ovodova, V.E. Vaskovsky, Y.S. Ovodov, The pectic substances of Zosteraceae. *Carbohydr. Res.* 6 (1968) 328.

[24] M.C. Jarvis, M.A. Hall, D.R. Threlfall, J. Friend, The polysaccharide structure of potato cell walls, chemical fractionation. *Planta* 152 (1981) 93.

[25] F. Dranca, M. Oroian, Extraction, purification and characterisation of pectin from alternative sources with potential technological applications. *Food Res. Int.* 113 (2018), 327–350.

[26] A. Noreen, J. Akram, I. Rasul, A. Mansha, N. Yaqoob et al., Pectins functionalised biomaterials; a new viable approach for biomedical applications: A review. *Int. J. Biol. Macromol.* 101 (2017) 254–272.

[27] K.C. Chang, N. Dhurandhar, X. You, A. Miyamoto, Cultivar/location and processing methods affect yield and quality of sunflower pectin. *J. Food Sci.* 59 (1994) 602.

[28] T. [Sakai, T. Sakamoto, J. Hallaert, E.J. Vandamme, Pectin, pectinase and protopectinase: Production, properties and applications. *Adv. Appl. Microbiol.* 39 (1993) 213.

[29] B. luke, C. Breuer, S. Keiko, Animal welfare: A central tenet of the association of zoos and aquariums. *New Phytolgist.* 201 (2013) 388.

[30] K. Chauhan, R. Kumar, M. Kumar, P. Sharma, G.S. Chauhan, Modified pectin-based polymers as green antiscalants for calcium sulfate scale inhibition. *Desalination* 30 (2012) 31–37.

[31] R. Geethanjali, A. Ali, F. Sabirneeza, S. Subhashini, Water-Soluble and biodegradable pectin-grafted polyacrylamide and pectin-grafted polyacrylic acid: Electrochemical investigation of corrosion-inhibition behaviour on mild steel in 3.5% NaCl media. *Ind. J. Mater. Sci.* (2014). doi.org/10.1155/2014/356075.

[32] R.K. Mishra, P.B. Sutar, J.P. Singhal, A.K. Banthia, Graft copolymerization of pectin with polyacrylamide. *Polym. Plast. Technol. Eng.* 46 (2007) 1079–1085.

[33] M.M. Fares, A.K. Maayta, M.M. Al-Qudah, Pectin as promising green corrosion inhibitor of aluminum in hydrochloric acid solution. *Corros. Sci.* 60 (2012) 112–117

[34] M.V. Fiori-Bimbi, P.E. Alvarez, H. Vaca, C.A. Gervasi, Corrosion inhibition of mild steel in hcl solution by pectin. *Corros. Sci.* (2014). doi:10.1016/j.corsci.2014.12.002.

[35] S.A. Umoren, I.B. Obot, A. Madhankumar, Z.M. Gasem, Performance evaluation of pectin as ecofriendly corrosion inhibitor for X60 pipeline steel in acid medium: Experimental and theoretical approaches. *Carbohydr. Polym.* (2015). doi:10.1016/j.carbpol.2015.02.036.

[36] N. Saidia, H. Elmsellema, M. Ramdania, A. Chetouania, K. Azzaouib, F. Yousfia, A. Aounitia, B. Hammouti, Using pectin extract as eco-friendly inhibitor for steel corrosion in 1 M HCl media. *Der PharmaChemica* 7(5) (2015) 87–94.

[37] M. Prabakaran, S. Ramesh, V. Periasamy, B. Sreedhar, The corrosion inhibition performance of pectin with propyl phosphonic acid and Zn2+ for corrosion control of carbon steel in aqueous solution. *Res. Chem. Interm.* 41 (2015) 4649–4671.

[38] A. Grassino, J. Halambek, S. Djaković, S.R. Brnčić, M. Dent, Z. Grabarić, Utilization of tomato peel waste from canning factory as a potential source for pectin production and application as tin corrosion inhibitor. *Food Hydrocolloids* 52 (2016) 265–274.

[39] M. Abdallah, A. Fawzy, H. Hawsawi, R.S. Abdel Hameed, S.S. Al-Juaid, Estimation of water-soluble polymers (poloxamer and pectin) as corrosion inhibitors for carbon steel in acidic medium. *Int. J. Electrochem. Sci.* 15 (2020) 8129–8144. doi:10.20964/2020.08.73.

14 Natural Gums as Corrosion Inhibitors

Monikandon Sukumaran
Annamalai University
Tamil Nadu Dr. J. Jayalalitha Fisheries University

Ravisankar Natarajamani and S. Poongothai
Annamalai University

Kesavan Devarayan
Tamil Nadu Dr. J. Jayalalitha Fisheries University

14.1 INTRODUCTION

Corrosion poses a widespread challenge, impacting the strength and durability of metals and alloys, which leads to significant economic losses and safety concerns across diverse industries. Conventional approaches to corrosion protection often involve synthetic inhibitors, which can be costly, environmentally harmful, and potentially hazardous to health [1]. In recent years, there has been a growing interest in exploring natural alternatives for corrosion inhibition, with natural gums emerging as promising candidates in this regard.

Natural gums are complex polysaccharides obtained from plant exudates, seeds, or other botanical sources. They possess a wide array of physical, chemical, and biological properties that make them appealing for various applications. Utilizing natural gums as corrosion inhibitors offers several advantages. Firstly, they are abundant, renewable, and cost-effective, providing a sustainable substitute for synthetic inhibitors. Moreover, natural gums are generally regarded as environmentally friendly and biocompatible, minimizing potential risks and ecological impacts.

Extensive studies have been conducted on the inhibitive properties of natural gums against corrosion, revealing their effectiveness in safeguarding metals and alloys from degradation. Natural gums exhibit inhibitory effects by forming a protective barrier on the metal surface, hindering the access of aggressive species, such as corrosive ions and oxygen, to the metal interface. Furthermore, their high molecular weight and complex chemical composition facilitate their adsorption onto the metal surface, enhancing the formation of a stable and durable film.

Several types of natural gums, including *Ficus benjamina* gum, *Anogessius leocarpus* gum, and *Ficus platyphylla* gum, among others, have been investigated for

DOI: 10.1201/9781003400059-14

their corrosion inhibition capabilities. These gums have shown promising results in inhibiting the corrosion of metals like aluminum and mild steel in various corrosive environments, such as sulfuric acid and hydrochloric acid solutions [2].

14.2   ANTI-CORROSION AGENTS

A corrosion inhibitor refers to a chemical compound that, when introduced in parts per million (ppm) to a corrosive environment, effectively reduces the corrosion rate [3]. These compounds have the ability to chemically and physically adsorb at the metal–solution interface, forming a protective layer that shields the metal from corrosive substances [4–8]. An ideal corrosion inhibitor should promptly bind to the metal surface through physiosorption or chemisorption processes [9–11].

Thus far, most industrially significant corrosion inhibitors have been synthesized using cost-effective raw materials or by selecting compounds with aromatic or long carbon chains that contain heteroatoms. However, considerable efforts have been devoted to discovering suitable natural alternatives as initial corrosion inhibitors for various corrosive conditions [12–15]. In acidic media, aldehydes, thioaldehydes, acetylenic chemicals, diverse natural alkaloids, nitrogen-based materials, and their derivatives have exhibited inhibitory properties. Generally, inhibitors of this kind slow down the reaction between metals and the corrosive medium. When choosing an inhibitor, several factors need to be considered, with the safety of the ecosystem and its inhabitants being of utmost importance. Organic inhibitors can be categorized into two groups: (1) synthetic organic inhibitors and (2) green inhibitors.

14.3   ORGANIC/SYNTHETIC INHIBITORS

Organic inhibitors commonly incorporate heteroatoms, such as O, N, S, and P, which contribute to their higher basicity and electron-donor capacity, enabling them to serve as effective corrosion inhibitors. These heteroatoms, specifically O, N, S, and P, serve as active sites for the adsorption process on the metal surface. In laboratory settings, synthetic inhibitors have been developed as robust alternatives to natural inhibitors (Figure 14.1).

The utilization of inhibitors represents one of the most effective approaches to prevent corrosion in metals. Currently, a wide range of inhibitors is employed, either selected from compounds with heteroatoms present in their aromatic or long-chain carbon systems or produced using cost-effective raw materials. However, it is important to note that the majority of these inhibitors have adverse effects on the environment. As a result, there has been a surge in research focused on discovering environmentally friendly corrosion inhibitors [15].

14.3   GREEN INHIBITORS

Green corrosion inhibitors are characterized by their non-toxic nature, absence of heavy metals, and positive environmental impact. These inhibitors are substances or combinations of compounds that are applied in minimal quantities to the exposed surfaces of metals in corrosive environments. They effectively impede or decelerate

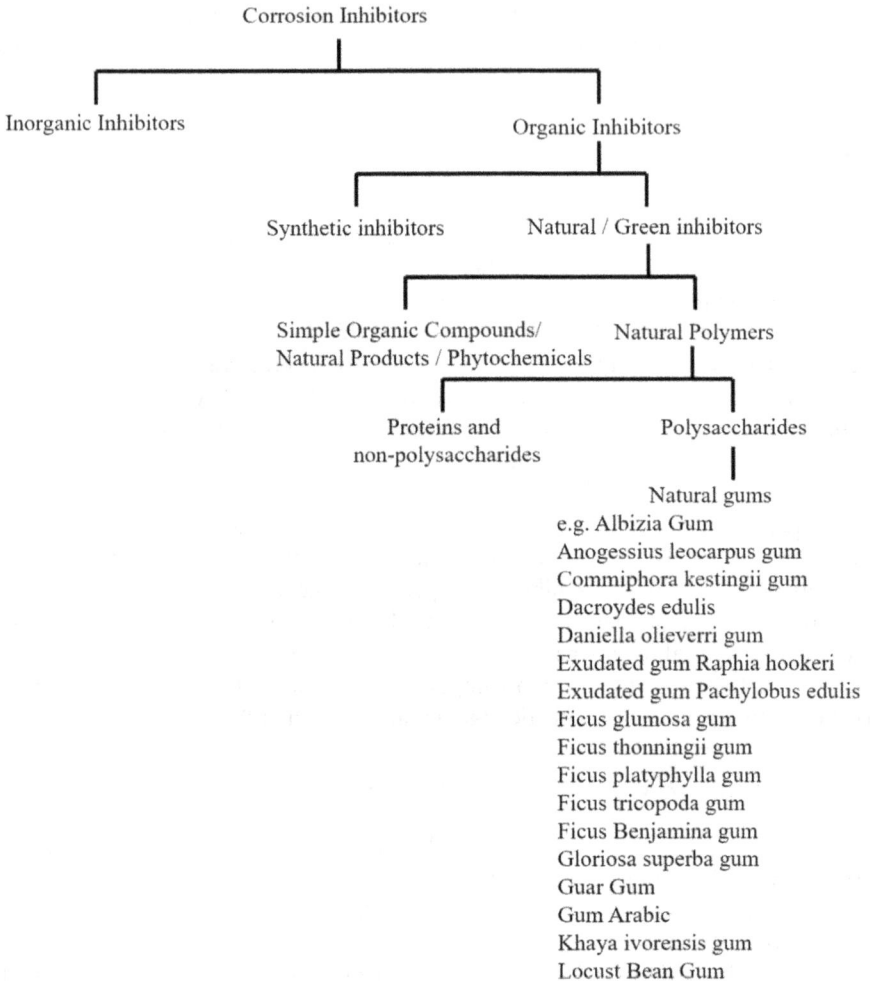

FIGURE 14.1   Classification of natural organic compounds-based inhibitors.

the corrosion process of the metal. Due to their ability to absorb substances, these inhibitors are also referred to as site-blocking elements or adsorption site blockers [1,16–18]. Green inhibitors, also known as eco-friendly inhibitors, demonstrate biocompatibility with the environment.

Gums derived from natural sources are effective corrosion inhibitors. These solid substances consist of polysaccharides, which are water-soluble and can form gels or jellies when immersed in water. They are insoluble in organic solvents such as alcohol, ether, hydrocarbons, and oils. Gums undergo hydrolysis, resulting in the formation of simple sugars like arabinose, galactose, mannose, and glucuronic acid. Some gums are obtained through exudation, primarily from tree stems but occasionally from other parts of plants. This exudation can occur due to plant injury or intentional tapping.

Apart from their applications in corrosion inhibition, gums find uses in various industries such as food, medicine, and technology. In the food and beverage sector, gums are used for thickening, stabilizing, emulsifying, and suspending purposes. Certain gums have dental and medical applications, while others serve as emulsifiers and binding agents in pharmaceutical formulations.

Studies have shown that gums possess significant corrosion inhibition properties [19]. The functional groups present in gums can form complexes with metal ions on metal surfaces. These complexes provide a protective layer, shielding the metal from corrosive substances. The oxygen and nitrogen atoms in various gums serve as adsorption centers. Most gums contain –COOH functional groups, which enhance electron transfer and facilitate adsorption-based inhibition. Furthermore, gums are considered non-toxic, environmentally friendly, and classified as "green" inhibitors.

The objective of the research discussed in this chapter is to summarize the use of different gums as environmentally friendly corrosion inhibitors (Table 14.1).

TABLE 14.1

Natural Gums and Their Application for Corrosion Inhibition of Different Metals in Various Environments

Inhibitor	Surface	Corrosive Environment	Analyses	References
Gum Arabic	MS & Al	H_2SO_4	WL, TM	[28]
Acacia var. *seyal*	MS	Drinking water	EIS, TP	[29]
Albizia zygia	MS	H_2SO_4	FTIR, TM	[20,21]
Anogessius leocarpus	MS	HCl	WL, GM	[22]
Commiphora kestingii	Al	H_2SO_4	WL, SEM	[30]
Commiphora pendunculata	Al	HCl	WL, TM	[31]
Ficus benjamin	Al	H_2SO_4	WL	[23]
Ficus glumosa	MS	H_2SO_4	WL, HE, SEM	[32]
Ficus platyphylla	MS	HCl	WL, FTIR, GCMS	[33]
Ficus thonningii	Al	HCl	WL, GM	[34]
Ficus tricopoda	Al	H_2SO_4	WL, GCMS	[35]
Gloriosa superba	Al	HCl	FTIR & GM	[36]
Guar gum	CS	H_2SO_4	WL, TP	[37]
Gum Arabic	Al	NaOH + KI	WL, HE	[38]
Khaya ivorensis	Ms	HCl	WL, GM	[39]
Pachylobus edulis	Al	Halide ions, HCl	WL	[40]
Pachylobus edulis with potassium halides	MS	H_2SO_4	HE, TM	[41]
Raphia hookeri	Al	HCl	WL, TM	[26]
Raphia hookeri with halides additive	Al	HCl	WL, GM	[27]

MS, mild steel; Al, aluminum, CS, carbon steel; WL, weight loss measurement; HE, hydrogen evolution; EIS, electrochemical impedance spectroscopy; TP, Tafel polarization; SEM, scanning electron microscopy; FTIR, Fourier-transform Infrared spectroscopy; GCMS, gas chromatography-mass spectrometry; GM, gasometric; TM, thermometric.

14.4 ALBIZIA GUM

Albizia gum, derived from trees in the *Albizia* genus, is available in various colors and sizes. It is highly water-soluble and has a large molecular size, making it intriguing for corrosion inhibition. Gas chromatography-mass spectrometry analysis has revealed the presence of nitrogen, sulfur, and oxygen in the gum's structure. Studies have shown that Albizia zygia gum is an effective inhibitor for mild steel corrosion in sulfuric acid solution [20]. The inhibition efficiency increases with concentration and decreases with temperature, and the mechanism is attributed to physical adsorption [21].

14.4.1 *Anogessius leocarpus* Gum

Anogessius leocarpus gum (*AL* gum) has been identified as an effective inhibitor for the corrosion of mild steel in hydrochloric acid solution containing potassium halide ions. Experimental methods, including weight loss and geometric analysis, have demonstrated that the inhibition efficiency increases with increasing concentration and decreases with temperature. The inhibition mechanism follows a physical adsorption process, conforming to the Langmuir adsorption isotherm [22].

14.4.2 *Commiphora kestingii* Gum

Commiphora kestingii gum, also known by various names such as Fula-fulfulde, Mbiji, and Nupe, is typically found in low-elevation dry woods from Togo to Nigeria. This gum has been found to be a corrosion inhibitor, and the inhibitory efficiency increases with temperature and concentration. Adsorption by Freundlich, Temkin, and Florry Huggins models has been reported [23].

14.4.3 *Dacroydes edulis* Gum

Dacroydes edulis gum, derived from the *Dacryodes edulis* tree, is an affordable and environmentally safe inhibitor for aluminum corrosion in an acidic medium. It has been tested using weight loss and thermometric tests, and the inhibition mechanism follows the Temkin adsorption isotherm [24].

14.4.4 *Daniella olieverri* Gum

Daniella olieverri gum (*DO* gum) is harvested from the fully grown stem of the *Daniella oliverra* plant during the dry season. This gum has been found to effectively inhibit the corrosion of mild steel in hydrochloric acid solution. Its properties, such as its acidic nature, solubility in water, and insolubility in acetone, chloroform, and ethanol, have been analyzed using Fourier Transform Infrared Spectroscopy (FTIR) and weight loss techniques [25].

14.4.5 *Raphia hookeri*

Exudated gum from *Raphia hookeri*, commonly known as Wine Palm, exhibits potent corrosion inhibition for aluminum in HCl solution. The inhibition efficiency increases with gum concentration and is significantly enhanced by the addition of halide ions. The adsorption process follows the Temkin adsorption isotherm. The Freundlich, Langmuir, and Temkin adsorption isotherms have also been found to be obeyed. The proposed mechanism involves physical adsorption [26].

14.4.6 PACHYLOBUS EDULIS

Pachylobus edulis, commonly known as African plum or pear safou, produces sticky exudates that contribute to the development of environmentally friendly corrosion inhibitors. Research has shown that the exudated gums from *Pachylobus edulis* (PE) can inhibit mild steel corrosion in H_2SO_4 solution when combined with potassium halide additives. The inhibitory effect was evaluated using hydrogen evolution and thermometric methods within a temperature range of 30°C–60°C. The data revealed that as the concentration of the gum increased, the corrosion rate decreased, but it increased with rising temperature. Potassium halides, specifically KI, KBr, and KCl, showed a synergistic effect in enhancing the inhibitory efficiency. The mechanism of inhibition followed the Temkin adsorption isotherm [41]. Furthermore, the weight loss method was used to observe the anti-corrosive effect of PE Gum with halide ions on aluminum corrosion in an acidic medium at temperatures between 30°C and 60°C. The results indicated that the inhibition effectiveness increased with higher inhibitor concentration and improved with the addition of halide ions, while it decreased with increasing temperature. The adsorption mechanism followed the Temkin adsorption isotherm, indicating physical adsorption [40].

14.4.7 FICUS GLUMOSA

Ficus glumosa, also known as the African rock fig, is a tree of small to medium size, typically reaching heights of 5–10 m and occasionally growing up to 24 m tall with a diameter of 50 cm. Through the use of weight loss, thermometric, and SEM techniques, it has been discovered that Ficus glumosa gum is an effective corrosion inhibitor for mild steel. The data obtained demonstrated that the inhibition efficiency increased with temperature and concentration. The inhibitory effect was attributed to chemical adsorption, following the Langmuir adsorption model [32].

14.4.8 FICUS THONNINGII

Ficus thonningii gum has been identified as a potent corrosion inhibitor for aluminum corrosion in hydrochloric acid solution using Gas Chromatography Mass Spectrometry (GC-MS) and Fourier transform infrared (FTIR) spectrophotometer techniques. The results revealed that the corrosion inhibition efficiency of the gum increased with the inhibitor concentration but decreased with increasing temperature, indicating physical adsorption. The adsorption behavior followed the Langmuir adsorption isotherm [34].

14.4.9 FICUS PLATYPHYLLA

The potential of *Ficus platyphylla* gum as a corrosion inhibitor for aluminum alloy in hydrochloric acid solution was investigated using weight loss and thermometric techniques. The results showed that the corrosion inhibition effectiveness of CP gum increased with concentration but decreased with rising temperature. The inhibition process was attributed to physical adsorption and adhered to the Langmuir adsorption isotherm [33].

14.4.10 *Ficus tricopoda*

Ficus tricopoda gum has been identified as a potent corrosion inhibitor for aluminum in acidic media using FTIR and GCMS techniques. The results indicated that FT gum exhibited initial physical adsorption followed by a chemical adsorption mechanism, suggesting the presence of both types of mechanisms. The adsorption behavior conformed to the Langmuir adsorption isotherm [35].

14.4.11 *Ficus benjamina*

Ficus benjamina gum, also known as the weeping fig, Benjamin's fig, or ficus tree, was investigated as a corrosion inhibitor for aluminum in sulfuric acid using weight loss and FTIR techniques. The results showed that the inhibition efficiency of *Ficus benjamina* gum increased with rising temperature and concentration, following the Frumkin and Dubinin Radushikevich adsorption model [30].

14.4.12 *Gloriosa superba*

Gloriosa superba gum has been identified as a corrosion inhibitor for aluminum in hydrochloric acid media using FTIR and GCMS techniques. The results revealed that Gloriosa superba gum exhibited initial physical adsorption followed by a chemical adsorption mechanism, indicating the presence of both types of mechanisms. The adsorption behavior followed the Langmuir adsorption isotherm [36].

14.5 GUAR GUM

Guar gum is a polysaccharide obtained from the endosperm of the seed of the legume plant *Cyamopsis tetragonolobus*. It consists of galactose and mannose sugars, with a linear chain of mannose residues linked at 1,4 positions and galactose residues linked at 1,6 positions for every other mannose, creating small side branches. Abdulla conducted a study on the corrosion inhibition potential of gums on carbon steel corrosion in 1 M H_2SO_4 solution. Weight loss and electrochemical methods were employed, and the results indicated that guar gum enhances the resistance to pitting corrosion of carbon steel in the presence of NaCl. The potentiodynamic polarization (PDP) method supported this finding, showing that guar gum acts as a mixed-type inhibitor. The data obtained fitted well with the Langmuir adsorption isotherm. The weight loss method also demonstrated that the inhibition efficiency increased with higher concentration [37].

14.6 GUM ARABIC

Gum Arabic is a widely used natural gum obtained from different *Acacia* tree species in the Leguminosae family. It has been investigated for its corrosion inhibition potential on aluminum in alkaline environments. Previous studies have suggested that the presence of arabinogalactan, oligosaccharides, polysaccharides, and glucoproteins in Gum Arabic contributes to its ability to prevent aluminum corrosion [34,35].

The effectiveness of Gum Arabic as a corrosion inhibitor for aluminum and mild steel in H_2SO_4 solution was evaluated using thermometric and weight loss methods. The results showed that the inhibition efficiency increased with higher inhibitor concentrations, following the Temkin adsorption isotherm for both aluminum and mild steel. Physical adsorption was observed for aluminum corrosion, while chemical adsorption was noted for mild steel corrosion. Thermodynamic tests indicated that Gum Arabic spontaneously adheres to the metal surface and exhibits superior corrosion inhibition for aluminum compared to mild steel [36].

The inhibitory action of gum exudates from *Acacia seyal* var. *seyal* was studied using potentiodynamic polarization and electrochemical impedance spectroscopy (EIS) methods. These gum exudates were found to be effective anodic inhibitors for drinking water. The results indicated that the percentage of inhibition increased with the concentration of the gum inhibitor, while the efficacy of inhibition remained unaffected by temperature changes. Chemisorption was believed to be the underlying mechanism [37].

Gum acacia has demonstrated effective corrosion inhibition for mild steel in HCl and H_2SO_4 solutions. The data obtained indicated that increasing the concentration of gum acacia reduced the corrosion rates of mild steel in both HCl and H_2SO_4. Gum acacia has been proposed as a green corrosion inhibitor for mild steel in acidic solutions, as evidenced by weight loss, hydrogen evolution, and polarization methods [38]. The results revealed that the inhibition efficiency increased with higher inhibitor concentrations, and the effect of inhibition was enhanced by halide additives. This mechanism followed the Temkin adsorption isotherm [39].

Through weight loss, gasometric, and thermometric techniques, it was demonstrated that *Khaya ivorensis* gum exudates (KI) serve as an efficient corrosion inhibitor for mild steel in a hydrochloric acid medium. The results showed that the inhibition efficiencies of the gum increased with increasing inhibitor concentrations and decreased with rising temperature. The inhibition mechanism was attributed to physical adsorption, and the Langmuir adsorption isotherm provided the best fit to describe exothermic spontaneous adsorption [39].

14.7 LOCUST BEAN GUM

Locust Bean gum, derived from the seeds of the carob tree, mainly consists of galactomannan-type polysaccharides with a galactose-to-mannose ratio of approximately 1:4. The corrosion suppression ability of Locust Bean Gum was investigated for carbon steel in H_2SO_4 solution. Electrochemical and potentiodynamic polarization methods were employed to test the inhibitory effect of carob gum on carbon steel in an acidic solution with the addition of NaCl [42].

14.8 PROSPECT OF INHIBITION MECHANISM THROUGH MOLECULAR MODELING

Previous reviews have highlighted the application of computer modeling in corrosion inhibitor research, particularly for organic corrosion inhibitors. However, limited research has been conducted on the use of quantum chemical modeling for natural gum components. Molecular modeling was employed to understand the corrosion control mechanism of primary components found in plant extracts. Specifically, the adsorption process of certain gum components derived from *Ferula asafoetida* and *Dorema ammoniacum* on mild steel corrosion in acidic environments was examined using quantum modeling. The calculations revealed important parameters and explained the differences in inhibitory effects between the two oleo gum resins. The active sites pertaining to adsorption of the gum molecules were evaluated using Frontier orbitals. The study demonstrated that the variations in constituent molecules of the two oleo gum resins accounted for the observed differences in their behavior [43].

14.9 CONCLUSION

In conclusion, the utilization of natural gums as corrosion inhibitors provides a promising alternative to traditional synthetic inhibitors. Their affordability, availability, sustainability, lack of hazards, and potential biodegradability make them appealing to various industries. Further research is needed to explore the inhibitory properties of other natural gums, evaluate their performance in different corrosive environments, and optimize their application methods. Embracing natural gums as corrosion inhibitors can contribute to the development of environmentally friendly and economically viable strategies for corrosion protection.

REFERENCES

[1] Devarayan K, Mayakrishnan G, Nagarajan S (2012) Green inhibitors for corrosion of metals: A review. *Chem Sci Rev Lett* 1, 1–8.
[2] Peter A, Obot IB, Sharma SK (2015) Use of natural gums as green corrosion inhibitors: An overview. *Int J Indus Chem* 6, 153–164.
[3] Riggs OL Jr (1973) Theoretical aspects of corrosion inhibitors. In: Nathan CC (Ed.), *NACE Corrosion Inhibition*. NACE, Houstan, pp. 7–27.
[4] Muthukumarasamy K, Pitchai S, Devarayan K, Nallathambi L (2022) *Thunbergia fragrans* extract as green inhibitor for mild steel in acidic medium. *Int J Corros Scale Inhib* 11(2), 686–694.
[5] Sukumaran M, Devarayan K, Marimuthu R (2022) Eco-friendly inhibitor for corrosion of TMT rod in marine environment. *Mater Today: Proc* 58, 898–901.
[6] Muthukumarasamy K, Pitchai S, Devarayan K, Nallathambi L (2020) Adsorption and corrosion inhibition performance of *Tunbergia fragrans* extract on mild steel in acid medium. *Mater Today: Proc* 33, 4054–4058.
[7] Rajeswari V, Devarayan K, Viswanathamurthi P. (2017) Expired pharmaceutical compounds as potential inhibitors for cast iron corrosion in acidic medium. *Res Chem Interm* 43, 3893–3913.

[8] Rose K, Sukumaran M, Devarayan K, Rajagopal K, Arumugam S (2016) Inhibition of steel corrosion in acid medium by ethanolic extract of *Cassia Fistula*, *Chem Sci Rev Lett* 5, 59–66.

[9] Rajeswari V, Kesavan D, Gopiraman M, Viswanathamurthi P (2013) Physicochemical studies of glucose, gellan gum, and hydroxypropyl cellulose-Inhibition of cast iron corrosion. *Carbohydr Polym* 95(1), 288–294.

[10] Kesavan D, Muthu Tamizh M, Sulochana N, Karvembu R (2012) 2-[(E)-{(1 S, 2 R)-1-Hydroxy-1-phenylpropan-2-ylimino} methyl] phenol for inhibition of acid corrosion of mild steel. *J Surf Deterg* 15, 751–756.

[11] Kesavan D, Tamizh MM, Gopiraman M, Sulochana N, Karvembu RX (2012) Physicochemical studies of 4-substituted N-(2-mercaptophenyl)-salicylideneimines: Corrosion inhibition of mild steel in an acid medium. *J Surf Deterg* 15, 567–576.

[12] Bouklah M, Hammouti B, Benhadda T, Benkadour M (2005) Thiophene derivatives as effective inhibitors for the corrosion of steel in 0.5 M H_2SO_4. *J Appl Electrochem* 35(11), 1095–1101.

[13] Fouda AS, Al-Sarawy AA, El-Katori EE (2006) Pyrazolone derivatives as corrosion inhibitors for C-steel HCl solution. *Desalination* 201, 1–13.

[14] Fiala A, Chibani A, Darchen A, Boulkamh DK (2007) Investigations of the inhibition of copper corrosion in nitric acid solutions by ketene dithioacetal derivatives. *Appl Surf Sci* 253(24), 9347–9356.

[15] Evans VR (1976) The corrosion and oxidation of metals (second supplementary volume), Monograph No. 00142444.

[16] Conway BE, Jerkiewicz G (Eds) (1994) In: *Proceedings of the Symposium on Electrochemistry and Materials Science of Cathodic Hydrogen Absorption and Adsorption*. The Electrochemical Society, Pennington.

[17] Jerkiewicz G, Borodzinski JJ, Chrzanowski W, Conway BE (1995) Examination of factors influencing promotion of H absorption into metals by site-blocking elements. *J Electrochem Soc* 142, 3755–3763.

[18] Henriquez-Roman JH, Sancy M, Paez MA, Padilla-Campos L, Zagal JH, Rangel CM, Thompson GE (2005) The influence of aniline and its derivatives on the corrosion behavior of copper in acid solution. *J Solid State Electrochem* 9(7), 504–511.

[19] Eddy NO, Gimba CE, Ameh PO, Ebenso EE (2011) GCMS studies on *Anogessus leocarpus* (Al) gum and their corrosion inhibition potential for mild steel in 0.1 M HCl. *Int J Electrochem Sci* 6, 5815–5829.

[20] Eddy NO, Abechi SE, Ameh PO, Ebenso EE (2013) GCMS, FTIR, SEM, physiochemical and rheological studies on *Albizia zygia* gum. *Walailak J Sci Technol* 10(3), 247–265

[21] Ameh PO (2014) Inhibitory action of albizia zygia gum on mild steel corrosion in acid medium. *Afr J Pure Appl Chem* 8(2), 37–46.

[22] Eddy NO, Ameh PO, Anduang OO (2012) Joint effect of *Anogessius gum* (AL gum) exudate an halide ions on the corrosion of mild steel in 0.1 M HCl. *Port Electrochim Acta*, 30(4), 235–245.

[23] Ameh PO (2014) Physicochemical characterization and inhibitive performance evaluation of commiphora kestingii gum exudate in acidic medium. *Int J Phys Sci* 9(8), 184–198.

[24] Umoren SA, Obot IB, Ebenso EE, Eddy NO (2008) Studies on the inhibitive effect of exudate gum from *Dacroydes edulis* on the acid corrosion of aluminium. *Port Electrochim Acta* 26, 199–209.

[25] Eddy NO, Ameh PO, Anduang OO, Ebenso EE (2012) Corrosion inhibition potential of *Daniella Olieverri* gum exudate for mild steel in acidic medium. *Int J Electrochem Sci* 7, 7425–7439.

[26] Umoren SA, Obot IB, Ebenso EE, Okafor PC (2009) The inhibition of aluminium corrosion in hydrochloric acid solution by exudate gum from *Raphia hookeri*. *Desalination* 247, 561–572.
[27] Umoren SA, Ebenso EE (2009) Studies of the anti-corrosive effect of *Raphia hookeri* exudate gum-halide mixtures for aluminium corrosion in acidic medium. *Pigment Resin Technol* 37(3, 173–182.
[28] Umoren SA (2008) Inhibition of aluminium and mild steel corrosion in acidic medium using gum arabic. *Cellulose* 15, 751–761.
[29] Buchweishaija I, Mhinzi GS (2008) Natural products as a source of environmentally friendly corrosion inhibitors: The case of gum exudates from *Acacia seyal* var. *Seyal. Electrochem Acta* 26(3), 257–265.
[30] Eddy NO, Ameh PO, Anduang OO (2014) Physicochemical characterization and corrosion inhibition potential of *Ficus benjamina* (FB) gum for aluminum in 0.1 M H2SO4. *Port Electrochim Acta* 32(3), 183–197
[31] Eddy NO, Ameh PO (2014) *Commiphora pedunculata* gum as a green inhibitor for the corrosion of aluminium alloy in 0.1 M HCl. *Res Chem Intermed* 40, 2641–2649.
[32] Ameh PO, Magaji L, Salihu T (2012) Corrosion inhibition and adsorption behaviour for mild steel by *Ficus glumosa* gum in H2SO4 solution. *Afr J Pure Appl Chem* 6(7), 100–106.
[33] Eddy NO, Ameh PO, Ebenso EE (2012) Chemical information from GCMS of Ficus *Platphylla gum* and its corrosion inhibition potential for mild steel in 0.1 M HCl. *Int J Electrochem Sci* 7, 5677–5691.
[34] Eddy NO, Ameh PO, Anduang OO, Emaikwu V (2014) Chemical information from GCMS and FTIR studies on *Ficus thonningii* gum and its potential as a corrosion inhibitor for aluminium in acidic medium. *Int J Chem Mater Environ Res* 1(1), 3–15.
[35] Eddy NO, Ameh PO, Gwarzo YM, Okop JI, Dodo NS (2013) Physicochemical study and corrosion inhibition potential of Ficus tricopoda for aluminium in acidic medium. *Port Electrochim Acta* 3(2), 79–93.
[36] Eddy NO, Ibok JU, Alobi ON, Sambo MM (2014) Adsorption and quantum chemical studies on the inhibition of the corrosion of aluminum in HCl by *Gloriosa superba* (GS) gum. *Chem Eng Comm* 201, 1360–1383.
[37] Mital HC, Adotey J (1973) Studies of *Albizia zygia* gum; stabilization of emulsions. *Pharm Acta Helv* 48, 412–419.
[38] Umoren SA (2009) Synergistic influence of gum arabic and iodide ion on the corrosion inhibition of aluminium in alkaline medium. *Port Electrochim Acta* 27(5), 565–577
[39] Ameh PO (2012) Adsorption and inhibitive properties of *Khaya ivorensis* gum for the corrosion of mild steel in HCl. *Int J Mod Chem* 2(1), 28–40.
[40] Umoren SA, Obot IB, Ebenso EE (2008) Corrosion inhibition of aluminium using exudate gum from *Pacchylobus edulis* in the presence of halide ions in HCl. *E J Chem* 5(2), 355–364.
[41] Umoren SA, Ekanem UF (2010) Inhibition of mild steel corrosion in H2SO4 using exudate gum from *Pachylobus edulis* and synergistic potassium halide additives. *Chem Eng Commun* 197, 1339–1356.
[42] Jano A, Lame A, Kokalari E (2012) Use of extracted green inhibitors as a friendly choice in corrosion protection of low alloy carbon steel. *Kem Ind* 61(11–12), 497–503.
[43] Behpour M, Ghoreishi SM, Khayatkashani M, Soltani N (2011) The effect of two oleogum resin exudate from *Ferula assafoetida* and *Dorema ammoniacum* on mild steel corrosion in acidic media. *Corros Sci* 53, 2489–2501.

15 Biopolymer Nanocomposites as Corrosion Inhibitors

Omar Dagdag
Gachon University

Rajesh Haldhar and Seong-Cheol Kim
Yeungnam University

Elyor Berdimurodov
National University of Uzbekistan

Hansang Kim
Gachon University

15.1 INTRODUCTION

Bio-polymeric nanomaterials have emerged as a favorable alternative to conventional materials in recent decades. This is as a result of its advantageous characteristics, which include adjustable surface chemistry, enhanced surface area, pore size, simple regeneration, and superior mechanical rigidity [1]. Polymeric materials are typically categorized into two groups: carbohydrate polymers and functionalized synthetic polymers. Carbohydrate biopolymers, such as chitosan, starch, and cellulose, consist of elongated chains of monosaccharides that are linked together with glycosidic bonds. Because they contain functional groups like amine, hydroxyl, and carboxyl, these types of biopolymers are frequently used to remove dyes and metal ions from solutions [2]. On the other hand, synthetic biopolymer adsorbents with carboxyl and amino groups may have better adsorption abilities. This is a result of the functional groups' interaction with the targeted contaminants and their capacity to bind to the polymeric matrix. Poly(anthranilic acid/4-nitroaniline/formaldehyde), poly(amidoamine-co-acrylic acid), poly(N-vinyl caprolactam-co-maleic acid), poly(styrene-co-maleic acid), poly(styrene-alt-maleic anhydride), along with poly(acrylonitrile-co-styrene) are just a few examples of the synthetic biopolymers that have already been reported [3,4]. According to research, these polymers are efficient in removing metal ions and colors from wastewater. The main cause of this is the chelating element included in the adsorbent structure.

15.2 BIOPOLYMER

Biopolymers are derived from organic matter and are made up of monomeric units that form linear or branched molecular structures. These monomeric units are made up of proteins' amino acids, saccharides, or nucleic acids (nucleotides) [5]. In comparison to conventional polymers, biopolymers are preferred because they are renewable, environmentally friendly, bioavailable, and biodegradable. Due to their higher economic worth and direct sourcing from the environment, they are frequently referred to as natural biodegradable polymers [6]. Biopolymers are used in various industries, including agriculture, medical dental implants, packaging, and catering items. A key factor in the manufacturing of biopolymers is the choice of substrate [7]. Consequently, biopolymers made from the trash and microorganisms have been produced and are viewed as a practical, affordable option [8].

15.2.1 SOURCES OF BIOPOLYMERS

A variety of naturally occurring biological materials, including plants, animals, microorganisms, and agricultural waste, can be used to create biopolymers. Biopolymers may be created chemically from monomeric components like sugars, oils, and amino acids that come from plants, including cotton, tapioca, banana, cassava, yams, sorghum, potatoes, wheat, rice, maize, and barley. You may also employ marine sources, including corals, sponges, fish, lobster, and shrimp, as well as animal sources like cattle. In addition, microbiological sources, including algae, fungus, and yeasts, are appropriate for the creation of biopolymers. Additionally, sources of biomass that are high in carbohydrates are used, including agro leftovers, crops, paper trash, wood waste, and green garbage. Additionally, triglyceride-containing vegetable oils from sources such as jojoba, rapeseed, safflower, castor, sunflower, soybean, and meadow foam oil are also used in the synthesis of biopolymers [9]. In particular, vegetable oils acquired from food producers are great substitutes for the creation of natural polymers [10] (Figure 15.1).

15.2.2 STRUCTURE OF BIOPOLYMERS

Evaluation criteria for biopolymers include their hydrophilicity, degradability, chirality, potential for chemical modification, and capacity to create semi-crystalline fiber morphologies. The creation of films, coatings, along with membranes is aided by research into the chemistry and structure of biopolymers. Glucopyranose rings take on a chair conformation to create cellulose, a natural polymer made of glucose monomers joined by β-1,4 glycosidic connections. Cellulose typically comprises 10,000–15,000 glucopyranose units in its polymer chain [12]. A natural homoglycan polymer made of both amylopectin and amylose units is starch. Amylopectin comprises glucopyranosyl groups linked by α-(1,4) and α-(1,6) glycosidic connections, whereas amylose is a linear molecule made up of glucopyranosyl groups linked by α-(1,4). Typically, starch has a molecular weight between 300 and 400 g/mol [13]. Through partial deacetylation of the latter, chitin, a biodegradable polymer made of D-glucosamine along with N-acetyl-D-glucosamine (D-GlcNAc) oligomers connected by 1,4 β-linkages, may be converted into chitosan. The linear polymeric

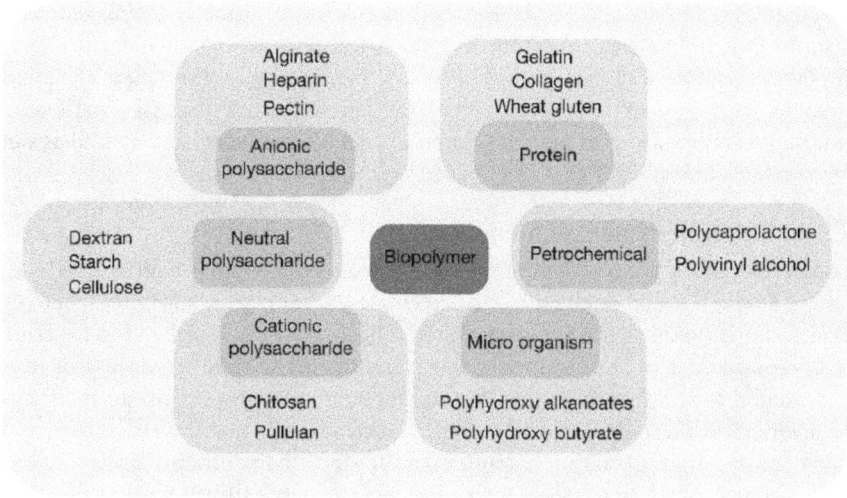

FIGURE 15.1 A visual depiction of various natural, renewable biopolymers classified according to their source type is presented [11].

chains of chitin contain protonated amino groups with a pKa value of 6.2–7.0, which leads to changes in the properties of chitosan, such as solubility, reactivity, adsorptivity, and biodegradability. The shells of insects and crustaceans are a source of chitin [14]. A polymer called chitin is made up of a chain of β-(1,4) GlcNAc units. Chitin exists in three forms, depending on where it comes from three types of chitins: α-, β-, and γ-chitins. The sugar chains in α-chitin are antiparallel and have reducing groups on the opposing sides. On the other hand, in β-chitin, sugar chains are parallel to one another and have reducing groups on the same surfaces. In γ-chitin, two sugar chains are parallel, and one is antiparallel to the other two [15].

A biodegradable polyester known as PCL is created chemically from caprolactone monomers that have ester bonds. The molecular weight of PCL, a semi-crystalline polymer, ranges from 3,000 to 80,000 g/mol [16]. PHAs (polyhydroxy alkanoates) are microorganism-produced polyesters that include tridecyl carbon groups linked to hydroxy fatty acids. This microbial polyester has a molecular weight between 50 and 1,000 kDa [17]. Pullulan is a microbial biopolymer that comprises linear chains of maltotriose units linked by α-(1,6) glycosidic linkages. The maltotriose units, in turn, consist of glucose monomers connected by α-(1,4) glycosidic linkages. Microbial pullulan has a molecular mass that typically ranges from 360 to 480 kDa [18]. Cellulose's inherent polymerization property can be utilized for the production of biopolymers on a large scale due to its non-degradable and renewable nature, making it a promising material for future applications [19].

15.2.3 Methods of Preparation

Biopolymers can be manufactured through either the fermentation of microorganisms using specific sources of carbon, nitrogen, minerals, and salts or through the polymerization of monomers. Biopolymers produced via fermentation are known as microbial biopolymers and are synthesized by microorganisms as a defense mechanism or storage material. Another technique for creating biopolymers is chemical polymerization of monomeric components that may be broken down by natural resources, enzymes, or microorganisms [20]. The following discussion will cover some of the commonly used methods for producing biopolymers from natural as well as synthetic sources.

15.2.3.1   Polymerization of Monomers

In order to create a microstructure, a monomer in a fluid condition must be allowed to move through a tubular channel at a certain flow rate. The light is thrown onto the stream as it is flowing, and the form of the reflected pulses is utilized to ascertain the microstructure of at least one unit in the monomeric fluid. A variety of reactions related to the functionalities of the reactants, including their steric effects, are involved in the polymerization phase. Alkanes only react in the presence of strong acids, while alkene units require easier steps than carbonyl groups, which require complicated processes. One of the most widely utilized natural polymers is polycaprolactone (PCL), which is primarily manufactured for its excellent miscibility with various polymers, disposability, and enhanced characteristics. PCL is made utilizing two processes: ring-opening polymerization of ε-caprolactone and polycondensation of a hydroxycarboxylic acid [21]. Before beginning the procedure, a number of factors were examined, including time, miscibility, percentage, and various catalysts. Comparative analysis revealed that the lactone-based ring-opening polymerization (ROP) route was the most appropriate and generated the required polymer [22].

15.2.3.2   Solvent-Based Extraction

Biopolymers were produced through the thermochemical modification process of biomass known as liquefaction, which involves the use of organic solvents. Various solvent systems are employed to pretreat and extract biopolymers from different types of biomasses [23]. Green solvents, which are considered alternatives, are used for the extraction process. These solvents come in a variety of forms, such as non-halogenated solvents, bio-derived solvents, deep eutectic solvents, ionic liquids, and expedited solvent systems [24]. Because of their special qualities, such as thermal stability, reusability, non-toxicity, low vapor pressure, and ionic liquids, they are used for the extraction of biopolymers. Ionic liquids like N,N-diisopropylethylamine (DIPEAAc and DIPEAP), 1-ethyl-3-methylimidazolium acetate (EMIMAc), 1-butyl-3-methylimidazolium chloride (BMIMCl), and 1-allyl-3-methylimidazolium bromide (AMIMBr) are examples of substances used to extract chitin, keratin, and collagen [23]. For the extraction of biopolymers such as glucan, xylan, lignin, chitin, cellulose, and marine polysaccharides, deep eutectic solvents are a mixture of two immiscible components [25]. Gamma valerolactone (GVL), a solvent generated from biomass, is used in the processing of biomass, including the extraction of carbohydrate polymers [26]. An accelerated solvent combination of ethanol and water is used to extract lignin from nutshells [27]. PHB is extracted using non-halogenated solvents such as ethyl acetate and butyl acetate [28]. In addition to the aforementioned solvents, neoteric solvents are also utilized for biopolymer extraction. These include mixed protic-aprotic ionic liquids, synthetic ionic liquids including BMIMCl, AMIMBr, and EMIMAc, and bio-based ionic liquids such as choline caprolate, choline caprylate, and choline acrylate [23].

15.2.3.3 Fermentation

Traditionally, microbes such as bacteria, fungi, and algae were ulilized to produce biopolymers through fermentation [29]. For example, Xanthan biopolymer was produced by inoculating *Xanthomonas campestris*, and the relative growth was measured using optical density, while the viscosity of the culture indicated the formation of the biopolymer product [30]. Different reactors were connected and carried out sequentially in place of batch plus continuous fermentation reactors. A comparison of the rotary disc reactor (RDR) with static fermentation found that the yield in the RDR was much higher than that in the static fermentation. Additionally, it was discovered that a faster rotor would have a negative effect on the culture's productivity and stability [31]. By transforming harmful pollutants into beneficial biopolymers such as PHAs and bacterial cellulose, industrial wastes may be used as fermentation feedstocks to synthesize biopolymers. The cascade method of recycling plastics has been proven to be beneficial in manufacturing biopolymers from microbial sources, and the waste streams may be treated with the right strains to create both PHAs and cellulose [32]. Similarly, crude glycerol was fermented with *Cupriavidus necator* and sugar cane molasses using *Bacillus megaterium* to create polyhydroxybutyrate [33].

15.2.3.4 Exo-Biopolymer Production

Usually advised are submerged cultures using a liquid medium for the generation of biopolymers from fungus. The growth of exo-biopolymer and mycelia in the bioreactor is optimized by adjusting the nitrogen, carbon, inorganic salt sources, pH, temperature, and agitation conditions. The dry weight of biomass is used to optimize the sources for time-dependent mycelial growth from *Paecilomyces japonica* [34]. Similarly, the production of exo-biopolymer from Paecilomyces tenuipes C240 fungi is achieved after optimization utilizing one factor at a time along with orthogonal matrix techniques. Numerous sugars, including fructose, sucrose, glucose, maltose, lactose, and xylose, are utilized as carbon sources, and yeast extract, tryptone, different kinds of peptone, and ammonium and nitrate salts are used as nitrogen sources. Additionally utilized are mineral supplies such as potassium nitrate, magnesium sulfate, and potassium bivalent salt [35]. When 19 various types of mushrooms are used to produce exo-biopolymers, it is shown that *Ganoderma lucidum* no. 1 with Phellinus linteus KCTC 6190 produces the most along with mycelial growth. Comparing several media, including potato malt peptone (PMP), yeast malt (YM), and mushroom complete medium (MCM), the PMP media had the highest exo-biopolymer production [36]. The fermentation of mycelia as well as exo-biopolymers depends on pH and temperature. Significant mycelial development in the Cordyceps militaris shake flask culture is visible after 7.5 days, and exo-biopolymer synthesis is seen after 9.5 days, as determined by measuring the relative dry weights [36].

15.2.3.5 Endo-Biopolymer Production

Similar to exo-biopolymers, endo-biopolymers (EBP) were made utilizing liquid submerged culture. To create EBP, potato dextrose agar (PDA) was added to Pleurotus eryngii under optimal acidic pH conditions. After 10 days, the mycelial pellet was centrifuged to separate the supernatant containing the EBP, followed by hot water extraction to remove the EBP. After being filtered, lyophilized, and kept for use in anticancer as well as immunomodulatory properties, the EBP was then washed with ethanol [37]. Anaerobic-aerobic fermentation systems were used to manufacture PHAs from organic wastes from palm oil mill effluent (POME) by adding various bacterial cultures. After the volatile fatty acids were removed using acidogenesis and acid polymerization, anaerobic fermentation with various bacteria for PHA synthesis in aerobic digesters was carried out. Similar methods were used by Eubacteria to create polyhydroxy octanoate (PHO), a valuable biopolyester, intracellularly. Due to the low melting temperature of this biopolyester, light composites were created [38]. Eight distinct mushrooms' effective endo-biopolymer synthesis was examined. Temperature and agitation are the ideal conditions needed for the manufacture of endo-biopolymers, with the rotary shaker being kept at 120 rpm and room temperature (25°C) throughout the manufacturing process. Three distinct mushrooms, *Pleurotus eryngii*, *Collybia confluens*, and *Ganoderma applanatum*, were used to make endo-biopolymers using MCM, which contains peptone, yeast extract, $MgSO_4$, KH_2PO_4, K_2HPO_4, and K_2HPO_4 [39].

Exo- and endo-biopolymers were extracted from the *Ganoderma lucidum* culture broth. After growing, the mycelia with supernatant were originally divided. The gathered mycelia were washed for the supernatant, and then the supernatant was precipitated. After dissolving the resulting supernatant, dialysis was used to get the EBP. The supernatant that was initially taken from the broth was precipitated and then washed to create a pure supernatant that produces exo-biopolymer during dialysis [40]. Both endo- and exo-biopolymers, which have strong therapeutic qualities and show potential for further research, are produced by fungi, primarily mushrooms.

15.2.3.6 Bulk Synthesis of Biopolymers

The cost efficiency of producing biopolymers depends on optimizing the feed for the bioreactor. A reduced kinetic model containing a mixture of different cultures was developed. It was discovered that the generation of PHB in a fed-batch reactor using *Ralstonia eutropha* requires a consistent feeding rate of both carbon and nitrogen. However, the use of dynamic feeding techniques during the growth of mixed cultures revealed a reduction in substrate inhibition with a doubling of biopolymer output. This suggests that the employment of various microorganisms and substrates results in maximum biopolymer synthesis. Besides pH, retention period, substrate concentration, and biomass feeding rate, other production-related factors are also taken into account. Additionally, by taking into account nutrient feeding standards along with dilution rates for high PHB production from biomass, a mathematical model relying on the genetic algorithm for batch culture may be effectively expanded to fed-batch cultivation [41].

The microbial synthesis of PHA from natural and mutant strains of *Pseudomonas putida* was studied mathematically in batch and fed-batch reactors. To compare the best-suited model, the process parameters were determined using the differential evolution technique. When compared to the wild strains, the impact of mutant strains was shown to maximize PHA synthesis while consuming less substrate. The carbon supply, dry cell mass, reactant (biomass), nitrogen source, and product (PHA) were the process characteristics taken into account during the building of the mathematical model. Ammonia was used as the rate-limiting substrate in the development of the kinetic equation for active biomass growth [41].

Biopolymers can be synthesized from a variety of sources, such as microbes, plants, and other natural renewable sources. Depending on the removal, separation, and purification necessary for their various uses, the extraction technique from these sources differs. As a result, some of the typical techniques for the industrial, pilot, and laboratory-scale synthesis of biopolymers have been reviewed.

15.3   BIOPOLYMER COMPOSITES

Biopolymer composites are substances comprised of two or more complete layers that create a resin and are strengthened with an appropriate filler, generally a natural fiber. According to their method of synthesis along with intended use, these composites have been categorized as green, hybrid, and textile composites, each having specific benefits and drawbacks. Due to their diverse qualities, such as biodegradability, biocompatibility, and inertness, cellulose fibers, which are utilized to create biopolymer nanocomposites, have a wide variety of uses [42]. Specifically, nanoindentation and thermogravimetric analysis (TGA) were used to evaluate nanocomposites made of chitosan and montmorillonite employing diluted acetic acid as a solvent. The presence of acetic acid deposits in the matrix had a significantly negative impact on the properties of nanocomposites, despite the matrix showing increased thermal stability, toughness, rigidity, and elastic modulus [43]. By standardizing and carrying out graft copolymerization of natural polymers caused by free radicals, green composites were created. Numerous process variables impacting the grafting were standardized throughout the graft-copolymer synthesis. When compared to the source natural polymers, natural composites made from raw polymers were shown to have better tensile along with mechanical characteristics [44]. The categorization of naturally occurring wood composites that can be found as direct sources is shown in Figure 15.2.

15.3.1   Methods of Preparation

Natural fibers and biopolymers have been combined in various ways to create new kinds of biopolymer composites that are reinforced with them. These composites are superior to biopolymer composites with metal-based fillers or nanofillers in terms of benefits. However, a detrimental impact on both mechanical and physical properties might be seen because of the low interfacial adhesion of the fiber with polymer matrix. By adding plasticizers or coupling agents that enhance the composites' adhesion and other qualities, these obstacles can be removed. The overall performance of fiber-polymer composites is improved by properly treating coupling agents and fiber to boost the adhesion to the polymer [45]. Based on the method of manufacture, biopolymer composites are categorized in Figure 15.3.

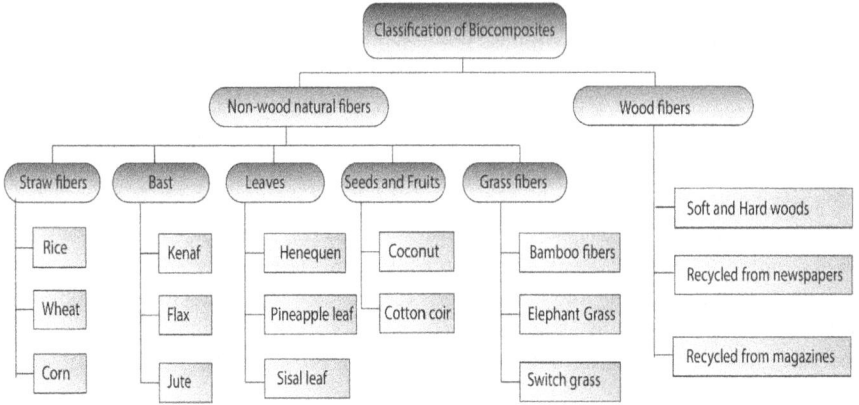

FIGURE 15.2 Natural biopolymer composites obtained from plants, including both wood and non-wood fibers, can be broadly classified [11].

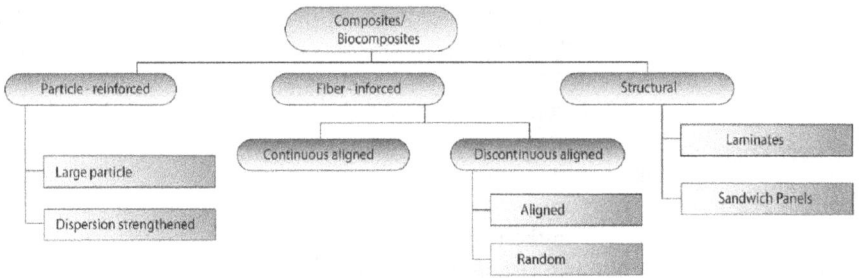

FIGURE 15.3 Biopolymer composites can be categorized according to the method of production [11].

15.3.1.1 Grafting

Due to the availability of low-cost, biodegradable filler materials, the synthesis of biopolymer composites has sped up. Lignocellulose, bacterial cellulose, nanocellulose, and, which are predominantly components derived from plants and petroleum, make up the majority of these reinforcing biofibers (Figure 15.3). However, the antagonistic nature of the hydrophobic polymeric matrix with the hydrophilic biofibers resulted in subpar performance. Following several attempts, it was discovered that the graft polymerization process improved the functionality of the grafted green composites. To enhance the thermal, physical, and chemical characteristics of the created composites, several graft polymerization techniques, such as ring-opening polymerization, grafting through coupling agent, and radical-induced grafting, were improved [46]. By means of polymerization, cellulose was added as reinforcement to green composites made of poly(3-hydroxybutyrate-co-3-hydroxyvalerate; PHB) and the copolymer. The biopolymer underwent grafting, and its copolymer, PHBV, which served as an interfacial coupling agent. The stress transfer among the interphases during the graft copolymerization process improved the mechanical characteristics, while the addition of cellulose made the material more flexible. This highlights the need for graft polymerization-based sustainable supplies of biopolymers for a variety of applications [47].

15.3.1.2 Molding

Biopolymer films were employed to create biopolymer composites utilizing a variety of molding methods, including compression molding, injection molding, sheet molding, and resin transfer molding. Due to the interaction stresses on the molded product formed from plant microfiber bundles, adding starch enhanced the bending strength. The stress-strain curve associated with Young's modulus, which was equivalent to that of metallic alloys, was utilized to examine the increase in bending strength brought about in the web-like network [48]. In the molding process, crucial factors including the impact of temperature, pressure, the number of plies, and tensile strength determine how strong the biopolymers will be. It was discovered that the tensile strength of poly lactic acid (PLA) biopolymer matrices reinforced with other fibers, including jute, flax, and cotton, had increased [49].

15.3.1.3 Extrusion and Injection

Utilizing cellulose fibers, a thermoplastic starch green composite was created utilizing the twin-screw injection technique and extrusion. Utilizing fiber nanofibrillation as well as distribution investigations, the dispersion of bleached wood with TEMPO-oxidized cellulose fibers across the starch biopolymer was investigated. The composites displayed enhanced composite characteristics with more moisture absorption and better fiber dispersion [50]. Employing the twin-screw injection technique, PLA was injected into two varieties of bamboo fibers, with the feeding rate serving as the primary conditional parameter that dictated the fiber bundle length. The mechanical properties of the biopolymer composite were affected by the screw speed, which changed the shear rate. The additional fibers improved PLA's heat stability and crystallinity as well [51]. For the creation of PLA biopolymer composites reinforced with sisal fibers, direct-injection molding and extrusion-injection molding techniques were investigated. In comparison to extrusion-injection molding, direct-injection molding had the greatest differences in fiber attrition and length. Direct-injection molding significantly enhanced the tensile, mechanical, and flexural characteristics with equivalent sisal fiber dispersion in both processes. Paulownia wood flour (PWF) was utilized similarly as reinforcement in PLA using extrusion techniques to maximize correlation on both biological and mechanical characteristics [52].

15.3.1.4 Pultrusion

In contrast to extrusion, which pushes the biopolymer material through the die or mold, pultrusion pulls the material through the mold. The generated biopolymer composite, which was created via the pultrusion process, has unidirectional fibers throughout and may be weaker along the cross-sectional direction. A number of plant-based biochar varieties with high specific surface area (SSA) were investigated in order to improve the flexural characteristics of the biochar composite. The biochar's enhanced porosity made it possible for the fiber and polymer to bond [53].

Pultrusion was used to create composite panels made of PE and Soy-PE resin, and it was discovered that these panels had higher tensile and flexural capabilities. The soy-based panels provided fresh opportunities for use in treatment, transportation, and other areas [54].

15.3.1.5 Compression Molding

Compression molding, also known as press molding, is a commonly used type of molding. To create a matrix with a homogeneous thickness, the biopolymer is heated to a high temperature between two plates and then a set force is applied to the plates' surfaces. The mat can also be piled with the woven plies as well as positioned once again between the plates. At the optimal temperature of 185°C, the tensile strength of PLA biopolymer composite was investigated using various test setups and found to be proportional to fiber strength. Similar to this, it was discovered that resorcinol-formaldehyde composites reinforced using fibers via *Saccaharum cilliare* had enhanced mechanical qualities with optimal cost effectiveness [49]. Red wine grape pomace (WGP) and white WGP biopolymer composite boards were found to have high break strength in the red WGP boards and maximum biodegradability in the white WGP boards. White WGP-based boards were totally destroyed in around 80% of cases, demonstrating their environmental inertness [55]. Extrusion and compression molding were used to create translucent films made of agar as well as soy-based protein biopolymers. By altering conformation, the inclusion of agar considerably reduced the solubility, but the compression molding procedure was able to increase the biocompatibility of the agar, filler, and soy protein biopolymer [56].

15.3.1.6 Infusion Molding

Vacuum is utilized during the infusion molding process to get the resin into the die or mold. Before the resin is applied to the mold, a vacuum is produced there, giving this procedure the titles "vacuum infusion" or "resin infusion process" [57]. Hemp fiber was added to a bio-resin made from epoxidized hemp oil using the infusion molding method. It was discovered that the mechanical and moisture-absorbing capabilities were equivalent to those of synthetic resin. Before posing, a full vacuum was kept at room temperature for around 24 hours to ensure a consistent thickness throughout the composite [58]. Even without any surface treatments, epoxy resin packed with short jute fibers demonstrated improved tensile strength, interfacial interaction, and strain at break [59].

15.3.1.7 Intercalation

The process of intercalation involves introducing a molecule or ion into host materials that are layered. The metal intercalation process is one form of intercalation, involving the mixing of metal, polymer, and fillers as well as the application of shear above the polymer's melting temperature. However, this method is only compatible with some biopolymers, such as PLA, and is incompatible with others, such as chitosan and PVA. Using shear mixing as well as magnetic stirring, nanofillers along with polymers are transformed into a colloidal solution in a different sort of inter-calation called solution intercalation. The intercalated network produced inside the polymeric chains is then visible after common solvents, including distilled water, ethanol, acetone, and dimethyl formamide, are evaporated. Utilizing the created polymeric chains, biopolymers like PLA alongside PCL may be intercalated to create nanocomposites [60].

15.3.1.8   Phase Separation

A solution-based method called solid-liquid phase separation involves crystallizing the solvent, which may then be purged using a vacuum. Phase separation techniques come in many different forms, including freezing, solvent-based extraction, dissolu-tion, gelation, and drying [45]. The thermally induced phase separation process or freeze-drying was used to create biodegradable porous foam bio-nanocomposites utilizing cellulose nanocrystals and PVA. This was accomplished by combining cel-lulose nanocrystals with PVA aqueous solutions in water at a 20:80 volume ratio. Using ImageJ and FTIR, the sample's functional groups and pore size may be deter-mined. The 3D PVA foam and 3D porous nanocomposite exhibited higher porosity, at 90% and 89%, respectively [61].

15.3.1.9   Filament Winding

An old and unusual procedure called filament winding involves wrapping a fila-mentous fiber across a revolving mandrel or shaft. To pull the fiber and create a composite, the mandrel's speed and orientation may be adjusted. This approach involved creating kenaf green composites, which showed good tensile qualities in torsion and split-disk tests [62]. This method's suitability for creating biopoly-mer composites depends on the fiber's tensile strength. Based on the position of the winding, the circumference of the mandrel, and the spaces between them, new techniques such as perimeter winding and cross winding have been devised. Recent advancements in robot techniques have made this cost-effective for input energy [63].

15.3.1.10   Melt Blending

This method entails melting any biopolymer along with mixing it with the necessary quantity of biopolymer composite to make composites. Vegetable oils were used to plasticize PLA, which improved the material's morphology, deformation, flexibility, tensile strength, and adherence to the plasticizer, fiber, and polymer [64]. At severe temperatures of 180°C and 200°C under ideal pressure and residual time conditions, two biopolymer composite systems based on PLLA with lignin as filler followed by poly ethylene glycol (PEG) as plasticizer were mixed and sandwiched. They were then vacuum-dried and chilled after molding, which revealed that the PEG addi-tion improved the flexibility and stiffness of the PLLA biopolymer composite [65]. Fumed silica with three distinct surface areas was used to make nanocomposites of PLA alongside PCL, which were then examined using thermal and microscopic methods. We carefully monitored the breakdown rate of these nanocomposites in compost at various temperatures. According to their rate of degradation, PCL composites decayed in the same compost at 40°C, whereas PLA nanocomposites did so at 58°C [66].

15.3.1.11   Solvent Casting

Any polymer that has been combined with a solvent can be molded using the solvent casting technique, and the biopolymer sticks to the mold once the solvent has been removed. Although mostly utilized for polymer composites, cellulose and fillers like lignin, including their nanofibrils, can also be used to create certain biopolymer composites, such as PHB. It was shown that the presence of lignin and cellulose nanofibrils reduced the size of PHA's pores. Similar to the previous example, the biopolymer composite of PCL made with coir fiber showed poor coir dispersion but increased tensile strength and decreased elongation at break [67]. The structure is removed from the mold that has solvent in it in this procedure, which is also known as salt leaching. This technique is well recognized for creating controlled porous composites, although being constrained by its mechanical characteristics [68].

15.3.1.12   Electrospinning

The Taylor cone, a key element for creating electrospun goods, is formed by the electrospinning procedure and depends on the voltage differential and biopolymer flow velocity. PVA and glucose oxidase were electrospun together to generate shuttle-shaped beads on the coated polymer, which allowed the encapsulation of enzymes with the highest degree of fiber consistency and repeatability [69]. Along with electrospinning, another approach called electrospraying was utilized to pull biopolymers such as PLLA-co-PCL, gelatin, and hydroxyapatite (HA). Biopolymer composites with the appropriate geometrical morphologies, such as fiber alignment, fiber density, pore size, and a large surface area, were made possible by electrospinning technologies [70]. When compared to the solvent casting approach, this method provided good zein particle dispersion in the gelation biopolymer, further demonstrating its superiority [71].

15.3.1.13   Film Stacking

As in injection molding, film stacking involves heating and compressing alternating layers of biopolymers to create biopolymer composite structures. Unreinforced PLLA fiber mats were made similarly to how paper is made by dispersing the fibers in water solutions [72]. The biodegradable PLLA mats have tensile characteristics comparable to those of glass fiber composites. The mechanical characteristics of flax and PHB biopolymer composites were improved when made utilizing film stacking as opposed to injection molding. The smaller areas lowered compression stress, which decreased permeability of the biopolymer composites and allowed them to support hydro-mechanical stresses. Overall, it was discovered that the film stacking procedure was less harsh than the standard injection molding method, maintaining the cohesiveness, structure, and composition of the applied biopolymers [73]. These are some of the principal techniques for creating biopolymer composites that are filled with metallic, organic, and natural materials. A schematic representation of the entire biopolymer composite production process is shown in Figure 15.4.

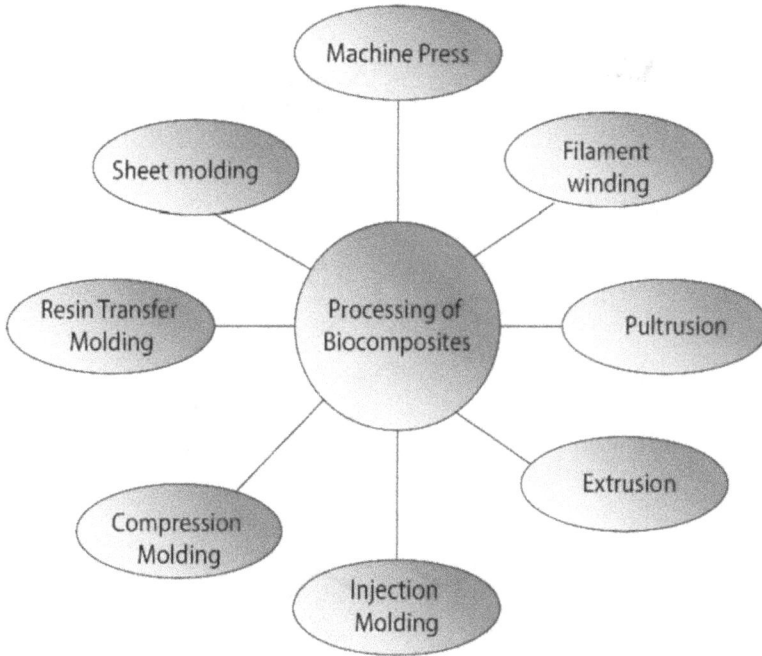

FIGURE 15.4   Various methods are employed in the production of biopolymer composites [11].

15.4   BIOPOLYMERS NANOCOMPOSITES AS CORROSION INHIBITORS

Due to their environmental friendliness and availability of numerous adsorption sites, biopolymers have attracted interest from all over the world as corrosion inhibitors for industrial metal substrates. However, it has been shown that most biopolymers act solely as weak inhibitors of corrosion. Numerous approaches have been investigated to get around this restriction, including copolymerization, the inclusion of chemicals that work in concert, and the insertion of inorganic nanoparticles (NPs) into the biopolymer matrix to improve the biopolymers' ability to suppress corrosion. The use of biopolymer composites along with nanocomposites as anticorrosion compounds can create a metal chelate that can shield metal surfaces from corrosive media's hostile ions.

Corrosion resistance may be achieved by using stable, ecologically benign, biodegradable, affordable, and renewable materials, notably carbohydrate polymers [74].

Chitosan, a naturally occurring carbohydrate polymer generated from waste-producing shrimp and crab shells, was mixed with green AgNPs to create a chitosan-Ag NPs nanocomposite (see Figure 15.5). Chitosan was used in the preparation of the Ag NPs as a capping and reduction agent, with their sizes varying from 3 to 6 nm. The resultant nanocomposite demonstrated a high degree of mixed-type inhibitor activity against mild steel corrosion in industrial chilling water, having an efficiency of around 97%–98%. Using WLM and electrochemical techniques, the nanocomposite's

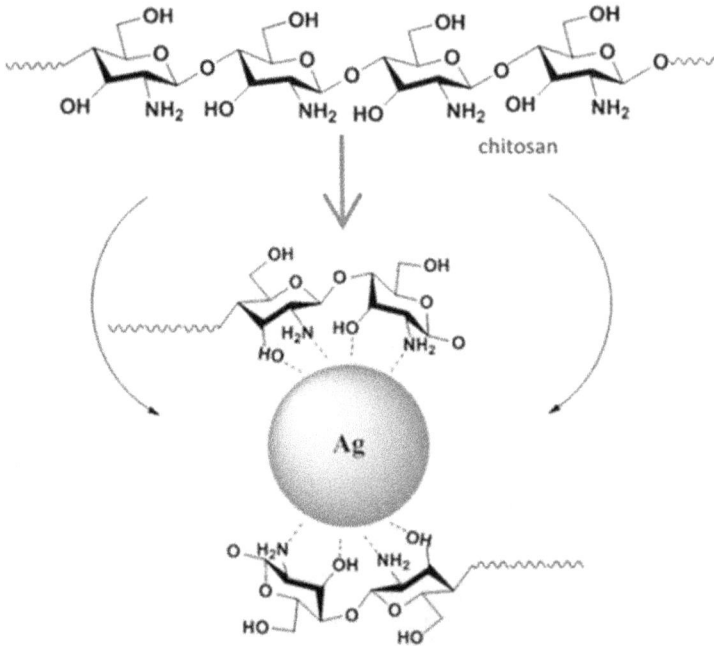

FIGURE 15.5   Schematic synthesis route for SNPs-CT NC samples [75].

ability to suppress corrosion when introduced to the corrosive medium was assessed. Measurements of adsorption showed that the nanocomposite's anticorrosion activity was caused by spontaneous physisorption. According to biological research, the nanocomposite had antimicrobial properties [75].

Chitosan (CHT) and green copper oxide (CuO), which was made using olive oil, were combined to form a nanocomposite (CHT-CuO NC) in Umoren et al. [76], which served as a corrosion inhibitor for X60 carbon steel. Using gravimetric along with electrochemical techniques, the nanocomposite's inhibitory effectiveness was found to be 90.35% when introduced to an HCl medium in a 5 wt.% concentration. An active site-blocking mechanism caused the inhibition, which prevented corrosion by creating a coating that was adsorbed on the carbon steel.

In a different study, carboxymethyl cellulose (CMC) and green AgNPs, which were made using honey as a reducing agent naturally, were combined to form a nanocomposite. In an acidic solution (5% H_2SO_4), this nanocomposite was applied to St37 steel as a corrosion inhibitor. With a maximum inhibitory efficiency of 96.37% when added to the corrosive medium, the CMC/AgNPs nanocomposite outperformed CMC alone, according to gravimetric, electrochemical, and surface assessment studies. The CMC/AgNPs nanocomposite displayed mixed-type inhibitor behavior, delaying both cathodic and anodic processes [77].

Primary aminated-modified cellulose (PAC), a natural carbohydrate polymer, was created by modifying cellulose to improve inhibition. Then, a variety of metal oxide NPs were combined with PAC to create nanocomposites, such as PAC/Fe_3O_4

NPs, PAC/CuO NPs, and PAC/NiO NPs. All three nanocomposites showed anticorrosion activity when tested on C-steel through the addition of the inhibitor to 1 M of HCl using mixed inhibitor features. The corrosion inhibition for PAC/CuO NP, PAC/Fe_3O_4 NPs, and PAC/NiO NPs at 250 ppm was 88.1%, 93.2%, and 96.1%, along with 98.6%, respectively [78].

In order to create a green nanocomposite of silver as well as gum Arabic, another natural polymer, honey was utilized as a reducing agent. Steel was employed with this nanocomposite as a corrosion inhibitor in acidic environments, such as 15% HCl and 15% H_2SO_4. When the nanocomposite was applied to the two mediums, gravity measurements and electrochemical techniques revealed that it operated via separate processes. When the nanocomposite was introduced to 15% H_2SO_4, it was shown to be a mixed-type inhibitor, whereas it was discovered to be an anodic inhibitor in the context of 15% HCl. The observed inhibition was brought on by the nanocomposite's ionic alongside neutral forms being adsorbed in its presence. AgNPs were discovered on the surface of the steel as Ag, Ag_2O, and AgO [79].

Utilizing cellulose and niacin as biopolymers was still another use. When the inhibitor was applied to the corrosive saline solutions (3.5% NaCl) in the corrosion medium, the nanocomposite demonstrated effectiveness in inhibiting copper. The effectiveness of the inhibition was examined using potential polarization, EIS methods, SEM, and EDS. The ethyl cellulose-niacin (NEC) campsite was the most effective of the composite's several formulations. NEC has a 94.6% effectiveness as a combined corrosion inhibitor [80].

15.5 CONCLUSION

Because of their biodegradability, recyclability, eco-friendliness, and other advantages, biopolymers are becoming more and more well-liked as an alternative to synthetic, petroleum-based polymers. However, compared to synthetic polymers, biopolymer materials' mechanical qualities are less strong, which restricts their application in a number of different industries. Different methods have been developed to overcome this issue, such as adding plasticizers, nanofilters, and coupling agents to biopolymer blends to enhance the structural and functional qualities of biopolymers. Currently, governments and businesses are spending more money on research to improve the effectiveness of biopolymers and biocomposites. The study of bio-based technology is a potential area for sustainable development, and it is expanding. The literature emphasizes the utilization of diverse environmentally friendly resources to make metal, including metal oxide NPs, as this chapter discusses. These green NPs can also be coupled with biopolymers to enhance their useful qualities for corrosion prevention applications.

REFERENCES

[1] D. Kai, M. J. Tan, P. L. Chee, Y. K. Chua, Y. L. Yap, and X. J. Loh, "Towards lignin-based functional materials in a sustainable world," *Green Chemistry*, vol. 18, pp. 1175–1200, 2016.

[2] D. S. Chaudhary, "Understanding amylose crystallinity in starch-clay nanocomposites," *Journal of Polymer Science Part B: Polymer Physics*, vol. 46, pp. 979–987, 2008.

[3] M. F. Moradali and B. H. Rehm, "Bacterial biopolymers: From pathogenesis to advanced materials," *Nature Reviews Microbiology*, vol. 18, pp. 195–210, 2020.

[4] S. H. Ko, D. Chandra, W. Ouyang, T. Kwon, P. Karande, and J. Han, "Nanofluidic device for continuous multiparameter quality assurance of biologics," *Nature Nanotechnology*, vol. 12, pp. 804–812, 2017.

[5] A. George, M. Sanjay, R. Srisuk, J. Parameswaranpillai, and S. Siengchin, "A comprehensive review on chemical properties and applications of biopolymers and their composites," *International Journal of Biological Macromolecules*, vol. 154, pp. 329–338, 2020.

[6] R. Rendón-Villalobos, A. Ortíz-Sánchez, E. Tovar-Sánchez, and E. Flores-Huicochea, "The role of biopolymers in obtaining environmentally friendly materials," In: P. Matheus (ed.), *Composites from Renewable and Sustainable Materials*, vol. 151, IntechOpen, Rijeka, 2016.

[7] A. M. Díez-Pascual, *Synthesis and Applications of Biopolymer Composites*. Vol. 20, MDPI, 2019, p. 2321.

[8] R. Francis, S. Sasikumar, and G. P. Gopalan, "Synthesis, structure, and properties of biopolymers (natural and synthetic)," *Polymer Composites*, pp. 11–107, 2013.

[9] V. Flaris and G. Singh, "Recent developments in biopolymers," *Journal of Vinyl and Additive Technology*, vol. 15, pp. 1–11, 2009.

[10] V. Sharma and P. P. Kundu, "Addition polymers from natural oils-A review," *Progress in Polymer Science*, vol. 31, pp. 983–1008, 2006.

[11] G. P. Udayakumar, S. Muthusamy, B. Selvaganesh, N. Sivarajasekar, K. Rambabu, S. Sivamani, et al., "Ecofriendly biopolymers and composites: Preparation and their applications in water-treatment," *Biotechnology Advances*, vol. 52, p. 107815, 2021.

[12] A. C. O'sullivan, "Cellulose: The structure slowly unravels," *Cellulose*, vol. 4, pp. 173–207, 1997.

[13] A. Buleon, P. Colonna, V. Planchot, and S. Ball, "Starch granules: Structure and biosynthesis," *International Journal of Biological Macromolecules*, vol. 23, pp. 85–112, 1998.

[14] G. Lodhi, Y.-S. Kim, J.-W. Hwang, S.-K. Kim, Y.-J. Jeon, J.-Y. Je, *et al.*, "Chitooligosaccharide and its derivatives: Preparation and biological applications," *BioMed Research International*, vol. 2014, 2014.

[15] E. Cabib, "Chitin: Structure, metabolism, and regulation of biosynthesis," In: W. Tanner, and F. A. Loewus (Eds.), *Plant Carbohydrates II: Extracellular Carbohydrates*, pp. 395–415, 1981.

[16] B. Azimi, P. Nourpanah, M. Rabiee, and S. Arbab, "Poly (ϵ-caprolactone) fiber: An overview," *Journal of Engineered Fibers and Fabrics*, vol. 9, 2014. doi:10.1177/155892501400900309.

[17] C. Reddy, R. Ghai, and V. C. Kalia, "Polyhydroxyalkanoates: An overview," *Bioresource Technology*, vol. 87, pp. 137–146, 2003.

[18] K. Sugumaran and V. Ponnusami, "Review on production, downstream processing and characterization of microbial pullulan," *Carbohydrate Polymers*, vol. 173, pp. 573–591, 2017.

[19] A. Kumar, G. Sharma, M. Naushad, H. Ala'a, A. García-Peñas, G. T. Mola, et al., "Bio-inspired and biomaterials-based hybrid photocatalysts for environmental detoxification: A review," *Chemical Engineering Journal*, vol. 382, p. 122937, 2020.

[20] S. Mohan, O. S. Oluwafemi, N. Kalarikkal, S. Thomas, and S. P. Songca, "Biopolymers-application in nanoscience and nanotechnology," *Recent Advances in Biopolymers*, vol. 1, pp. 47–66, 2016.

[21] G. P. Udayakumar, G. Kirthikaa, S. Muthusamy, B. Ramakrishnan, and N. Sivarajasekar, "Comparison and evaluation of electrospun nanofiber membrane for the clarification of grape juice," In: *Sustainable Development in Energy and Environment: Select Proceedings of ICSDEE 2019*, pp. 77–92, 2020. Singapore: Springer.

[22] M. Labet and W. Thielemans, "Synthesis of polycaprolactone: A review," *Chemical Society Reviews*, vol. 38, pp. 3484–3504, 2009.

[23] H. Mahmood and M. Moniruzzaman, "Recent advances of using ionic liquids for bio-polymer extraction and processing," *Biotechnology Journal*, vol. 14, p. 1900072, 2019.

[24] Y. Gu and F. Jérôme, "Bio-based solvents: An emerging generation of fluids for the design of eco-efficient processes in catalysis and organic chemistry," *Chemical Society Reviews*, vol. 42, pp. 9550–9570, 2013.

[25] Y. T. Tan, A. S. M. Chua, and G. C. Ngoh, "Deep eutectic solvent for lignocellulosic biomass fractionation and the subsequent conversion to bio-based products-A review," *Bioresource Technology*, vol. 297, p. 122522, 2020.

[26] K. Prasad and M. Sharma, "Green solvents for the dissolution and processing of biopoly-mers," *Current Opinion in Green and Sustainable Chemistry*, vol. 18, pp. 72–78, 2019.

[27] A. P. Klein, E. S. Beach, J. W. Emerson, and J. B. Zimmerman, "Accelerated solvent extraction of lignin from *Aleurites moluccana* (Candlenut) nutshells," *Journal of Agricultural and Food Chemistry*, vol. 58, pp. 10045–10048, 2010.

[28] A. Aramvash, N. Gholami-Banadkuki, F. Moazzeni-Zavareh, and S. Hajizadeh-Turchi, "An environmentally friendly and efficient method for extraction of PHB biopolymer with non-halogenated solvents," *Journal of Microbiology and Biotechnology*, vol. 25, pp. 1936–1943, 2015.

[29] I. Chang, M. Jeon, and G.-C. Cho, "Application of microbial biopolymers as an alternative construction binder for earth buildings in underdeveloped countries," *International Journal of Polymer Science*, vol. 2015, pp. 1–9, 2015.

[30] R. Silman and P. Rogovin, "Continuous fermentation to produce xanthan biopolymer," *Biotechnology and Bioengineering*, vol. 12, pp. 75–84, 1970.

[31] N. Pa'e, K. A. Zahan, and I. I. Muhamad, "Production of biopolymer from Acetobacter xylinum using different fermentation methods," *International Journal of Engineering and Technology*, vol. 11, pp. 90–98, 2011.

[32] H. Al Battashi, S. Al-Kindi, V. K. Gupta, and N. Sivakumar, "Polyhydroxyalkanoate (PHA) production using volatile fatty acids derived from the anaerobic digestion of waste paper," *Journal of Polymers and the Environment*, vol. 29, pp. 250–259, 2021.

[33] H. Al-Battashi, N. Annamalai, S. Al-Kindi, A. S. Nair, S. Al-Bahry, J. P. Verma, *et al.*, "Production of bioplastic (poly-3-hydroxybutyrate) using waste paper as a feedstock: Optimization of enzymatic hydrolysis and fermentation employing Burkholderia sac-chari," *Journal of Cleaner Production*, vol. 214, pp. 236–247, 2019.

[34] J.-T. Bae, J. Sinha, J.-P. Park, C.-H. Song, and J.-W. Yun, "Optimization of submerged culture conditions for exo-biopolymer production by *Paecilomyces japonica*," *Journal of Microbiolog and Biotechnology*, vol. 10, pp. 482–487, 2000.

[35] C.-P. Xu, S.-W. Kim, H.-J. Hwang, J.-W. Choi, and J.-W. Yun, "Optimization of sub-merged culture conditions for mycelial growth and exo-biopolymer production by Paecilomyces tenuipes C240," *Process Biochemistry*, vol. 38, pp. 1025–1030, 2003.

[36] S. Kim, H. Hwang, J. Park, Y. Cho, C. Song, and J. Yun, "Mycelial growth and exo-bio-polymer production by submerged culture of various edible mushrooms under different media," *Letters in Applied Microbiology*, vol. 34, pp. 56–61, 2002.

[37] Y.-T. Jeong, S.-C. Jeong, Y.-A. Gu, R. Islam, and C.-H. Song, "Antitumor and immu-nomodulating activities of endo-biopolymers obtained from a submerged culture of Pleurotus eryngii," *Food Science and Biotechnology*, vol. 19, pp. 399–404, 2010.

[38] Salmiati, Z. Ujang, M. Salim, M. Md Din, and M. Ahmad, "Intracellular biopolymer productions using mixed microbial cultures from fermented POME," *Water Science and Technology*, vol. 56, pp. 179–185, 2007.

[39] Y.-T. Jeong, B.-K. Yang, C.-R. Li, and C.-H. Song, "Anti-tumor effects of exo-and endo-biopolymers produced from submerged cultures of three different mushrooms," *Mycobiology*, vol. 36, pp. 106–109, 2008.

[40] B.-K. Yang, Y.-A. Gu, Y.-T. Jeong, and C.-H. Song, "Anti-complementary activities of exo-and endo-biopolymer produced by submerged mycelial culture of eight different mushrooms," *Mycobiology*, vol. 35, pp. 145–149, 2007.

[41] S. Y. Lai, P. C. Kuo, W. Wu, M. F. Jang, and Y. S. Chou, "Biopolymer production in a fed-batch reactor using optimal feeding strategies," *Journal of Chemical Technology & Biotechnology*, vol. 88, pp. 2054–2061, 2013.

[42] M. J. John and S. Thomas, "Biofibres and biocomposites," *Carbohydrate Polymers*, vol. 71, pp. 343–364, 2008.

[43] S. Wang, L. Shen, Y. Tong, L. Chen, I. Phang, P. Lim, *et al.*, "Biopolymer chitosan/montmorillonite nanocomposites: Preparation and characterization," *Polymer Degradation and Stability*, vol. 90, pp. 123–131, 2005.

[44] V. K. Thakur, M. K. Thakur, and R. K. Gupta, "Graft copolymers of natural fibers for green composites," *Carbohydrate Polymers*, vol. 104, pp. 87–93, 2014.

[45] L. Cardon, K. Ragaert, R. De Santis, and A. Gloria, "Design and fabrication methods for biocomposites," In: *Biomedical Composites*. Elsevier, pp. 17–36, 2017.

[46] L. Wei and A. G. McDonald, "A review on grafting of biofibers for biocomposites," *Materials*, vol. 9, p. 303, 2016.

[47] L. Wei, N. M. Stark, and A. G. McDonald, "Interfacial improvements in biocomposites based on poly (3-hydroxybutyrate) and poly (3-hydroxybutyrate-co-3-hydroxyvalerate) bioplastics reinforced and grafted with α-cellulose fibers," *Green Chemistry*, vol. 17, pp. 4800–4814, 2015.

[48] H. Yano and S. Nakahara, "Bio-composites produced from plant microfiber bundles with a nanometer unit web-like network," *Journal of Materials Science*, vol. 39, pp. 1635–1638, 2004.

[49] A. Rubio-López, A. Olmedo, A. Díaz-Álvarez, and C. Santiuste, "Manufacture of compression moulded PLA based biocomposites: A parametric study," *Composite Structures*, vol. 131, pp. 995–1000, 2015.

[50] M. Hietala, P. Rollo, K. Kekäläinen, and K. Oksman, "Extrusion processing of green biocomposites: Compounding, fibrillation efficiency, and fiber dispersion," *Journal of Applied Polymer Science*, vol. 131, 2014.

[51] G. Gamon, P. Evon, and L. Rigal, "Twin-screw extrusion impact on natural fibre morphology and material properties in poly (lactic acid) based biocomposites," *Industrial Crops and Products*, vol. 46, pp. 173–185, 2013.

[52] S. Chaitanya and I. Singh, "Processing of PLA/sisal fiber biocomposites using direct- and extrusion-injection molding," *Materials and Manufacturing Processes*, vol. 32, pp. 468–474, 2017.

[53] L. K. Bowlby, G. C. Saha, and M. T. Afzal, "Flexural strength behavior in pultruded GFRP composites reinforced with high specific-surface-area biochar particles synthesized via microwave pyrolysis," *Composites Part A: Applied Science and Manufacturing*, vol. 110, pp. 190–196, 2018.

[54] R. R. Vuppalapati, K. Chandrashekhara, and T. P. Schuman, "Manufacturing and performance evaluation of core-filled, pultruded bio-composite panels," In: *Proceedings of the 3rd Annual ISC Research Symposium, ISCRS 2009*, 2009.

[55] Y. Jiang, J. Simonsen, and Y. Zhao, "Compression-molded biocomposite boards from red and white wine grape pomaces," *Journal of Applied Polymer Science*, vol. 119, pp. 2834–2846, 2011.

[56] T. Garrido, A. Etxabide, P. Guerrero, and K. De la Caba, "Characterization of agar/soy protein biocomposite films: Effect of agar on the extruded pellets and compression moulded films," *Carbohydrate Polymers*, vol. 151, pp. 408–416, 2016.

[57] S. Phillips, P. Kuo, C. Demaria, L. Lessarda, N. Yan, P. Hubert, *et al.*, "Development of multi-scale biocomposites from flax, nanocellulose and epoxy by resin infusion," In: *NIPMMP Conference*. Montreal, Quebec, 2013.

[58] G. Francucci, N. W. Manthey, F. Cardona, and T. Aravinthan, "Processing and characterization of 100% hemp-based biocomposites obtained by vacuum infusion," *Journal of Composite Materials*, vol. 48, pp. 1323–1335, 2014.

[59] P. Valášek and M. Müller, "Tensile characteristics of epoxy/jute biocomposites prepared by vacuum infusion," In: J. Machado, F. Soares, and G. Veiga (Eds.), *Innovation, Engineering and Entrepreneurship*, HELIX 2018. Lecture Notes in Electrical Engineering, vol. 505, pp. 574–580, 2019. Cham: Springer. https://doi.org/10.1007/978-3-319-91334-6_78

[60] B. Sharma, P. Malik, and P. Jain, "Biopolymer reinforced nanocomposites: A comprehensive review," *Materials Today Communications*, vol. 16, pp. 353–363, 2018.

[61] A. Kumar, Y. S Negi, N. K Bhardwaj, and V. Choudhary, "Synthesis and characterization of cellulose nanocrystals/PVA based bionanocomposite," *Advanced Materials Letters*, vol. 4, pp. 626–631, 2013.

[62] S. Misri, M. Ishak, S. Sapuan, and Z. Leman, "Filament winding process for Kenaf fibre reinforced polymer composites," In: M. Salit, M. Jawaid, N. Yusoff, and M. Hoque (Eds.), *Manufacturing of Natural Fibre Reinforced Polymer Composites*, pp. 369–383, 2015. Cham: Springer. https://doi.org/10.1007/978-3-319-07944-8_18

[63] N. Minsch, F. Herrmann, T. Gereke, A. Nocke, and C. Cherif, "Analysis of filament winding processes and potential equipment technologies," *Procedia CIRP*, vol. 66, pp. 125–130, 2017.

[64] B. W. Chieng, N. A. Ibrahim, Y. Y. Then, and Y. Y. Loo, "Epoxidized vegetable oils plasticized poly (lactic acid) biocomposites: Mechanical, thermal and morphology properties," *Molecules*, vol. 19, pp. 16024–16038, 2014.

[65] M. A. Rahman, D. De Santis, G. Spagnoli, G. Ramorino, M. Penco, V. T. Phuong, et al., "Biocomposites based on lignin and plasticized poly (L-lactic acid)," *Journal of Applied Polymer Science*, vol. 129, pp. 202–214, 2013.

[66] K. Fukushima, D. Tabuani, C. Abbate, M. Arena, and P. Rizzarelli, "Preparation, characterization and biodegradation of biopolymer nanocomposites based on fumed silica," *European Polymer Journal*, vol. 47, pp. 139–152, 2011.

[67] H. C. Obasi, A. A. Chaudhry, K. Ijaz, H. Akhtar, and M. H. Malik, "Development of biocomposites from coir fibre and poly (caprolactone) by solvent casting technique," *Polymer Bulletin*, vol. 75, pp. 1775–1787, 2018.

[68] A. Prasad, M. R. Sankar, and V. Katiyar, "State of art on solvent casting particulate leaching method for orthopedic scaffoldsfabrication," *Materials Today: Proceedings*, vol. 4, pp. 898–907, 2017.

[69] J. D. Schiffman and C. L. Schauer, "A review: Electrospinning of biopolymer nanofibers and their applications," *Polymer Reviews*, vol. 48, pp. 317–352, 2008.

[70] A. Rogina, "Electrospinning process: Versatile preparation method for biodegradable and natural polymers and biocomposite systems applied in tissue engineering and drug delivery," *Applied Surface Science*, vol. 296, pp. 221–230, 2014.

[71] L. Deng, X. Kang, Y. Liu, F. Feng, and H. Zhang, "Characterization of gelatin/zein films fabricated by electrospinning vs solvent casting," *Food Hydrocolloids*, vol. 74, pp. 324–332, 2018.

[72] A. Le Duigou, P. Davies, and C. Baley, "Seawater ageing of flax/poly (lactic acid) biocomposites," *Polymer Degradation and Stability*, vol. 94, pp. 1151–1162, 2009.

[73] P. Ouagne, L. Bizet, C. Baley, and J. Bréard, "Analysis of the film-stacking processing parameters for PLLA/flax fiber biocomposites," *Journal of Composite Materials*, vol. 44, pp. 1201–1215, 2010.

[74] C. Verma and M. Quraishi, "Carbohydrate polymer-metal nanocomposites as advanced anticorrosive materials: A perspective," *International Journal of Corrosion and Scale Inhibition*, vol. 11, pp. 507–523, 2022.

[75] H. Fetouh, A. Hefnawy, A. Attia, and E. Ali, "Facile and low-cost green synthesis of eco-friendly chitosan-silver nanocomposite as novel and promising corrosion inhibitor for mild steel in chilled water circuits," *Journal of Molecular Liquids*, vol. 319, p. 114355, 2020.

[76] P. S. Umoren, D. Kavaz, and S. A. Umoren, "Corrosion inhibition evaluation of chitosan-CuO nanocomposite for carbon steel in 5% HCl solution and effect of KI addition," *Sustainability*, vol. 14, p. 7981, 2022.

[77] M. M. Solomon, H. Gerengi, and S. A. Umoren, "Carboxymethyl cellulose/silver nanoparticles composite: Synthesis, characterization and application as a benign corrosion inhibitor for St37 steel in 15% H2SO4 medium," *ACS Applied Materials & Interfaces*, vol. 9, pp. 6376–6389, 2017.

[78] M. Gouda and H. M. A. El-Lateef, "Novel cellulose derivatives containing metal (Cu, Fe, Ni) oxide nanoparticles as eco-friendly corrosion inhibitors for c-steel in acidic chloride solutions," *Molecules*, vol. 26, p. 7006, 2021.

[79] M. M. Solomon, H. Gerengi, S. A. Umoren, N. B. Essien, U. B. Essien, and E. Kaya, "Gum Arabic-silver nanoparticles composite as a green anticorrosive formulation for steel corrosion in strong acid media," *Carbohydrate Polymers*, vol. 181, pp. 43–55, 2018.

[80] M. S. Hasanin and S. A. Al Kiey, "Environmentally benign corrosion inhibitors based on cellulose niacin nano-composite for corrosion of copper in sodium chloride solutions," *International Journal of Biological Macromolecules*, vol. 161, pp. 345–354, 2020.

16 Cellulose-Composites as Corrosion Inhibitors

Abhinay Thakur
Lovely Professional University

Ashish Kumar
Government of Bihar

16.1 INTRODUCTION

Corrosion is a natural process that occurs when metals react with the environment, causing degradation and deterioration of metallic structures. Its adverse impact extends to various industries worldwide [1]. Economically, corrosion costs the global economy billions of dollars each year. In the United States alone, corrosion's direct cost is estimated to be around $276 billion annually, equivalent to approximately 3.1% of the country's GDP. The consequences of corrosion are particularly evident in infrastructure, where it is responsible for approximately 20% of bridge failures in the US [2–4]. The transportation sector also suffers from corrosion-related expenses, with airlines worldwide spending around $4 billion annually on maintenance and repairs. In the maritime industry, corrosion costs account for approximately 10% of total operational expenses for shipping companies. The energy industry faces significant challenges, as corrosion in the oil and gas sector can result in leaks, spills, and environmental damage, costing billions of dollars in maintenance, repairs, and remediation efforts [5,6]. The manufacturing sector bears a global cost of over $300 billion annually due to corrosion, encompassing equipment replacement, maintenance, and productivity losses [7]. Moreover, corrosion-related incidents can harm the environment, such as oil or chemical spills from corroded pipelines, endangering aquatic life and ecosystems. To combat corrosion and mitigate its detrimental effects, various methods of corrosion protection have been developed and employed. These include the application of protective coatings, cathodic protection, sacrificial anodes, and the use of corrosion inhibitors. Among these methods, corrosion inhibitors have gained considerable attention due to their ability to impede or slow down the corrosion process [8,9].

Corrosion inhibitors are substances that, when added to the corrosive environment or applied to the metal surface, reduce the corrosion rate and extend the lifespan of metallic structures. They achieve this by forming a protective barrier on the metal surface, inhibiting the electrochemical reactions that drive corrosion [10]. Corrosion inhibitors can be organic or inorganic compounds, and they work through various mechanisms, including adsorption, film formation, and alteration of the electrolyte chemistry. The significance of corrosion inhibition lies in its potential to mitigate the economic, structural, and safety risks associated with corrosion. By incorporating corrosion inhibitors into maintenance and protection strategies, industries can reduce the frequency of repairs and replacements, lower maintenance costs, and enhance the lifespan of assets. Furthermore, corrosion inhibition contributes to the overall safety and reliability of structures, minimizing the risk of failures and accidents [11]. While traditional corrosion inhibitors have been effective in many cases, they often come with their own set of limitations and concerns. Some inhibitors contain toxic or hazardous substances, posing risks to human health and the environment. For example, chromates, which have been widely used as corrosion inhibitors, are known to be carcinogenic and environmentally harmful. Additionally, some inhibitors may have limited durability or adverse effects on the mechanical properties of metals. Given the growing emphasis on sustainability and environmental responsibility, there is a clear need for corrosion inhibitors that are not only effective but also sustainable and eco-friendly. The development of such inhibitors aligns with the principles of green chemistry and promotes the use of renewable resources, non-toxic materials, and environmentally benign processes. Sustainable corrosion inhibitors can reduce the ecological impact associated with corrosion protection practices while ensuring long-term performance and safety [12,13].

In recent years, there has been a growing interest in utilizing cellulose-based composites as corrosion inhibitors. The aim is to enhance the inherent corrosion inhibition properties of cellulose by incorporating various reinforcements and fillers into cellulose matrices. Vijayan P et al. [14] focused on utilizing cellulose nanofibers (CNF) to facilitate the release of healing agents in epoxy coatings. In their study, epoxy monomer and amine curing agents were immobilized onto CNF. The resulting materials, namely epoxy immobilized CNF (EiCNF) and amine curing agent immobilized CNF (AiCNF), were subjected to characterization using techniques such as Fourier-transform infrared spectroscopy (FTIR), thermogravimetric analysis (TGA), and scanning electron microscopy (SEM). Through these analyses, the researchers elucidated the mechanism and nature of interaction between EiCNF and AiCNF. It was observed that there was a chemical interaction between the amine curing agent and CNF, while the interaction between the epoxy monomer and CNF was purely physical. Furthermore, the researchers conducted a preliminary investigation into the self-healing ability of epoxy coatings incorporated with both EiCNF and AiCNF. The presence of the dual healing agents supported on CNF demonstrated effective imparting of self-healing ability to the epoxy coatings. The addition of these materials allows for the improvement of mechanical strength, barrier properties, and overall corrosion resistance of the composites. Similarly, Zhu et al. [15] prepared aminated hydroxyethyl cellulose-induced biomimetic hydroxyapatite coatings on the AZ31 magnesium alloy. The objective was to enhance the cytocompatibility and corrosion resistance of the magnesium alloy, aiming to develop a biodegradable medical material. The researchers successfully prepared a biomimetic hydroxyapatite coating on the surface of the AZ31 magnesium alloy using a sol-gel spin-coating method and biomimetic mineralization, with the aminated hydroxyethyl cellulose (AHEC) playing a crucial role. Figure 16.1 illustrates the surface morphology of different substrates: AZ31 (Figure 16.1a), AHEC-coated AZ31 (Figure 16.1b), and HA/AHEC-coated AZ31 (Figure 16.1c). The AHEC-coated AZ31 exhibits a uniform

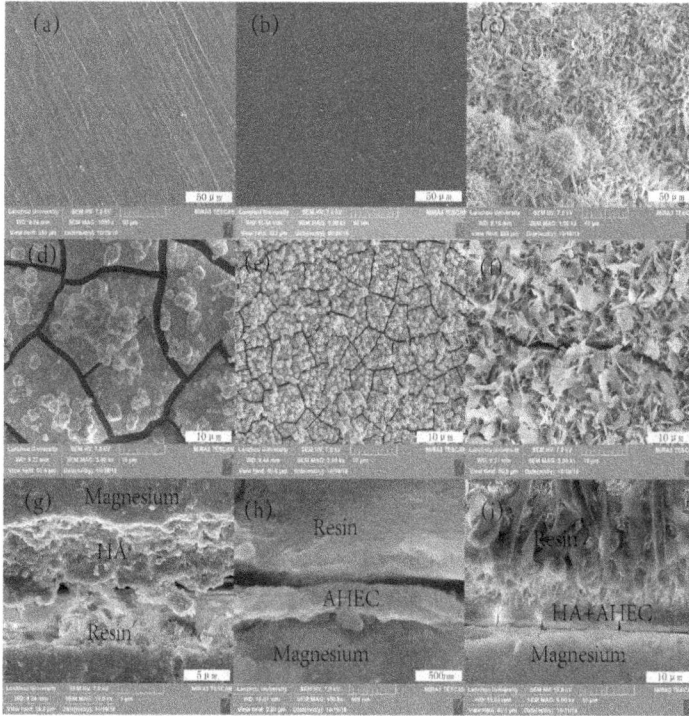

FIGURE 16.1 SEM images of AZ31, aminated hydroxyethyl cellulose (AHEC)/AZ31, and hydroxyapatite/aminated hydroxyethyl cellulose (HA/AHEC)/AZ31 specimens. The images show the specimens before (a–c) and after (d–f) immersion in simulated body fluid (SBF) for 7 days. Additionally, the cross-sections of HA/AZ31, AHEC/AZ31, and HA/AHEC/AZ31 are presented in (g), (h), and (i), respectively. Adapted with permission from Ref. [15]. Distributed under CCBY 4.0.

and dense structure, contrasting with the untreated AZ31, which shows scratches. The surface of the HA/AHEC-coated AZ31 displays a flower-like and homogeneous structure composed of porous and consistent flake-like crystals. This porous structure is favorable for bone cell biology reactions, such as bone apatite formation and osteoblast proliferation. The porous structure is attributed to hydrogen release. After immersion in simulated body fluid (SBF) for 7 days, the surface morphologies of the samples are depicted in Figure 16.1d–f. The uncoated AZ31 samples exhibit wide and deep cracks and pits, with minimal HA presence. The AHEC-coated samples show smaller cracks and pits with dense and spherical HA particles, indicating that AHEC induces HA formation in SB. However, the HA/AHEC-coated AZ31 exhibits the fewest cracks and pits with a narrower and shallower structure. These cracks are attributed to water loss of corrosion products and surface shrinkage. Thus, it can be inferred that the HA/AHEC double coating enhances the corrosion resistance of AZ31 and provides protection against SBF corrosion. The cross-section morphologies of the HA, AHEC, and HA/AHEC coatings are presented in Figure 16.1g–i.

No distinct dividing line is observed in the HA/AHEC-coated AZ31, suggesting that the AHEC coating induces the formation of the HA coating on the surface of the AZ31 magnesium alloy.

Wang et al. [16] synthesized a series of poly (methyl methacrylate) and poly (butyl acrylate) (PMMA-co-PBA) copolymer latexes through mini emulsion polymerization, incorporating 1-octyl-3-methylimidazolium hexafluorophosphate (C8mimPF6), cellulose nanocrystals (CNCs), and C8mimPF6-CNCs. These newly developed composites were applied as coatings on mild steel panels, and their anti-corrosion performance was evaluated by immersing the coated samples in a 3.5 wt.% sodium chloride (NaCl) solution for a specified duration. The results demonstrated that the C8mimPF6-CNCs sample exhibited synergistic effects, leading to the highest coating resistance, charge transfer resistance, and corrosion inhibition efficiency, as well as the lowest diffusion coefficient and corrosion rate. The Nyquist impedance plots of the coated samples immersed in a 3.5 wt.% NaCl solution for 96 hours are depicted in Figure 16.2. These plots, obtained from electrochemical impedance spectroscopy (EIS) tests, provide valuable insights into the different aspects of the coating. In Figure 16.2a, the bottom left region at high frequency corresponds to defects and pinholes in the coating, while the upper right region at low frequency represents the interface between the coating and the mild steel surface. The shrinking semi-circles observed during the immersion test indicate a decrease in the anti-corrosion performance over time. Furthermore, the initial linear segment of the curve shown in Figure 16.2c is attributed to the pseudo-two-time constant phenomenon resulting from the penetration of electrolytes through the coating defects.

The choice of reinforcements and fillers in cellulose-based composites is crucial and depends on the desired properties and targeted application. Nanofillers, such as nanocellulose, clay nanoparticles, carbon nanotubes, or graphene, have shown promising results in enhancing the mechanical strength and barrier properties of cellulose-composites. These nanofillers have high aspect ratios and large surface areas, which promote strong interfacial interactions with cellulose and improve the composite's mechanical integrity and resistance to environmental factors. Other reinforcing materials, such as fibers or coatings, can also be incorporated into cellulose matrices to further enhance the corrosion inhibition properties. Fibers, such as glass fibers or natural fibers like sisal or jute, provide additional mechanical reinforcement to the composite, improving its structural stability and resistance to external forces. Coatings, on the other hand, can act as protective layers, enhancing the barrier properties of the composite and preventing corrosive species from reaching the metal surface.

The synergistic effects between cellulose and the reinforcing materials in these composites contribute to their enhanced corrosion resistance. The presence of cellulose in the composite provides inherent properties like the ability to form protective layers on the metal surface and inhibit the diffusion of corrosive species [17–19]. The reinforcing materials, on the other hand, improve the mechanical strength, barrier properties, and stability of the composite, further enhancing its corrosion inhibition performance. By tailoring the composition, structure, and fabrication parameters of cellulose-based composites, researchers can optimize their corrosion inhibition properties for specific applications. This allows for the development of tailored

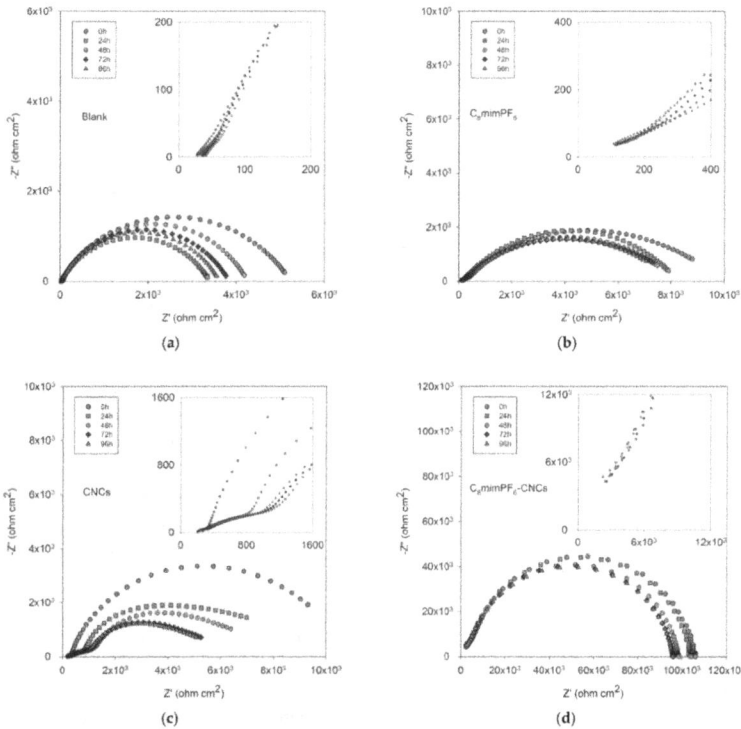

FIGURE 16.2 Nyquist impedance plots of the coated samples after being immersed in a 3.5 wt.% NaCl solution for 96 hours. The plots include the following: (a) Blank sample, representing the neat PMMA-co-PBA coating; (b) C8mimPF6 sample, representing the PMMA-co-PBA coating with 10 wt.% C8mimPF6; (c) CNCs sample, representing the PMMA-co-PBA coating with 0.5 wt.% CNCs; and (d) C8mimPF6-CNCs sample, representing the PMMA-co-PBA coating with 10 wt.% C8mimPF6 and 0.5 wt.% CNCs. Adapted with permission from Ref. [16]. Distributed under CCBY 4.0.

corrosion protection solutions that are not only effective but also sustainable and eco-friendly [20]. The utilization of cellulose-based composites as corrosion inhibitors represents a significant advancement in the field of corrosion science and offers promising alternatives to conventional inhibitors with improved performance and reduced environmental impact. Table 16.1 showcases some cellulose-based composites that have been used as corrosion inhibitors [21–31].

The objective of this chapter is to explore the use of cellulose-composites as corrosion inhibitors, focusing on their potential as sustainable and eco-friendly alternatives. Cellulose, as a renewable and abundant biopolymer, offers unique advantages in terms of its availability, renewability, non-toxicity, and biodegradability. Cellulose-composites, formed by incorporating reinforcing fillers into cellulose matrices, have shown promise in enhancing the corrosion inhibition properties of cellulose. By investigating the fabrication techniques, mechanisms of corrosion inhibition, characterization methods, challenges, and future perspectives of cellulose-composites as

TABLE 16.1

Some Cellulose-Based Composites That Have Been Used as Corrosion Inhibitors

Cellulose Composite	Corrosion Inhibition Mechanism	Application
Cellulose nanocrystals (CNC) reinforced with graphene oxide (GO)	Barrier protection and adsorption of corrosive species	Coatings for metals, protective films, and anti-corrosive coatings
Cellulose nanofibrils (CNF) reinforced with carbon nanotubes (CNT)	Barrier protection and enhanced mechanical strength	Coatings, films, and surface treatments for corrosion protection
Cellulose acetate (CA) nanofibers embedded with cerium oxide (CeO$_2$) nanoparticles	Inhibition of corrosion reactions and self-healing properties	Surface coatings, nanocomposite films, and anti-corrosive coatings
Cellulose-based hydrogels loaded with zinc oxide (ZnO) nanoparticles	Release of corrosion inhibitors and formation of protective films	Self-healing coatings, protective coatings, and corrosion-resistant materials
Cellulose nanofiber (CNF) composites with silane-functionalized GO	Barrier protection and enhanced mechanical properties	Coatings, films, and composite materials for corrosion resistance
Cellulose nanocrystals (CNC) combined with metal nanoparticles (e.g., silver, gold)	Synergistic effects of cellulose and metal nanoparticles	Antimicrobial coatings, protective films, and corrosion-resistant materials
Cellulose nanofibrils (CNF) reinforced with zinc oxide (ZnO) nanoparticles	Inhibition of corrosion reactions and UV protection	Coatings for outdoor applications, corrosion-resistant materials
CA nanocomposites with layered double hydroxides (LDH)	Barrier protection and controlled release of corrosion inhibitors	Smart coatings, self-healing films, and corrosion-resistant materials
Cellulose-based aerogels incorporated with ionic liquids	Enhanced corrosion resistance and self-healing properties	Coatings, encapsulation, and corrosion protection of sensitive materials
Cellulose nanofiber (CNF) composites with polyaniline (PANI)	Conductive coatings and inhibition of corrosion reactions	Corrosion protection for electronic devices, coatings for metal substrates
Cellulose nanocrystals (CNC) reinforced with clay nanoparticles	Barrier protection and improved mechanical properties	Coatings, films, and surface treatments for corrosion resistance
CA nanofibers embedded with zinc phosphate (Zn$_3$(PO$_4$)$_2$) nanoparticles	Inhibition of corrosion reactions and phosphate release	Protective coatings, self-healing materials, and corrosion-resistant films
Cellulose nanofiber (CNF) composites with titanium dioxide (TiO$_2$) nanoparticles	Photocatalytic corrosion inhibition and UV protection	Self-cleaning coatings, outdoor applications, and corrosion-resistant materials
Cellulose-based hydrogels loaded with manganese dioxide (MnO$_2$) nanoparticles	Oxygen reduction reaction and barrier protection	Coatings for metal surfaces, corrosion-resistant materials
Cellulose nanocrystals (CNC) combined with polymeric corrosion inhibitors	Controlled release of inhibitors and barrier protection	Coatings, films, and corrosion protection of sensitive materials

corrosion inhibitors, this chapter aims to contribute to the understanding and development of sustainable corrosion protection strategies. It will highlight the potential of cellulose-composites to address the limitations of traditional inhibitors and provide insights into their applications in various industries. Ultimately, this chapter seeks to promote the use of sustainable and eco-friendly corrosion inhibitors for the long-term protection of metallic structures.

16.2   CELLULOSE AS A CORROSION INHIBITOR

16.2.1   OVERVIEW OF CELLULOSE

Cellulose, a naturally occurring biopolymer, is a crucial component responsible for the structural integrity of plants. Composed of glucose units arranged in long chains, cellulose exhibits remarkable mechanical strength, making it a highly valuable material. As the most abundant biopolymer on Earth, cellulose is readily available in various plant sources such as wood, cotton, hemp, and biomass materials [32–34]. Wood, a primary source of cellulose, finds extensive use in industries including construction, paper production, and textiles. The global production of paper and paperboard heavily relies on wood pulp derived from cellulose. Cotton, known for its high purity of cellulose, is a significant contributor to the textile industry. In 2020, global cotton production reached approximately 25 million metric tons, emphasizing the widespread use of cellulose-based fibers in textiles [35]. Beyond these traditional applications, ongoing research explores the potential of cellulose in emerging fields. In biomedical engineering, cellulose-based materials are being investigated for applications like tissue engineering, drug delivery systems, and wound healing. The energy storage sector explores cellulose-derived materials for supercapacitors and batteries. Additionally, cellulose-based materials show promise in environmental remediation due to their ability to adsorb pollutants. These applications highlight the versatility and potential of cellulose in addressing diverse challenges and advancing sustainable solutions. As researchers delve deeper into its properties and capabilities, cellulose continues to garner attention as a valuable and versatile biopolymer.

16.2.2   INHERENT CORROSION RESISTANCE PROPERTIES OF CELLULOSE

The hydrophilic nature of cellulose, with its strong affinity for water molecules, can be illustrated with the example of cellulose-based coatings used for corrosion protection. When a cellulose-based coating is applied to a metal surface, the hydroxyl groups present in cellulose interact with water molecules in the environment [36,37]. This interaction leads to the formation of a hydrate film on the metal surface, effectively creating a barrier between the metal and the corrosive agents. For instance, cellulose nanocrystals (CNCs) have been used to develop environmentally friendly coatings for corrosion protection. CNCs have a high density of hydroxyl groups, making them highly hydrophilic. When CNCs are incorporated into a coating formulation, they form a network structure that absorbs and retains water, creating a hydrate film on the metal surface. This film acts as a barrier against corrosive agents such as oxygen, moisture, and ions, thereby inhibiting corrosion.

In addition to the formation of hydrate films, the ability of cellulose to form complex networks and structures can be demonstrated through cellulose-based composites [38–45]. For example, cellulose fibers can be combined with reinforcing materials like carbon nanotubes or graphene oxide to create composite materials with improved corrosion resistance properties. The entanglement of cellulose fibers and the incorporation of reinforcing materials create a dense network within the composite, impeding the diffusion of corrosive species toward the metal surface [46]. This barrier effect enhances the corrosion inhibition performance of the composite material. The inherent stability and durability of cellulose can be highlighted through the use of cellulose-based films for corrosion protection. Cellulose films have been developed by depositing cellulose onto metal surfaces using various techniques such as spin-coating or dip-coating. These films exhibit excellent resistance to UV radiation, moisture, and temperature fluctuations, ensuring long-term protection of the metal against corrosion. The stability and durability of cellulose films make them suitable for outdoor applications where metals are exposed to harsh environmental conditions [47,48].

The formation of a hydrate film is essential for inhibiting corrosion because it effectively blocks the diffusion of corrosive species to the metal surface. By creating a moisture-rich environment, cellulose hampers the transport of oxygen, moisture, and ions, which are essential for corrosion reactions. As a result, the metal is shielded from the corrosive environment, reducing the likelihood of corrosion initiation and propagation [49–51]. Moreover, cellulose's ability to form complex networks and structures contributes to its corrosion resistance. The long chains of cellulose can entangle and create a dense network, hindering the movement and diffusion of corrosive species toward the metal surface. This barrier effect further slows down the corrosion process, providing additional protection to the metal [52–55]. Another advantage of cellulose is its inherent stability and durability. Cellulose is a stable biopolymer that can withstand a wide range of environmental conditions. It exhibits excellent resistance to UV radiation, moisture, and temperature fluctuations, making it a reliable material for long-term corrosion protection. The stability of cellulose ensures that its protective properties remain intact over extended periods, contributing to the longevity of cellulose-based corrosion inhibition strategies.

16.3 CELLULOSE-BASED COMPOSITES FOR CORROSION INHIBITION

16.3.1 Types of Cellulose-Based Composites

Cellulose-based composites offer a wide range of possibilities for corrosion inhibition applications, thanks to the incorporation of various reinforcements and fillers. One example of a cellulose-based composite is cellulose fiber-reinforced composites. Cellulose fibers, sourced from wood or plant residues, are combined with a matrix material, such as epoxy or polyurethane. The cellulose fibers act as a reinforcing phase, enhancing the mechanical strength and barrier properties of the composite [56,57]. These composites exhibit improved corrosion resistance due to the efficient dispersion and alignment of cellulose fibers, which create a dense network that hinders the diffusion of corrosive species toward the metal surface. Another type of cellulose-based composite is cellulose nanoparticle-reinforced composites. Cellulose nanoparticles, derived from the breakdown of cellulose fibers, are incorporated into a matrix material. These nanoparticles possess unique properties, including a high aspect ratio, large surface area, and excellent dispersibility [58]. When added to a composite, cellulose nanoparticles enhance its mechanical strength, barrier properties, and corrosion resistance. For example, the incorporation of cellulose nanoparticles into a polymer matrix can create a composite with improved resistance to corrosion-induced degradation, as the nanoparticles form a protective barrier on the metal surface.

Cellulose-based nanocomposites can also be created by incorporating nanofillers into a cellulose matrix. These nanofillers can include carbon nanotubes, graphene oxide, or metal oxide nanoparticles. By introducing these nanofillers, the composite gains additional functionalities that enhance its corrosion inhibition properties [59,60]. For instance, the inclusion of metal oxide nanoparticles can provide sacrificial protection, forming a protective oxide layer on the metal surface. Carbon-based nanofillers, on the other hand, improve the electrical conductivity and barrier properties of the composite, preventing the formation of localized corrosion cells.

16.3.2   Reinforcements and Fillers in Cellulose-Composites

The performance of cellulose-based composites for corrosion inhibition relies on the choice of reinforcements and fillers incorporated into the cellulose matrix [61–63]. The selection of these materials depends on their compatibility with cellulose and their ability to enhance the mechanical, barrier, and corrosion resistance properties of the composite. Various reinforcements can be used in cellulose-composites, including natural fibers, such as jute, sisal, or bamboo, and synthetic fibers, such as glass or carbon fibers. These reinforcements improve the mechanical strength, stiffness, and toughness of the composites, making them more resistant to mechanical stress and corrosion-induced damage. For example, in a cellulose-based composite reinforced with natural fibers such as jute or sisal, the long and aligned fibers provide enhanced mechanical strength and resistance to corrosion-induced damage. These composites can be used in applications where high mechanical performance and corrosion resistance are required, such as in the automotive or construction industries. Similarly, cellulose-composites reinforced with synthetic fibers like glass or carbon fibers exhibit excellent mechanical properties, including high tensile strength and stiffness. These composites can withstand harsh environmental conditions and provide long-lasting corrosion protection. They are commonly used in marine applications, where corrosion resistance is critical due to exposure to saltwater and moisture [64].

The incorporation of nanoparticles as fillers in cellulose-composites further enhances their corrosion inhibition properties. For instance, the addition of zinc oxide nanoparticles to a cellulose matrix can provide self-healing properties, as the nanoparticles release zinc ions when the composite is damaged, forming a protective layer on the metal surface and preventing further corrosion. These composites are particularly useful in industries such as oil and gas, where metal structures are exposed to aggressive environments. Carbon-based fillers, such as carbon nanotubes or graphene oxide, offer multiple benefits to cellulose-composites. They improve electrical conductivity, allowing for effective cathodic protection and reducing the risk of galvanic corrosion. Moreover, carbon-based fillers enhance the barrier properties of the composites, preventing the diffusion of corrosive species [65]. These composites find applications in sectors such as aerospace, where lightweight materials with excellent corrosion resistance are highly desirable. For example, metal oxide nanoparticles, such as zinc oxide or cerium oxide, can provide additional corrosion protection by acting as sacrificial anodes or by forming a passivation layer on the metal surface. Carbon-based fillers, such as carbon nanotubes or graphene oxide, can improve the electrical conductivity and barrier properties of the composites, inhibiting corrosion and preventing the formation of localized corrosion cells. The proper dispersion and alignment of reinforcements and fillers within the cellulose matrix are crucial for achieving optimal performance. Surface modifications and functionalization techniques can be employed to enhance the interfacial adhesion between cellulose and the reinforcing materials, ensuring efficient stress transfer and improved mechanical properties.

16.3.3 Processing/Fabrication Techniques for Cellulose-Composites

Fabrication techniques play a crucial role in the development of cellulose-composites by ensuring the proper dispersion and alignment of reinforcements and fillers within the cellulose matrix. These techniques determine the final structure, properties, and performance of the composites. Several fabrication techniques are commonly employed for cellulose-composites, including solution casting, melt compounding, extrusion, compression molding, and others. Each technique has its advantages and limitations, and the choice depends on factors such as the desired composite structure, matrix material, reinforcements, and fillers, as well as the intended application of the composite.

16.3.3.1 Solution Casting

Solution casting is a versatile and widely used fabrication technique for cellulose-composites. It involves the dissolution of cellulose and the desired reinforcing materials or fillers in a solvent to create a homogeneous solution. The solution is then cast into a mold or onto a substrate, and the solvent is subsequently evaporated or dried, resulting in the formation of a solid composite. One of the key advantages of solution casting is its ability to provide precise control over the composition and dispersion of the components in the composite. The process allows for the incorporation of various types of reinforcements and fillers, such as fibers, nanoparticles, or other additives, into the cellulose matrix. This versatility enables the tailoring of composite properties to meet specific requirements for different applications. Solution casting is particularly suitable for the production of thin films, coatings, or membranes. Thin films find application in areas such as packaging, electronics, and sensors, where a high surface area-to-volume ratio is desired. Coatings can provide corrosion protection, barrier properties, or surface modification functionalities. Membranes are used in separation processes, water purification, or biomedical applications.

Solution casting offers the advantage of producing these thin and uniform layers with ease. The control over the composition and dispersion achieved through solution casting allows for the optimization of the properties of the resulting cellulose-composites. For example, the reinforcement loading, the ratio of cellulose to reinforcing materials, and the distribution of fillers can be precisely adjusted. This control enables the enhancement of mechanical strength, stiffness, and other properties of the composites, depending on the specific requirements of the application. By tailoring the composition and dispersion, solution casting enables the creation of composites with improved resistance to corrosion, wear, or impact. To illustrate the application of solution casting, let's consider the development of a cellulose-based composite film for corrosion protection. In this case, cellulose would serve as the matrix material, while reinforcing materials such as nanoparticles or fibers would be incorporated to enhance the barrier properties and corrosion resistance of the film.

First, cellulose is dissolved in a suitable solvent, such as N-methylmorpholine N-oxide (NMMO), to form a cellulose solution [66,67]. The reinforcing material, such as zinc oxide nanoparticles, is then dispersed into the cellulose solution with the help of ultrasonication or mechanical stirring. The dispersion process ensures a uniform distribution of the nanoparticles within the cellulose matrix. Once the homogeneous cellulose-reinforcement solution is prepared, it is poured onto a flat substrate or cast into a mold, depending on the desired film thickness and shape. The solution is spread evenly to achieve a uniform film thickness. The solvent is then evaporated under controlled conditions, such as low temperature and reduced pressure, allowing the film to solidify. The resulting cellulose-composite film possesses enhanced corrosion resistance properties due to the presence of zinc oxide nanoparticles. The nanoparticles act as a barrier to the diffusion of corrosive species, preventing them from reaching the metal surface and initiating corrosion. The film can be applied to metal surfaces as a protective coating, providing long-term corrosion inhibition.

Although solution casting offers numerous advantages, it also has certain limitations. One limitation is scalability, as the technique is more suitable for laboratory-scale production rather than large-scale manufacturing. The process parameters need to be carefully controlled to ensure reproducibility and uniformity of the resulting composites. Additionally, achieving a high reinforcement loading can be challenging, as it may lead to difficulties in dispersing the reinforcements uniformly within the cellulose matrix.

16.3.3.2 Melt Compounding

Melt compounding is a widely employed fabrication technique for the production of cellulose-composites. It involves blending cellulose, reinforcing materials, and fillers in a molten state. The components are mixed using various techniques such as extrusion, injection molding, or hot pressing. The molten mixture is then cooled and solidified to form a composite material with enhanced properties. One of the key advantages of melt compounding is its suitability for thermoplastic matrices. Thermoplastics, such as polyethylene, polypropylene, or polystyrene, exhibit the characteristic of becoming soft and moldable when heated and solidifed upon cooling. This behavior allows for the processing of the molten mixture during compounding and subsequent shaping.

Melt compounding offers several benefits for the fabrication of cellulose-composites. Firstly, it enables large-scale production, making it suitable for industrial applications. The continuous processing nature of techniques like extrusion allows for the efficient production of composite materials in a continuous form, such as sheets, profiles, or fibers. Another advantage of melt compounding is the potential to incorporate high levels of reinforcements or fillers. The high temperatures and shear forces applied during processing facilitate the dispersion and alignment of these components within the cellulose matrix [22]. This uniform distribution enhances the mechanical properties of the composite, including strength, stiffness, and impact resistance. The increased loading of reinforcements or fillers can result in composites with enhanced functionality and performance. To illustrate the application of melt compounding, let's consider the development of a cellulose-based composite using extrusion. In this case, cellulose fibers and carbon nanotubes (CNTs) will be incorporated to enhance the mechanical properties and electrical conductivity of the composite. Initially, cellulose fibers are blended with a thermoplastic matrix, such as polypropylene, to form a mixture. The CNTs are then added to the blend, either as a separate component or pre-dispersed in a carrier medium. The mixture is fed into an extruder, where it undergoes heating and mixing. The high temperature and shear forces within the extruder melt the thermoplastic matrix, allowing for the dispersion of cellulose fibers and CNTs throughout the molten matrix.

As the molten mixture is forced through a die, it takes on the desired shape, such as a rod, sheet, or filament. Upon exiting the die, the material cools and solidifies, resulting in a cellulose composite with well-dispersed cellulose fibers and aligned CNTs within the thermoplastic matrix. The resulting composite exhibits improved mechanical properties, such as higher tensile strength and modulus, as well as enhanced electrical conductivity due to the presence of CNTs. While melt compounding offers significant advantages, it is not without limitations. One limitation is the requirement of high processing temperatures. The high temperatures necessary for melting the thermoplastic matrix can potentially lead to thermal degradation of cellulose or sensitive additives incorporated in the composite. Therefore, careful selection of processing parameters, including temperature and residence time, is essential to minimize thermal degradation and preserve the integrity of the cellulose and other components. Additionally, melt compounding may pose challenges in achieving a uniform dispersion of reinforcements or fillers within the matrix. The choice of processing conditions, such as temperature, screw speed, and residence time, needs to be optimized to ensure effective dispersion and alignment of the components. Insufficient dispersion can result in reduced mechanical properties and compromised performance of the composite.

16.3.3.3 Extrusion

Extrusion is a widely used fabrication technique for the production of cellulose-composites. It involves the process of forcing a molten cellulose matrix through a die, which contains incorporated reinforcements or fillers. The extruded material is then cooled and solidified to form a continuous composite profile with the desired shape. One of the key advantages of extrusion is its ability to produce cellulose-composites with complex shapes and continuous lengths. The extrusion process allows for the formation of profiles, such as rods, tubes, or sheets, with consistent cross-sectional geometries. This makes extrusion suitable for a wide range of applications, including construction materials, automotive components, and packaging. The extrusion technique also offers precise control over the distribution and alignment of reinforcements or fillers within the cellulose matrix. The incorporation of reinforcing materials, such as fibers or nanoparticles, improves the mechanical properties of the composite, including strength, stiffness, and impact resistance. By controlling the die design and processing parameters, such as temperature, pressure, and shear rate, the orientation and distribution of reinforcements can be tailored to meet specific performance requirements [23,24].

To illustrate the application of extrusion, let's consider the production of a cellulose composite with wood fibers as reinforcements. Initially, cellulose is melted and mixed with wood fibers, either in a pre-mixed form or by feeding the fibers into the extruder along with the cellulose matrix. The mixture is then fed into the extruder, where it undergoes heating and melting. As the molten cellulose-fiber mixture passes through the die, it takes on the shape of the die opening and solidifies upon exiting the die. The resulting composite profile contains uniformly dispersed wood fibers within the cellulose matrix, providing enhanced mechanical properties and a natural appearance due to the presence of the fibers. The composite profile can be further processed, such as cutting or surface finishing, to meet specific application requirements. Extrusion can also be combined with other techniques to produce reinforced fibers or filaments. For example, in fiber spinning, the extrusion process is modified to allow for the continuous formation of fibers. In this case, the molten cellulose matrix is forced through a spinneret, which contains multiple small openings. As the extruded material exits the spinneret, it solidifies into individual fibers. These fibers can be collected, wound, and further processed to create cellulose-based reinforced textiles or composite materials with enhanced strength and durability.

While extrusion offers significant advantages, it is not without limitations. One limitation is the potential for the degradation of cellulose at high processing temperatures. Cellulose is susceptible to thermal degradation, which can lead to a reduction in its molecular weight and changes in its chemical structure [27,28]. Therefore, careful control of processing parameters, such as temperature and residence time, is essential to minimize thermal degradation and preserve the integrity of the cellulose matrix. Another challenge in extrusion is achieving a high reinforcement loading and uniform dispersion of nanoparticles within the cellulose matrix. The viscosity of the molten cellulose matrix and the distribution and flow behavior of the reinforcements can affect their dispersion and alignment. Achieving a high degree of dispersion and uniform distribution is crucial to maximizing the mechanical and functional properties of the composite.

16.3.3.4 Compression Molding

Compression molding is a widely used fabrication technique for the production of cellulose-composites. It involves the application of heat and pressure to a mixture of cellulose, reinforcing materials, and fillers in a mold to consolidate the composite into a solid shape. Compression molding offers several advantages in terms of controlled thickness, high density, and the ability to produce complex shapes. The process of compression molding begins with the preparation of the mixture or composite formulation. Cellulose is typically combined with reinforcing materials, such as fibers or particles, and fillers, which can enhance the mechanical and functional properties of the composite. The mixture is preheated to soften the cellulose matrix and facilitate the flow of the material during compression. Once the mixture reaches the desired temperature, it is placed into a mold cavity. The mold is then closed, and pressure is applied to the mixture using a hydraulic press or mechanical press. The pressure helps to distribute the composite materials evenly within the mold cavity, ensuring good interfacial bonding between the cellulose matrix and the reinforcements or fillers. Simultaneously, the heat aids in the melting and flow of the cellulose matrix, allowing it to fill the mold cavities completely.

The applied pressure and heat are maintained for a specific period to allow for the consolidation and curing of the composite. This ensures the formation of a solid, uniform, and dimensionally stable product. The duration of the compression molding process depends on factors such as the composition of the mixture, the desired thickness, and the curing characteristics of the materials used. To illustrate the application of compression molding, let's consider the production of a cellulose composite with glass fibers as reinforcements. In this case, cellulose is mixed with glass fibers and any additional fillers or additives, such as flame retardants or colorants. The mixture is then preheated to soften the cellulose matrix and enhance its flowability. Next, the preheated mixture is placed into a mold cavity that corresponds to the desired shape of the final product. The mold is closed, and pressure is applied to the mixture using a press. The applied pressure distributes the cellulose matrix and glass fibers evenly within the mold cavity, ensuring proper interfacial bonding. The pressure and heat are maintained for a specific period to allow for curing and consolidation of the composite. During this time, the cellulose matrix solidifies, and the glass fibers become embedded within it, creating a strong and durable composite material. After the curing period, the mold is opened, and the finished cellulose-composite product is removed.

Compression molding offers several advantages in the fabrication of cellulose-composites. One of the key advantages is the ability to achieve controlled thickness and high density in the final product. The pressure applied during compression ensures uniform distribution of the materials and consolidation of the composite, resulting in a product with consistent thickness and high-density characteristics. Another advantage of compression molding is the versatility in producing complex shapes. The mold cavity can be designed to match intricate geometries, enabling the production of cellulose-composites with intricate details, such as textured surfaces or intricate patterns. This makes compression molding suitable for a wide range of applications, including automotive components, electrical enclosures, and consumer goods. Compression molding also allows for the incorporation of a wide range of reinforcements and fillers in the cellulose matrix. Depending on the desired properties of the composite, various types of reinforcing materials, such as fibers, particles, or flakes, can be added to enhance strength, stiffness, and impact resistance. Fillers can be incorporated to modify other properties, such as thermal conductivity or flame retardancy.

Despite its advantages, compression molding does have some limitations. One limitation is the longer processing time compared to other fabrication techniques.

The combination of heat and pressure, along with the curing time, can result in longer production cycles. This may be a consideration for high-volume production where shorter cycle times are desired. Another limitation is the suitability of compression molding for continuous production. Due to the nature of the process and the time required for curing, compression molding is more commonly used for batch production or the production of smaller quantities. Continuous production may be challenging due to the need for rapid mold opening and closure, as well as shorter curing times.

16.3.3.5 Electrospinning

Electrospinning is a versatile and widely used technique for the fabrication of nanofibrous cellulose-composites. It involves the use of high voltage to create a charged jet of a polymer solution containing cellulose and other desired components. This charged jet is then collected on a grounded collector, resulting in the formation of a non-woven mat of ultrafine fibers. Electrospinning offers numerous advantages, including a high surface area-to-volume ratio, controlled fiber morphology, and the ability to incorporate various types of reinforcements and fillers. The electrospinning process begins with the preparation of a polymer solution that includes cellulose as the main component. The cellulose can be in the form of cellulose acetate, cellulose esters, or cellulose derivatives. The polymer solution is typically prepared by dissolving the cellulose and any other desired additives, such as reinforcing materials or fillers, in a suitable solvent. The choice of solvent depends on the solubility of cellulose and the desired properties of the final composite. Once the polymer solution is prepared, it is loaded into a syringe or a reservoir connected to a high-voltage power supply. The solution is then dispensed through a needle or a spinneret, which is connected to the power supply. The high voltage creates an electric field that induces a charge on the droplet of the polymer solution at the tip of the needle. The repulsion between the charges in the droplet causes it to elongate and form a thin jet. As the jet travels toward the grounded collector, the solvent evaporates, resulting in the solidification of the polymer fibers. The fibers are randomly deposited on the collector, forming a non-woven mat or a fibrous membrane. The morphology and structure of the fibers can be controlled by adjusting various parameters, including the applied voltage, the flow rate of the solution, and the distance between the needle and the collector.

The electrospinning process offers several advantages for the fabrication of cellulose-composites. One of the key advantages is the high surface area-to-volume ratio of the resulting nanofibers. The ultrafine diameter of the fibers, typically in the range of hundreds of nanometers to a few micrometers, provides a large surface area for interactions with the surrounding environment. This high surface area facilitates enhanced properties, such as increased mechanical strength, improved barrier properties, and enhanced drug-loading capacity in drug delivery applications. Another advantage of electrospinning is the ability to control the morphology and structure of the fibers. By adjusting the process parameters, such as the voltage, solution concentration, and collection distance, it is possible to achieve different fiber diameters, orientations, and alignments. This control over fiber morphology allows for the tailoring of the mechanical, optical, and functional properties of the cellulose-composites.

Furthermore, electrospinning offers versatility in incorporating various types of reinforcements and fillers into the cellulose matrix. For example, inorganic nanoparticles, such as metal oxides or carbon-based materials, can be dispersed within the polymer solution before electrospinning. These nanoparticles can provide additional functionalities, such as antimicrobial properties, electrical conductivity, or catalytic activity, to the resulting cellulose-composites.

To illustrate the application of electrospinning in cellulose-composites, let's consider the production of a nanofibrous membrane with cellulose nanocrystals (CNCs) as reinforcements. In this case, cellulose is dissolved in a suitable solvent, such as an ionic liquid or a mixture of organic solvents. The CNCs are dispersed within the cellulose solution to enhance the mechanical properties of the composite. The cellulose-CNC solution is loaded into a syringe, and a high voltage is applied to the syringe needle. As the solution is ejected from the needle, the electrostatic forces induce the formation of a charged jet. The jet undergoes stretching and thinning due to the repulsion between the charges, resulting in the formation of nanofibers. The nanofibers are collected on a grounded collector, forming a nanofibrous membrane. The resulting cellulose-CNC nanofibrous membrane exhibits enhanced mechanical properties, such as increased tensile strength and modulus, compared to pure cellulose membranes. The CNCs, being highly stiff and reinforcing, effectively reinforce the cellulose matrix, improving its load-bearing capacity. The nanofibrous structure of the membrane provides a high surface area, making it suitable for applications in filtration, separation, and biomedical devices.

Despite its advantages, electrospinning also has some limitations that need to be considered. One limitation is the potential use of toxic solvents during the electrospinning process. Some solvents used for cellulose dissolution or to enhance the spinnability of the solution may be hazardous to human health or the environment. Therefore, careful solvent selection and proper handling procedures are necessary to ensure safety. Another limitation is the relatively low production rate of electrospinning. The process typically operates at a slower speed due to the time required for solvent evaporation and fiber deposition. This makes it challenging to scale up the production of electrospun cellulose-composites for large-scale applications [29,30].

16.3.3.6 3D Printing

3D printing, also known as additive manufacturing, is a transformative fabrication technique that has gained significant attention in recent years. It enables the creation of complex three-dimensional objects by depositing materials layer by layer based on digital designs. While initially developed for plastics and metals, 3D printing has expanded to include a wide range of materials, including cellulose-composites. In the context of cellulose-composites, 3D printing offers several advantages and opportunities. One of the key advantages is the ability to customize the printed objects according to specific requirements. This customization capability is particularly valuable in fields such as healthcare, where personalized medical devices or implants can be fabricated with tailored geometries and properties. By leveraging the design flexibility of 3D printing, cellulose-composites can be precisely engineered to meet the desired functionalities and performance criteria. Another advantage of 3D printing is the ability to fabricate complex geometries that are challenging or even impossible to achieve with traditional manufacturing techniques. Traditional methods, such as machining or molding, often have limitations in terms of geometrical complexity. In contrast, 3D printing enables the fabrication of intricate structures, including internal cavities, overhangs, and intricate lattice-like architectures. This capability opens up new possibilities for the design of lightweight and optimized cellulose-composites with enhanced mechanical properties. Furthermore, 3D printing allows for the incorporation of multiple materials into a single printed object. This capability is particularly relevant for cellulose-composites, as it enables the combination of cellulose with other reinforcing materials or fillers. By incorporating materials such as nanocellulose, carbon nanotubes, or metal nanoparticles, the mechanical, electrical, or thermal properties of the cellulose-composites can be further enhanced. For example, the addition of nanocellulose can significantly increase the stiffness and strength of the printed objects, while the incorporation of conductive materials can enable the development of cellulose-based electronics.

To illustrate the application of 3D printing in cellulose-composites, let's consider the fabrication of a cellulose-based scaffold for tissue engineering. Tissue engineering aims to create functional tissues or organs by combining cells with supportive scaffolds. In this case, cellulose and a biocompatible polymer, such as polycaprolactone (PCL), can be blended to form a composite ink suitable for 3D printing. The cellulose provides structural support and biocompatibility, while the PCL enhances the mechanical properties and printability of the scaffold. The 3D printing process begins with the preparation of the composite ink. Cellulose and PCL are dissolved in a compatible solvent and mixed thoroughly to ensure a homogeneous blend. The ink is then loaded into a syringe or a deposition system attached to the 3D printer. The printer follows a digital design, depositing the ink layer by layer to create the desired scaffold structure. During printing, the solvent evaporates, allowing the cellulose and PCL to solidify and bond together, forming a continuous scaffold structure. The printing parameters, such as printing speed, layer thickness, and temperature, can be optimized to ensure good printability and mechanical integrity of the scaffold. Post-processing steps, such as drying or annealing, may be required to further enhance the mechanical properties of the printed object. The resulting cellulose-PCL scaffold exhibits excellent biocompatibility and mechanical properties, making it suitable for tissue engineering applications. The cellulose component provides a bioactive and biomimetic environment for cell attachment and growth. The 3D-printed scaffold can be designed with specific porosity and pore size, allowing for nutrient and waste exchange within the tissue construct. Furthermore, the mechanical properties of the scaffold can be tailored to match the target tissue, ensuring proper support and functionality.

Despite the advantages, there are challenges and limitations associated with 3D printing of cellulose-composites that need to be addressed. One challenge is the compatibility of the materials used in the printing process. Cellulose-based inks or filaments must be carefully formulated to ensure proper rheological properties and printability. The viscosity, shear thinning behavior, and drying kinetics of the ink need to be optimized to achieve accurate deposition and bonding between layers. Another challenge is the limited mechanical properties of the printed cellulose-composites. Compared to traditional manufacturing methods, 3D printing often results in lower mechanical strength and stiffness due to the layer-by-layer deposition process. The anisotropic nature of the printed objects, with weaker interlayer bonding, can lead to reduced mechanical performance in certain directions. Strategies such as the incorporation of reinforcing fibers, optimization of printing parameters, or post-printing treatments (e.g., annealing) can be employed to enhance the mechanical properties of the printed cellulose-composites. Furthermore, the scalability of 3D printing for cellulose-composites is another aspect that needs consideration. While 3D printing offers high design flexibility and customization, it may still face limitations in terms of production volume and speed. The time required for printing complex structures or large objects can be significantly longer compared to traditional manufacturing methods. The development of faster printing techniques, such as continuous printing or multi-nozzle systems, can help address these limitations.

16.3.4 Corrosion Inhibition Efficiency of Cellulose-Composites

Cellulose-composites offer enhanced corrosion inhibition efficiency compared to pure cellulose due to the synergistic effects between cellulose and the incorporated reinforcements or fillers. The improved mechanical and barrier properties of the composites contribute to their corrosion resistance performance. The mechanical reinforcement provided by cellulose fibers or nanoparticles enhances the strength and integrity of the composite, making it more resistant to mechanical damage that can promote corrosion initiation. Additionally, the presence of reinforcements can improve the toughness and ductility of the composite, preventing crack propagation and enhancing its overall durability in corrosive environments.

The barrier properties of cellulose-composites are improved by the inclusion of reinforcements and fillers, which hinder the diffusion of corrosive species toward the metal surface. The dense network structure created by cellulose fibers or nanoparticles acts as a physical barrier, reducing the permeability of corrosive agents and slowing down the corrosion process. Furthermore, the incorporation of specific fillers, such as metal oxide nanoparticles or carbon-based materials, can provide additional corrosion protection mechanisms. Metal oxide nanoparticles can release inhibiting ions, form protective films, or act as sacrificial anodes, while carbon-based materials can provide barrier properties and prevent the formation of localized corrosion cells.

The corrosion inhibition efficiency of cellulose-composites can be evaluated through various techniques, including electrochemical tests, such as potentiodynamic polarization and EIS, and accelerated corrosion tests, such as salt spray or immersion tests. These evaluations assess parameters such as corrosion current density, polarization resistance, and the formation of corrosion products. For instance, in an experiment, Gan et al. [36] synthesized a novel inhibitor called borated aminated cellulose citrate (BACC) using citric acid, bagasse cellulose, diethanolamine, and boric acid through mechanical activation-assisted solid-phase reaction (MASPR) technology. The study revealed that BACC primarily acted as a mixed-type inhibitor in the simulated cooling water, with a greater impact on the anodic reaction. Furthermore, the inhibition efficiency of BACC increased with higher BACC concentrations. BACC also exhibited excellent scale inhibition properties in the simulated cooling water. The weight loss and electrochemical tests confirmed the inhibitory effects of

BACC on A3 steel corrosion in the simulated cooling water system. With a BACC concentration of 200 mg/L, the inhibition efficiency reached 88.33% in the simulated cooling water. Polarization curves indicated that BACC acted as a mixed-type inhibitor and mainly influenced the anodic reaction. The thermodynamic parameter ΔG^0 ads suggested that the adsorption of BACC on the A3 steel surface followed the Langmuir adsorption isotherm and involved chemisorption. SEM and X-ray photoelectron spectroscopy (XPS) analyses further supported the formation of a protective film composed of adsorbed BACC molecules on the A3 steel surface, guarding against corrosive attack. Additionally, BACC demonstrated a promising scale inhibition effect in the simulated cooling water, achieving an inhibition rate of 91.57% with a BACC concentration of 150 mg/L at 70°C. SEM and X-ray diffraction (XRD) results indicated that BACC influenced the particle size and surface morphology of the $CaCO_3$ scale deposits. The scale inhibition mechanism of BACC appeared to involve hindering the formation of $CaCO_3$ scale deposits through threshold inhibition and modification of microcrystal growth. These comprehensive findings establish BACC as a potential and effective inhibitor for corrosion and scale control in cooling water systems.

Similarly, Azani et al. [7] compared the effects of cellulose nanocrystal (CNC) fillers on the chemical, mechanical, and corrosion properties of epoxy-Zn protective coatings for mild steel in a 3.5% NaCl solution. The researchers investigated the potential of oil palm frond cellulose nanocrystals (OPF-CNC) as nanofillers to enhance the barrier protection of epoxy-Zn coatings and reduce corrosion. Epoxy-Zn coatings with varying ratios of OPF-CNC (0.5–10.0 wt.%) were prepared and analyzed. ATR analysis revealed specific peaks at 3,450 and 1,030 cm^{-1}, indicating the formation of a crosslinking network between the epoxy group and the OPF-CNC group. The inclusion of 0.5 wt.% OPF-CNC significantly increased the hardness of the epoxy-Zn coating by 36% compared to a commercial coating. The anti-corrosion performance of the coatings exposed to 3.5 wt.% NaCl solution was evaluated using electrochemical measurements, salt spray tests, and AFM analyses. OCP measurements confirmed that the mild steel coated with 0.5 wt.% OPF-CNC exhibited the least susceptibility to corrosion, providing the highest level of corrosion protection. Impressively, the coated mild steel demonstrated remarkable corrosion inhibition even under high-temperature conditions, achieving performance levels of 93%–99%. After a 24-hour salt spray test, the coated mild steel exhibited minimal rust formation and no blistering. The improved corrosion resistance was supported by AFM analysis, which showed low surface roughness after a 7-day immersion period.

Arukalam et al. [5] studied the corrosion inhibition and adsorption behavior of ethyl hydroxyethyl cellulose on mild steel corrosion. The study utilized weight loss measurement, EIS, polarization, and quantum chemical calculation techniques to analyze the corrosion inhibition of mild steel in a 1.0 M H_2SO_4 solution with ethyl hydroxyethyl cellulose. The results demonstrated that the corrosion reaction was inhibited by EHEC in the acidic environment, and the effectiveness of inhibition increased with the concentration of EHEC. Additionally, the presence of iodide ions further enhanced the inhibition efficiency through a synergistic effect. The adsorption of EHEC on the corroding metal surface was confirmed through impedance analysis, which followed a modified Langmuir isotherm. The researchers also employed density functional theory to analyze the adsorption characteristics of EHEC at a molecular level, considering the frontier molecular orbitals (HOMO and LUMO) as well as the local reactivity of the EHEC molecule. Figure 16.3 displays the geometry-optimized structures of EHEC, as well as the HOMO and LUMO orbitals, Fukui functions, and the total electron density. Within the EHEC molecule, the HOMO orbital exhibits saturation around the aromatic nucleus, which corresponds to the region of the highest electron density. This region is often susceptible to electrophilic attacks and serves as the active center with a strong bonding potential to the metal surface. On the other hand, the LUMO orbital is saturated around the ethoxy function and signifies the site where nucleophilic attacks take place.

Sangeetha et al. [67] studied the corrosion-mitigating potential of AHEC upon mild steel in 1 M HCl. The concentration of AHEC positively influenced the inhibition efficiency, with a maximum of 93% achieved at room temperature for 900 ppm of AHEC. Notably, AHEC exhibited higher inhibition efficiency compared to other cellulose derivatives tested thus far. However, as the temperature increased from 303 to 323 K, the protective ability of AHEC diminished. Polarization studies indicated that AHEC acted as a mixed-type inhibitor. The adsorption of the inhibitor on the mild steel surface followed the Frumkin adsorption isotherm, predominantly through physical interactions. Impedance plots revealed that the charge transfer resistance

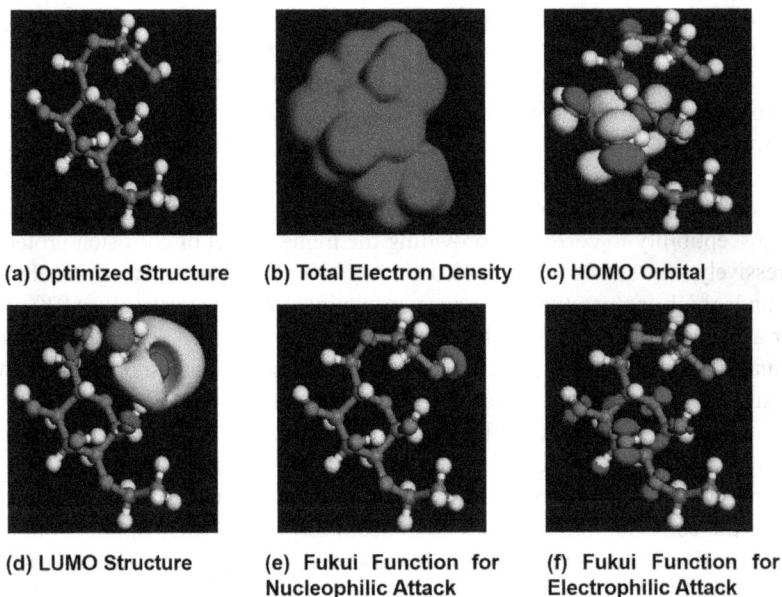

(a) Optimized Structure   (b) Total Electron Density   (c) HOMO Orbital

(d) LUMO Structure   (e) Fukui Function for   (f) Fukui Function for
                               Nucleophilic Attack   Electrophilic Attack

FIGURE 16.3   Electronic properties of ethyl hydroxyethyl cellulose (EHEC) [C, gray; H, white; O, red]. Adapted with permission from Ref. [5]. Distributed under CCBY 4.0.

increased with higher inhibitor concentrations. The formation of a protective layer by the inhibitor on the mild steel surface was confirmed through SEM and AFM studies.

Similarly, Youssef et al. [68] focused on the cost-effective and straightforward synthesis of a biodegradable and biocompatible hydrogel, namely chitosan/hydroxyl ethyl cellulose/polyaniline loaded with graphene oxide doped by silver nanoparticles (CS/HEC/PAni/GO@Ag) bionanocomposite, for energy storage applications. SEM analysis confirms the compatibility of chitosan, hydroxyl ethyl cellulose, and polyaniline, as well as the uniform distribution of GO@Ag-NPs within the bionanocomposite hydrogels. XRD provides evidence of the structural presence of GO@Ag-NPs in the matrix. Increasing the content of GO@Ag-NPs slightly enhances the swelling percentage and antibacterial activity. The combination of chitosan and cellulose further promotes the biodegradation of the fabricated bionanocomposites, which is augmented by the addition of graphene oxide (GO). Furthermore, the incorporation of 5% GO@Ag-NPs significantly improves the dc-conductivity of the hydrogels, increasing it by approximately 25 times from 3.37×10^{-3} to 8.53×10^{-2} S/cm. These fabricated hydrogels possess affordability, environmental friendliness, high capacitance, and permittivity, enabling them to efficiently store electrical energy.

Gouda et al. [69] addressed the issue of corrosion, which is a significant problem for alloys and metals, causing damage to buildings, machinery, and industrial systems such as petroleum refineries. The CeO_2-nanoparticle-loaded CMC was synthesized using a simple method, and its structural configuration was characterized using FE-SEM/EDS, TEM, FT-IR, and thermal analyses. The corrosion protection performance of mild steel coated with CeO_2-CMC films in 1.0 M HCl solutions was evaluated using EOCP-time, EIS, and PDP techniques. In addition, the relationship between the structure of the coating films and their corrosion protection was analyzed through DFT calculation and MC simulation. The experimental results and computational analysis confirmed that the CeO_2-CMC-coated films exhibited high corrosion resistance. To analyze the chemical composition of the composite particles, EDS-SEM analysis was conducted on CMC and CMC-CeO_2, as depicted in Figure 16.4a and b, respectively. In the case of CMC prior to the addition of CeO_2 particles, the atomic percentages of O and C were found to be 58.99% and 41.01%, respectively. Following the incorporation of CeO_2 particles, the atomic percentages of O, C, and Ce were observed to be 63.41%, 35.21%, and 1.38%, respectively. The decrease in carbon content and the increase in oxygen content suggest a chemical transformation that involves the inclusion of CeO_2 oxide.

Additionally, the presence of ceria in the coatings improved the protection capacity, with an increase of up to 3% achieving 98.4% efficiency. The DFT calculation and MC simulation further supported the influence of the chemical composition of the coated films on their corrosion protection, aligning with the experimental observations. Figure 16.5 clearly shows that the HOMO level of the CeO_2-CMC molecule is located on the nitrogen and oxygen atoms, as depicted in Figure 16.5. These specific sites are favorable for electrophilic attacks on the surface of mild steel (MS). Additionally, the energy gap (ΔE) plays a crucial role in enhancing the corrosion protection capability of the inhibitor molecule. A smaller ΔE value leads to an increased protection ability. The CeO_2-CMC molecule exhibits a lower ΔE

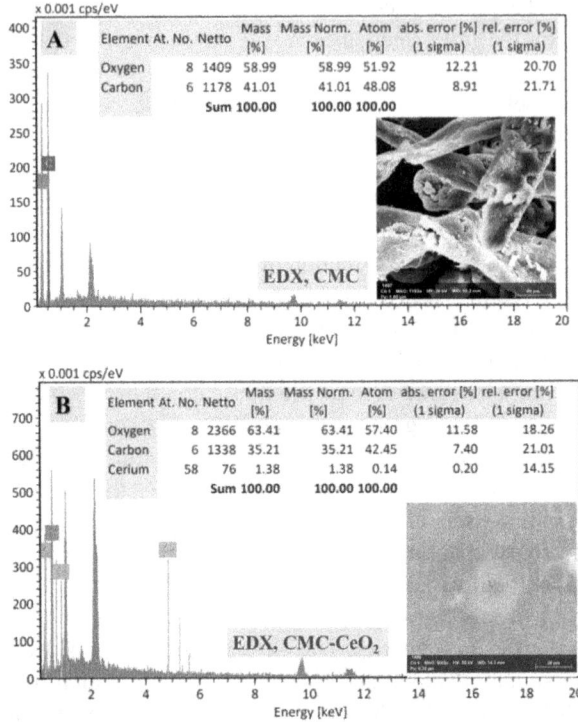

FIGURE 16.4   EDS of CMC (a) and CMC-CeO2 (b), and the inset is the chemical elemental analysis based on EDS peak analysis in addition to the EDS investigated area. Adapted with permission from Ref. [69]. Distributed under CCBY 4.0.

value (0.18 eV) compared to the CeO_2 and CMC molecules. This indicates a strong tendency for CeO_2-CMC molecules to be adsorbed onto the steel interface.

16.4   MECHANISMS OF CORROSION INHIBITION BY CELLULOSE-COMPOSITES

16.4.1   Interactions Between Cellulose and Metal Surfaces

The interactions between cellulose and metal surfaces play a crucial role in the corrosion inhibition properties of cellulose-based composites. Understanding these interactions is essential for designing effective corrosion protection strategies and enhancing the performance of cellulose-composites. When cellulose comes into contact with a metal surface, several types of interactions can occur. One of the primary interactions is the physical adsorption of cellulose molecules onto the metal surface. This adsorption is driven by intermolecular forces, such as van der Waals forces and hydrogen bonding, between the hydroxyl groups on cellulose and the metal surface. The physical adsorption of cellulose creates a protective layer on the metal surface, inhibiting the direct contact between the metal and corrosive agents in the

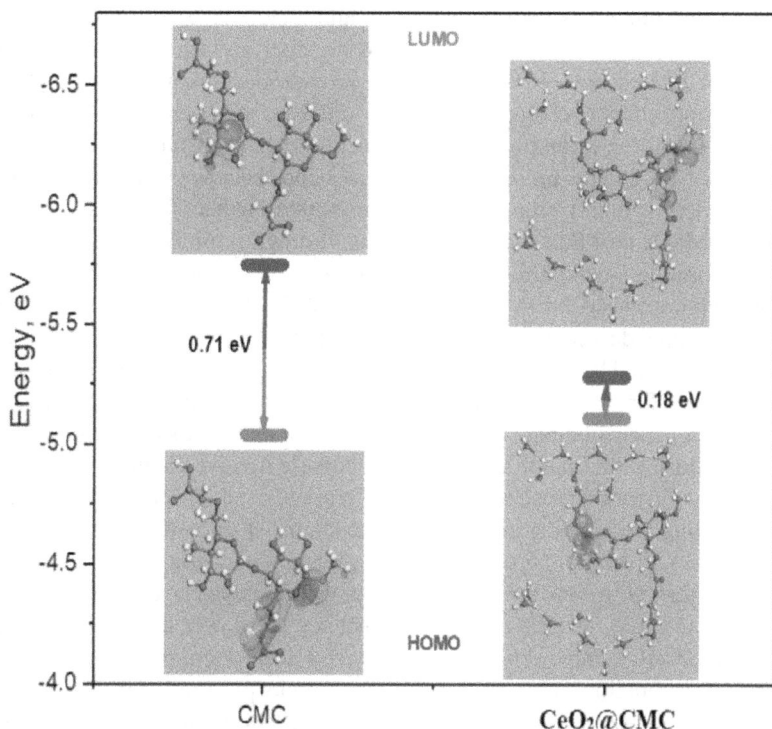

FIGURE 16.5 Energy diagram of the Frontier molecular orbitals for the CMC and CeO2-CMC molecules utilizing the DFT method. Adapted with permission from Ref. [69]. Distributed under CCBY 4.0.

environment. Chemical interactions between cellulose and metal surfaces can also occur. Cellulose contains reactive functional groups, such as hydroxyl and carboxyl groups, which can form chemical bonds with metal atoms or oxides on the metal surface. For example, cellulose can chelate metal ions, forming stable complexes that inhibit corrosion. These chemical interactions provide additional protection to the metal surface and enhance the corrosion inhibition properties of cellulose.

The formation of a hydrate film on the metal surface is another important interaction between cellulose and metals. Due to its hydrophilicity, cellulose has a strong affinity for water molecules. When cellulose comes into contact with a metal surface, it can absorb and retain water, forming a hydrate film. This film acts as a physical barrier, preventing the diffusion of corrosive species to the metal surface and reducing the corrosion rate. The interactions between cellulose and metal surfaces can be influenced by various factors. The surface characteristics of the metal, such as roughness, cleanliness, and oxide layer composition, can affect the adsorption and bonding of cellulose molecules. The chemical composition and structure of cellulose, including the degree of polymerization, crystallinity, and functional groups, also play a role in the interactions. Additionally, environmental factors, such as temperature, humidity, and pH, can influence the interactions between cellulose and metal surfaces. The interactions between cellulose and metal surfaces have been studied using various characterization techniques. Surface analysis techniques, such as SEM, atomic force microscopy (AFM), and XPS, provide insights into the morphology, topography, and chemical composition of the cellulose-metal interfaces. These techniques can reveal the formation of protective films, the distribution of cellulose on the metal surface, and the presence of chemical bonds between cellulose and metals. Wang et al. [16] synthesized a series of poly (methyl methacrylate) and poly (butyl acrylate) (PMMA-co-PBA) copolymer latexes through mini emulsion polymerization, incorporating 1-octyl-3-methylimidazolium hexafluorophosphate (C8mimPF6), cellulose nanocrystals (CNCs), and C8mimPF6-CNCs. These newly developed composites were applied as coatings on mild steel panels, and their anti-corrosion performance was evaluated by immersing the coated samples in a 3.5 wt.% sodium chloride (NaCl) solution for a specified duration. Based on the aforementioned results, the synergistic anti-corrosion mechanism of C8mimPF6 and CNCs can be summarized as follows. During the initial stage of immersion, corrosive elements like oxygen, moisture, and Cl⁻ ions diffuse through the micropores and defects of the copolymer coating, leading to oxidation and reduction reactions at the interface with the mild steel. These reactions can induce pitting, delamination of coatings, and failure of the protective effect. However, the introduction of CNCs enhances the barrier effect of the coating, resulting in higher coating resistance and lower diffusion coefficient. Although the CNC sample shows increased water absorption, most of the absorbed water moisture remains within the coating, limiting its penetration through the coating layer. This behavior is similar to other nanofillers, such as nanosilica, which obstruct the diffusion of water moisture and corrosive ions. Despite the improved barrier effect, some water moisture and corrosive ions can still pass through the coating, accumulating at the interface and potentially causing corrosion. As the coating degrades, encapsulated C8mimPF6 is released, and the C8mim⁺ cations replace Na⁺ ions, effectively blocking corrosion sites on the mild steel surface. In summary, CNCs reinforce the barrier effect of the coating, while C8mimPF6 inhibits corrosion at the coating-mild steel interface. As a result, the C8mimPF6-CNCs composite coating demonstrates superior protection compared to other samples, exhibiting the highest corrosion inhibition efficiency, lowest diffusion coefficient, and corrosion rate as shown in Figure 16.6.

Understanding the interactions between cellulose and metal surfaces is crucial for the development and optimization of cellulose-based composites for corrosion inhibition. By tailoring the properties of cellulose, such as its degree of polymerization, surface functionalization, and compatibility with metal surfaces, the interactions can be optimized to enhance the corrosion protection properties. Furthermore, the incorporation of reinforcements and fillers into cellulose matrices can modify the interactions and improve the overall performance of cellulose-composites in inhibiting corrosion.

16.4.2 ROLE OF FUNCTIONAL GROUPS IN CORROSION INHIBITION

Functional groups play a crucial role in the corrosion inhibition properties of organic compounds, including cellulose-based materials. These functional groups can interact with metal surfaces, alter the electrochemical processes involved in

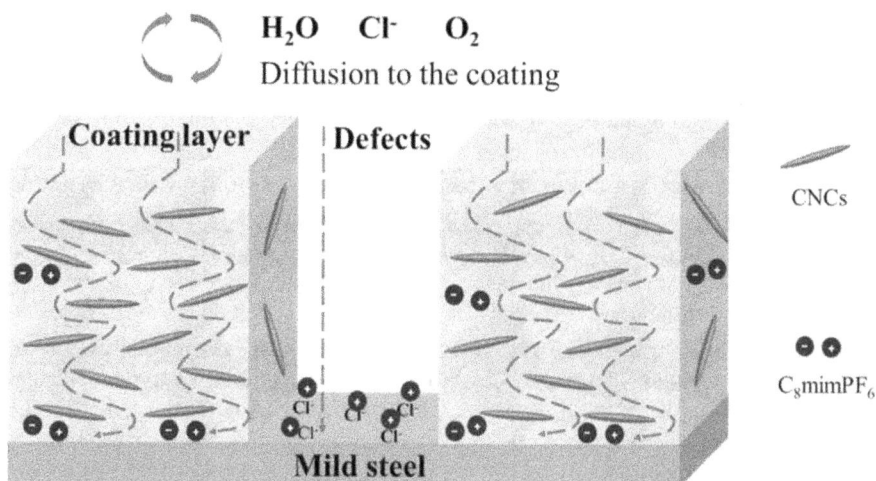

corrosion, and form protective films that hinder the access of corrosive species to the metal surface. Understanding the role of functional groups is essential for designing effective corrosion inhibitors and optimizing the performance of cellulose-based composites for corrosion protection. Hydroxyl (-OH) groups are one of the most common functional groups found in cellulose and many other organic compounds. These groups are highly polar and can form hydrogen bonds with metal atoms or oxides on the metal surface. Hydroxyl groups are capable of physical adsorption onto metal surfaces, creating a protective layer that inhibits the access of corrosive agents to the metal. The presence of hydroxyl groups also enhances the hydrophilicity of cellulose, allowing it to form hydrate films and retain moisture, which further slows down the corrosion process.

Carboxyl (-COOH) groups are another important functional group involved in corrosion inhibition. Carboxyl groups can form coordination complexes with metal ions, chelating them and inhibiting their corrosive activity. The coordination between carboxyl groups and metal ions can result in the formation of stable protective films on the metal surface, preventing further corrosion. Carboxyl groups also have acidic properties, which can contribute to the formation of passivation layers on metal surfaces, reducing the corrosion rate. Amino (-NH$_2$) groups are functional groups that can act as corrosion inhibitors through their ability to coordinate with metal ions and form protective films. Amino groups can also act as proton acceptors or donors, altering the pH of the surrounding environment and influencing the corrosion reactions. They can also promote the formation of passivation layers on metal surfaces, reducing the corrosion rate.

In addition to these common functional groups, other functional groups present in cellulose or incorporated into cellulose-based composites can contribute to corrosion inhibition. For example, sulfonic acid (-SO$_3$H) groups have been used as functional groups in cellulose-based materials for corrosion protection. Sulfonic acid groups can donate protons and lower the pH of the metal surface, promoting passivation and inhibiting corrosion. They can also form coordination complexes with metal ions, providing further protection. The role of functional groups in corrosion inhibition can be further enhanced by modifying their chemical structure and introducing specific substituents. For example, introducing electron-withdrawing groups, such as nitro (-NO_2) or cyano (-CN) groups, can enhance the inhibiting effect by increasing the electron density on the metal surface and promoting the formation of protective films. On the other hand, electron-donating groups, such as alkyl (-CH_3) or alkoxy (-OR) groups, can enhance the adsorption properties of the functional groups onto the metal surface. It is important to note that the effectiveness of functional groups in corrosion inhibition depends on their compatibility with the metal surface, the nature of the corrosive environment, and the overall composition of the cellulose-based composite. The concentration, distribution, and accessibility of the functional groups within the composite also play a significant role in their corrosion inhibition properties.

In conclusion, functional groups in cellulose-based materials play a crucial role in corrosion inhibition by interacting with metal surfaces, altering electrochemical processes, and forming protective films. Hydroxyl, carboxyl, amino, sulfonic acid, and other functional groups contribute to the corrosion inhibition properties of cellulose-based composites. The chemical structure and substituents of these functional groups can be tailored to enhance their inhibiting effects. Understanding the role of functional groups is essential for designing effective corrosion inhibitors and optimizing the performance of cellulose-based composites in various corrosion protection applications [70–72].

16.4.3 INTERFACIAL INTERACTIONS IN CELLULOSE-COMPOSITE SYSTEMS

Interfacial interactions play a critical role in the performance and properties of cellulose-composite systems. These interactions occur between cellulose and the reinforcing materials or fillers incorporated into the composite, and they strongly influence the mechanical, thermal, and chemical behavior of the composite. Understanding the interfacial interactions is essential for tailoring the properties of cellulose-composites and optimizing their performance in various applications. In cellulose-composite systems, the interfacial interactions can be categorized into three main types: physical interactions, chemical interactions, and mechanical interactions.

- Physical interactions involve the physical bonding or entanglement between cellulose and the reinforcing materials or fillers. In cellulose fiber-reinforced composites, for example, the cellulose fibers can mechanically interlock with the matrix material, enhancing the load transfer between the fibers and the matrix. This interlocking mechanism improves the composite's mechanical strength and stiffness, preventing fiber debonding and improving the overall structural integrity. Physical interactions also contribute to the barrier properties of cellulose-composites by hindering the diffusion of corrosive species through the composite structure.
- Chemical interactions occur when chemical bonds form between cellulose and the reinforcing materials or fillers. These chemical bonds can be

covalent, ionic, or hydrogen bonds, depending on the nature of the interacting groups. For example, in cellulose nanoparticle-reinforced composites, cellulose nanoparticles can form hydrogen bonds with the matrix material, leading to strong interfacial adhesion and improved mechanical properties. Chemical interactions at the interface can enhance the overall performance of cellulose-composites by improving the compatibility between the components and facilitating stress transfer across the interface.

- Mechanical interactions refer to the mechanical forces acting at the interface between cellulose and the reinforcing materials or fillers. These forces can arise from differences in thermal expansion coefficients, resulting in thermal stresses at the interface. When the composite is subjected to thermal cycling, these stresses can lead to interfacial delamination or debonding, compromising the mechanical performance of the composite. To mitigate such issues, the selection of compatible materials with similar thermal expansion coefficients is crucial to ensure good interfacial adhesion and minimize the occurrence of mechanical failures.

The interfacial interactions in cellulose-composite systems can be further influenced by surface modifications and treatments. Surface modifications involve altering the surface characteristics of cellulose or the reinforcing materials/fillers to enhance their compatibility and interfacial adhesion. For example, cellulose fibers can be chemically treated to introduce functional groups that can react with the matrix material, improving the bonding strength at the interface. Similarly, surface modifications of reinforcing materials or fillers can involve the addition of coupling agents or surface treatments to promote chemical bonding or enhance the mechanical interlocking between the components. The interfacial interactions in cellulose-composites can be studied and characterized through various analytical techniques. Surface analysis techniques, such as SEM and AFM, can provide valuable insights into the morphology, roughness, and interfacial bonding of the composite. Additionally, spectroscopic techniques, such as Fourier-transform infrared spectroscopy (FTIR) and XPS, can be employed to investigate the chemical interactions and bonding at the interface. The optimization of interfacial interactions in cellulose-composites is crucial for achieving desirable properties and performance. Proper control of processing parameters, including temperature, pressure, and curing time, can influence the interfacial adhesion and bonding strength. Additionally, the selection of suitable reinforcing materials or fillers with compatible surface properties and chemical structures is essential to ensure strong interfacial interactions.

16.5   ADVANTAGES AND LIMITATIONS OF CELLULOSE AS A CORROSION INHIBITOR

16.5.1   ADVANTAGES OF CELLULOSE AS A CORROSION INHIBITOR

- **Abundant and renewable**: One of the significant advantages of cellulose as a corrosion inhibitor is its abundance and renewability. Cellulose is the most abundant biopolymer on Earth and can be sourced from various plant materials, such as wood, cotton, and hemp. Its availability in large quantities makes it a cost-effective and sustainable option for corrosion protection.

- **Biodegradable and environmentally friendly**: Cellulose is a biodegradable material, which means it can be broken down by natural processes without causing harm to the environment. This makes it an environmentally friendly alternative to synthetic corrosion inhibitors, which often have non-biodegradable components and can persist in the environment for extended periods. Cellulose-based corrosion inhibitors minimize the environmental impact and promote sustainability.

- **Non-toxic**: Cellulose is non-toxic and poses minimal health risks compared to many synthetic corrosion inhibitors. This is particularly beneficial in applications where corrosion protection is required in environments that may come into contact with food, water, or human/animal tissues. The non-toxic nature of cellulose makes it suitable for use in industries such as food processing, water treatment, and biomedical applications.

- **Versatility**: Cellulose can be modified and tailored to suit different corrosion protection needs. It can be processed into various forms, such as fibers, films, coatings, and composites, allowing for versatile applications. Cellulose can also be chemically modified to introduce specific functional groups that enhance its corrosion inhibition properties. This versatility makes cellulose adaptable to different industries and corrosion prevention requirements.

- **Barrier formation**: Cellulose has the ability to form hydrate films and create dense networks that act as physical barriers, preventing direct contact between the metal surface and corrosive agents in the environment. These barrier formations inhibit the diffusion of moisture, oxygen, and ions, which are essential for corrosion reactions. The barrier effect provided by cellulose contributes to its corrosion resistance and helps prolong the lifespan of metal structures.

- **Stability and durability**: Cellulose is a stable material that can withstand harsh environmental conditions, including exposure to UV radiation, moisture, and temperature fluctuations. It retains its structural integrity over time, ensuring long-term corrosion protection. Cellulose-based corrosion inhibitors exhibit good durability, making them suitable for applications that require extended protection against corrosion.

16.5.2 LIMITATIONS OF CELLULOSE AS A CORROSION INHIBITOR

- **Water sensitivity**: Cellulose has a strong affinity for water molecules and can absorb and retain moisture. While this hydrophilic property can contribute to the formation of hydrate films and enhance corrosion inhibition, it can also make cellulose susceptible to degradation in the presence of high moisture levels. Excessive exposure to water can lead to swelling, dimensional changes, and a decrease in mechanical properties, affecting the long-term effectiveness of cellulose-based corrosion inhibitors.

- **Limited solubility**: Native cellulose has limited solubility in common organic solvents, which can limit its processability and the formation of homogeneous coatings or films. This challenge can be addressed through chemical modifications or by using specific solvents or solvent systems that are compatible with cellulose. However, additional processing steps or modifications may be required to enhance the solubility and application potential of cellulose as a corrosion inhibitor.
- **Mechanical weakness**: Native cellulose exhibits relatively low mechanical strength and stiffness compared to synthetic polymers or metals. This mechanical weakness can limit the application of cellulose as a standalone corrosion inhibitor in high-stress environments or where mechanical properties are crucial. However, cellulose can be reinforced with other materials, such as fibers or nanoparticles, to enhance its mechanical performance and overcome this limitation.
- **Sensitivity to microbial attack**: Cellulose can be susceptible to microbial attack, particularly by certain bacteria and fungi that have enzymes capable of breaking down cellulose. Microbial degradation can weaken the protective barrier formed by cellulose and compromise its corrosion inhibition properties. To mitigate this limitation, appropriate measures such as biocides or additives can be employed to protect cellulose-based corrosion inhibitors from microbial attack.
- **Compatibility with metal surfaces**: The compatibility between cellulose and different metal surfaces can vary. While cellulose generally exhibits good compatibility with many metals, certain metals or alloys may require surface pre-treatment or modifications to ensure strong adhesion and effective corrosion inhibition. The interfacial interactions between cellulose and metal surfaces need to be carefully considered to maximize the corrosion protection provided by cellulose-based inhibitors.
- **Limited temperature resistance**: Cellulose has a relatively low thermal stability and can degrade at high temperatures. This limits its application as a corrosion inhibitor in environments where elevated temperatures are present, such as in high-temperature industrial processes. However, cellulose can be combined with heat-resistant additives or incorporated into composite systems to improve its temperature resistance and expand its application range.

16.6 CHALLENGES AND FUTURE PERSPECTIVES

16.6.1 Optimization of Cellulose-Composite Formulations

Optimization of cellulose-composite formulations is a critical aspect in harnessing the full potential of cellulose-based materials for various applications. Cellulose-composites, which combine cellulose with reinforcing materials and fillers, offer improved mechanical, barrier, and corrosion resistance properties. The optimization process involves carefully selecting the components, their proportions, and the processing techniques to achieve the desired properties and performance of the composite.

16.6.1.1 Selection of Components

The first step in the optimization process is the selection of suitable components for the cellulose-composite formulation. This includes choosing the type of cellulose, such as cellulose fibers or nanoparticles, and the reinforcing materials and fillers. The choice of components depends on the desired properties of the composite, the target application, and the compatibility between the components.

- **Cellulose**: Different sources of cellulose, such as wood, cotton, or agricultural residues, can be used. Each source has unique properties that influence the performance of the composite. For example, wood-based cellulose fibers may provide better mechanical strength, while cotton-based cellulose fibers may offer improved flexibility and moisture absorption properties. The selection of cellulose also depends on its availability, cost, and sustainability.
- **Reinforcing materials**: Various reinforcing materials, including natural fibers (e.g., jute, sisal), synthetic fibers (e.g., glass, carbon), and nanoparticles (e.g., metal oxides, carbon nanotubes), can be incorporated into the cellulose matrix. The choice of reinforcing material depends on the desired mechanical properties, such as tensile strength, stiffness, and toughness. Each reinforcing material has its own advantages and limitations, and the selection should be based on the specific requirements of the application.
- **Fillers**: Fillers, such as nanoparticles or microparticles, can be added to further enhance the properties of cellulose-composites. For example, metal oxide nanoparticles can provide additional corrosion protection, while carbon-based fillers can improve electrical conductivity or barrier properties. The selection of fillers depends on the desired functionalities and the compatibility with the cellulose matrix.

16.6.1.2 Proportions of Components

The proportions of cellulose, reinforcing materials, and fillers in the composite formulation significantly impact the final properties of the composite. The optimization process involves determining the optimal ratio or concentration of each component to achieve the desired mechanical, barrier, and corrosion resistance properties. The proportions can be determined through experimental trials or modeling techniques. Various characterization techniques, such as microscopy, spectroscopy, and mechanical testing, can be employed to evaluate the effect of different component ratios on the composite properties. The aim is to find the optimum balance that maximizes the synergistic effects between cellulose and the reinforcing materials while ensuring good dispersion and interfacial adhesion between the components.

16.6.1.2.1 Processing Techniques

The choice of processing technique is crucial in optimizing cellulose-composite formulations. Different processing techniques, such as solution casting, melt compounding, extrusion, or compression molding, can be employed based on the type of matrix material and the desired composite structure.

- **Solution casting**: This technique involves dissolving cellulose and other components in a solvent, followed by casting the solution into a mold and drying it to form a solid composite. Solution casting allows for precise control over the composition and dispersion of the components. It is particularly suitable for small-scale production and research purposes.
- **Melt compounding**: Melt compounding involves mixing cellulose, reinforcing materials, and fillers in a molten state, followed by cooling and solidification. This technique is commonly used for thermoplastic matrix materials. It enables large-scale production of cellulose-composites and is well-suited for continuous manufacturing processes.
- **Extrusion**: In extrusion, a molten cellulose matrix with incorporated reinforcements or fillers is forced through a die, resulting in a continuous composite profile. This technique is advantageous for producing cellulose-composites with complex shapes and continuous lengths. Extrusion provides good control over the alignment and distribution of the reinforcing materials, thereby enhancing the mechanical properties of the composite.
- **Compression molding**: Compression molding involves applying heat and pressure to a mixture of cellulose, reinforcing materials, and fillers in a mold to consolidate the composite. It is suitable for producing cellulose-composites with controlled thickness and high density. Compression molding allows for the fabrication of complex shapes and provides good control over the interfacial adhesion between the components.

16.6.1.3   Characterization and Evaluation

After the optimization of cellulose-composite formulations, thorough characterization and evaluation are essential to assess the properties and performance of the composite. Various techniques can be employed to evaluate the mechanical, barrier, and corrosion resistance properties.

- **Mechanical testing**: Tensile, flexural, and impact testing can be performed to assess the mechanical properties, such as strength, stiffness, toughness, and ductility, of the cellulose-composites. These tests provide information about the composite's structural integrity and its ability to withstand mechanical stress and deformation.
- **Barrier property evaluation**: The barrier properties of cellulose-composites can be evaluated by measuring parameters such as water vapor transmission rate (WVTR) and oxygen permeability. These tests assess the ability of the composite to hinder the diffusion of moisture and corrosive species, thereby reducing the corrosion rate.
- **Corrosion resistance testing**: Various corrosion resistance tests, such as electrochemical tests (e.g., potentiodynamic polarization, EIS) and accelerated corrosion tests (e.g., salt spray, immersion tests), can be conducted to evaluate the corrosion inhibition efficiency of the cellulose-composites. These tests assess parameters such as corrosion current density, polarization resistance, and the formation of corrosion products.

- **Durability and long-term performance**: The long-term stability and performance of cellulose-composites need to be evaluated to ensure their suitability for practical applications. Factors such as moisture absorption, thermal stability, UV resistance, and resistance to microbial attack can impact the durability and effectiveness of the composites. Long-term exposure tests or accelerated aging tests can be conducted to assess the composite's performance under realistic conditions.

16.6.2 Exploration of Novel Cellulose Sources

Cellulose, as the most abundant biopolymer on Earth, can be sourced from various plant materials such as wood, cotton, and hemp. However, there is ongoing research and exploration into novel cellulose sources to expand the availability and utilization of this versatile material. The exploration of novel cellulose sources offers several advantages, including sustainable and renewable options, improved properties, and potential applications in various industries.

- **Agricultural residues**: One promising area of exploration is the utilization of agricultural residues as a source of cellulose. Agricultural residues include materials such as rice straw, wheat straw, corn husks, and sugarcane bagasse. These residues are generated in large quantities worldwide and are often considered as waste or burned, leading to environmental issues. However, they contain a significant amount of cellulose that can be extracted and utilized for various applications. By converting agricultural residues into cellulose-based materials, it is possible to transform waste streams into valuable resources and contribute to a more sustainable and circular economy.
- **Algae**: Algae are another potential source of cellulose that is currently being explored. Algae are microorganisms that grow in water and can be cultivated in large-scale photobioreactors or open ponds. They have a high growth rate and can be harvested multiple times per year, making them a renewable and fast-growing source of cellulose. Algae-derived cellulose has unique properties and can be used in applications such as biofuels, bioplastics, and biocomposites. Furthermore, algae cultivation has the potential to mitigate carbon dioxide emissions and contribute to environmental sustainability.
- **Bacterial and microbial cellulose**: Bacterial cellulose (BC) and microbial cellulose are produced by certain bacteria, such as Acetobacter xylinum. Unlike plant-based cellulose, BC is produced through a fermentation process in which bacteria synthesize and secrete cellulose fibrils. BC has a highly pure and uniform structure, excellent mechanical properties, and a high degree of polymerization, making it suitable for a wide range of applications. BC can be used in medical and biomedical fields, such as wound dressings, tissue engineering scaffolds, and drug delivery systems. The exploration of bacterial and microbial cellulose as a novel source offers unique properties and opportunities for innovation.

- **Marine resources**: Marine resources, such as seaweed and other marine algae, are being explored as potential sources of cellulose. Seaweeds are abundant in coastal regions and offer several advantages, including rapid growth rates, low nutrient requirements, and the ability to absorb carbon dioxide. Seaweed-derived cellulose has the potential to be used in applications such as food packaging, textiles, and biomedical materials. The utilization of marine resources for cellulose production not only diversifies the sources of cellulose but also promotes the sustainable use of marine ecosystems.
- **Genetically modified plants**: Genetic engineering techniques can be employed to modify plants to produce cellulose with enhanced properties. By manipulating the genes responsible for cellulose synthesis, it is possible to engineer plants that produce cellulose with desired characteristics, such as improved mechanical strength, increased crystallinity, or altered chemical composition. Genetically modified plants can also be designed to produce cellulose with specific functionalities, such as self-healing capabilities or responsiveness to external stimuli. The exploration of genetically modified plants offers opportunities for tailoring cellulose properties for specific applications.
- **Waste biomass**: Waste biomass from various industries, such as pulp and paper, agriculture, and forestry, can be explored as a source of cellulose. By utilizing waste biomass, it is possible to reduce waste generation, lower environmental impact, and create value-added products. Waste biomass can undergo extraction processes to obtain cellulose and then be processed into cellulose-based materials. The utilization of waste biomass as a cellulose source contributes to the circular economy concept and promotes sustainability.

The exploration of novel cellulose sources presents both opportunities and challenges. One challenge is the development of efficient extraction methods for cellulose from these sources. Different sources may require specific extraction techniques to obtain high-quality cellulose. Additionally, the scalability of production processes and the economic viability of utilizing novel cellulose sources need to be considered [15]. Furthermore, the properties and performance of cellulose derived from novel sources may differ from traditional cellulose sources. It is essential to characterize and understand the properties of cellulose obtained from different sources to optimize its utilization and tailor it for specific applications.

16.6.3 LONG-TERM STABILITY AND PERFORMANCE ENHANCEMENT

Long-term stability and performance enhancement are crucial factors in the development and utilization of cellulose-based materials. The durability and reliability of these materials over extended periods of time are essential for their successful implementation in various applications. Therefore, efforts are being made to improve the long-term stability and enhance the performance of cellulose-based materials through various strategies, including surface modifications, incorporation of additives, and optimization of processing techniques.

16.6.3.1   Surface Modifications

Surface modifications play a vital role in improving the long-term stability and performance of cellulose-based materials. By modifying the surface properties, it is possible to enhance the material's resistance to degradation, moisture absorption, and chemical attack. Surface modifications can be achieved through various techniques, including chemical treatments, coating applications, and grafting of functional groups.

- Chemical treatments involve the use of reagents to alter the surface properties of cellulose. For example, acetylation, esterification, or etherification can be used to introduce hydrophobic groups onto the cellulose surface, reducing its affinity for water and enhancing moisture resistance. These surface modifications can also improve the material's resistance to microbial attacks and biological degradation.
- Coating applications involve the deposition of protective layers onto the cellulose surface. These coatings can provide a physical barrier against environmental factors, such as UV radiation, moisture, and chemical substances. Examples of protective coatings include polymer films, metal oxide coatings, or nanocomposite coatings. These coatings can improve the material's resistance to degradation, enhance mechanical properties, and provide additional functionalities, such as antimicrobial properties or self-cleaning capabilities.
- Grafting of functional groups involves the attachment of specific chemical moieties onto the cellulose surface [73]. This technique allows for the introduction of desired properties, such as improved adhesion, enhanced compatibility with other materials, or responsiveness to external stimuli. Functional groups can be grafted onto cellulose through various methods, including chemical reactions, plasma treatments, or radiation-induced grafting.

16.6.3.2   Incorporation of Additives

The incorporation of additives is another approach to enhance the long-term stability and performance of cellulose-based materials. Additives can be introduced during the processing of cellulose-composites or incorporated into the cellulose matrix through blending or impregnation techniques. These additives can provide additional functionalities and improve the material's resistance to degradation, mechanical properties, and barrier properties.

- Antioxidants and stabilizers are commonly used additives to improve the long-term stability of cellulose-based materials. They inhibit or delay the degradation processes, such as oxidation or hydrolysis, which can lead to the deterioration of the material over time. Examples of antioxidants include hindered phenols, phosphites, or thioethers. These additives scavenge free radicals and prevent chain scission reactions, thereby enhancing the material's resistance to oxidative degradation.
- UV stabilizers are additives specifically designed to protect cellulose-based materials from UV radiation-induced degradation. They absorb or reflect

UV radiation, preventing its penetration into the material and reducing the degradation caused by UV exposure. UV stabilizers can be organic or inorganic compounds, such as benzotriazoles, benzophenones, or metal oxide nanoparticles.

- Flame retardants can be incorporated into cellulose-based materials to enhance their fire resistance. Flame retardants act by reducing the flammability and slowing down the combustion process. They can be divided into different classes, including halogenated compounds, phosphorus-based compounds, or intumescent systems. The incorporation of flame retardants improves the material's resistance to ignition and propagation of flames, ensuring its long-term safety.
- Nanoparticles, such as clay nanoparticles, carbon nanotubes, or graphene oxide, can be added to cellulose-composites to enhance their mechanical properties, barrier properties, and resistance to degradation. These nanoparticles can reinforce the cellulose matrix, improve the interfacial adhesion, and hinder the diffusion of corrosive species or moisture. The incorporation of nanoparticles can significantly enhance the long-term stability and performance of cellulose-based materials.

16.6.3.3 Optimization of Processing Techniques

Optimizing the processing techniques is essential for achieving long-term stability and performance enhancement of cellulose-based materials. The choice of processing technique, parameters, and conditions can influence the material's structure, morphology, and properties. By optimizing these factors, it is possible to improve the material's resistance to degradation, mechanical properties, and barrier properties. Processing techniques, such as compression molding, extrusion, solution casting, or electrospinning, can be optimized to control the orientation, alignment, and dispersion of cellulose fibers or nanoparticles within the matrix. Proper dispersion and alignment of reinforcements are crucial for achieving optimal mechanical properties, barrier properties, and resistance to degradation. The optimization of processing parameters, such as temperature, pressure, shear rate, or residence time, can ensure the uniform distribution of cellulose and additives, promote interfacial adhesion, and minimize defects or voids within the material. Furthermore, the optimization of drying and curing processes is crucial to prevent the occurrence of defects, such as cracks or warpages, and ensure the material's dimensional stability. Proper drying and curing conditions allow for the complete removal of solvents or moisture, leading to a stable and robust material structure [68,74,75].

16.6.3.4 Characterization and Testing

Characterization and testing techniques are essential for evaluating the long-term stability and performance of cellulose-based materials. These techniques provide insights into the material's structural, morphological, mechanical, and barrier properties, as well as its resistance to degradation. Several techniques such as SEM, transmission electron microscopy (TEM), XRD, Fourier-transform infrared spectroscopy (FTIR), or thermogravimetric analysis (TGA) can be used to analyze the material's structure, morphology, and chemical composition. Mechanical testing, such as tensile testing or flexural testing, allows for the evaluation of the material's mechanical properties, including strength, stiffness, and toughness. Barrier property testing, such as water vapor permeability or gas permeability measurements, assesses the material's resistance to moisture or gas transmission. Accelerated aging tests, such as accelerated weathering or accelerated corrosion tests, can simulate the long-term degradation processes and assess the material's durability under harsh conditions. By employing a combination of characterization and testing techniques, it is possible to gain a comprehensive understanding of the material's performance and make informed decisions regarding its optimization and improvement.

16.6.4 POTENTIAL APPLICATIONS IN INDUSTRIES

Cellulose-based materials offer a wide range of potential applications in various industries due to their unique properties, eco-friendliness, and biodegradability. The versatility of cellulose allows for its utilization in numerous fields, ranging from packaging and textiles to electronics and biomedical applications. The following are some of the key industries where cellulose-based materials have significant potential:

- **Packaging industry**: The packaging industry is one of the largest consumers of cellulose-based materials. Cellulose can be processed into films, coatings, or composites to create biodegradable and renewable packaging materials. These materials provide excellent barrier properties against moisture, gases, and UV radiation, ensuring the protection and preservation of packaged goods. Cellulose-based packaging materials can replace conventional petroleum-based plastics, reducing environmental pollution and waste.
- **Textile industry**: Cellulose fibers, such as cotton, viscose, or lyocell, have long been used in the textile industry. The demand for sustainable and eco-friendly textiles has led to an increased interest in cellulose-based fibers. Cellulose fibers offer excellent moisture absorption, breathability, and comfort properties. They can be used in the production of clothing, home textiles, and non-woven materials. Additionally, advancements in nanocellulose technology have led to the development of cellulose-based nanofibers, which exhibit exceptional strength, flexibility, and functional properties. These nanofibers have the potential to enhance the performance of textiles, including improved tensile strength, abrasion resistance, and antibacterial properties.
- **Construction and building materials**: Cellulose-based materials have promising applications in the construction industry. Cellulose fibers can be used as reinforcement in cementitious materials, such as concrete or mortar, to improve their mechanical properties and durability. The addition of cellulose fibers enhances the flexural and impact strength, crack resistance, and reduces shrinkage and permeability of the cementitious matrix. Moreover, cellulose-based insulation materials, such as cellulose fibers or foam, provide effective thermal and acoustic insulation for buildings.
- **Automotive and aerospace industries**: The automotive and aerospace industries are increasingly exploring the use of cellulose-based materials

to replace conventional petroleum-based materials. Cellulose-composites can be used in interior components, such as door panels, dashboards, or seat backs, to reduce the weight of vehicles and enhance fuel efficiency. Cellulose fibers can also be incorporated into polymer matrices to improve the mechanical properties, thermal stability, and fire resistance of automotive and aerospace components. Furthermore, the low density and biodegradability of cellulose-based materials make them attractive for sustainable manufacturing in these industries.

- **Electronics and energy storage**: Cellulose-based materials have potential applications in the electronics and energy storage sectors. Cellulose nanofibers can be used as a sustainable alternative to conventional carbon-based materials in energy storage devices, such as supercapacitors or lithium-ion batteries. The high surface area, electrical conductivity, and mechanical properties of cellulose nanofibers contribute to the improved performance of energy storage systems. Moreover, cellulose-based materials can be utilized in flexible electronics, printed circuit boards, or sensor devices due to their excellent dielectric properties, thermal stability, and biocompatibility.

- **Biomedical and pharmaceutical industries**: Cellulose-based materials find wide applications in the biomedical and pharmaceutical sectors. Cellulose nanocellulose hydrogels can be utilized as scaffolds for tissue engineering and regenerative medicine due to their biocompatibility, biodegradability, and structural similarity to the extracellular matrix. Cellulose fibers can be processed into wound dressings, drug delivery systems, or medical implants due to their excellent moisture absorption, mechanical strength, and antimicrobial properties. Additionally, the use of cellulose-based materials in pharmaceutical formulations, such as tablets or capsules, can improve drug stability, controlled release, and patient compliance.

- **Environmental remediation**: Cellulose-based materials have the potential to contribute to environmental remediation efforts. Cellulose fibers or nanoparticles can be used as adsorbents for the removal of pollutants, heavy metals, or dyes from wastewater. The large surface area and hydrophilic nature of cellulose facilitate the adsorption and sequestration of contaminants, contributing to water purification processes. Furthermore, cellulose-based materials can be utilized for oil spill cleanup, air filtration, or soil remediation applications.

Despite the numerous potential applications of cellulose-based materials, there are certain limitations and challenges that need to be addressed. These include:

- **Cost and scalability**: The cost of producing cellulose-based materials, especially those derived from nanocellulose, can be relatively high compared to traditional materials. Developing cost-effective production methods and scaling up the manufacturing processes are necessary to make cellulose-based materials economically viable for large-scale industrial applications.

- **Processing and compatibility**: Cellulose-based materials often require specialized processing techniques to achieve desired properties and performance. Compatibility with existing processing equipment, such as extruders or injection molding machines, may need to be addressed. Additionally, the interfacial compatibility between cellulose and other components in composite systems, such as polymers or additives, can influence the material's performance and stability.

- **Durability and long-term stability**: Ensuring the long-term stability and durability of cellulose-based materials is crucial for their practical applications. Factors such as moisture absorption, degradation under harsh conditions, or susceptibility to microbial attack can impact the material's performance and lifespan. Strategies for improving the long-term stability, such as surface modifications, incorporation of additives, or optimization of processing techniques, need to be explored.
- **Regulatory and certification requirements:** Cellulose-based materials need to meet regulatory and certification requirements in various industries, such as food packaging, medical devices, or automotive components. Compliance with safety, health, and environmental standards is essential to gain acceptance and widespread adoption of cellulose-based materials.

16.7 CONCLUSION

In conclusion, cellulose-composites have emerged as a promising class of materials for corrosion inhibition due to their incorporation of cellulose, a renewable and biodegradable polymer. These composites offer numerous advantages, including enhanced mechanical properties, improved barrier performance, and increased corrosion resistance. The synergistic effects between cellulose and the incorporated reinforcements or fillers further contribute to their corrosion inhibition efficiency. The mechanical reinforcement provided by cellulose fibers or nanoparticles is a key advantage of cellulose-composites. These reinforcements enhance the strength, toughness, and durability of the composites, preventing crack propagation and mechanical damage that can accelerate corrosion. This makes cellulose-composites suitable for structural applications where corrosion resistance is critical. The barrier properties of cellulose-composites play a vital role in corrosion inhibition. The incorporation of reinforcements and fillers creates a dense network structure that hinders the diffusion of corrosive species toward the metal surface, thus reducing the rate of corrosion. In addition, specific fillers, such as metal oxide nanoparticles or carbon-based materials, provide additional corrosion protection mechanisms, such as sacrificial anodes or the formation of protective films. These barrier properties make cellulose-composites effective in preventing the interaction between metals and corrosive environments.

Cellulose-composites also offer advantages in terms of sustainability and environmental impact. Cellulose, derived from renewable sources such as plants, can be easily obtained and processed into various forms. The biodegradability of cellulose-composites reduces the environmental burden associated with the disposal of conventional corrosion inhibitors. Furthermore, their use contributes to the reduction of greenhouse gas emissions and supports the transition toward a more sustainable and circular economy. However, there are challenges and limitations that need to be addressed to optimize and implement cellulose-composites as corrosion inhibitors. The formulation of cellulose-composites requires careful consideration of factors such as the selection of reinforcements and fillers, their compatibility with cellulose, and the processing techniques employed. Optimization of these formulations is crucial to achieve the desired properties and performance of the composites. Additionally, long-term stability studies and performance enhancement strategies, including surface modifications and the incorporation of additives, need to be investigated to ensure the durability and reliability of cellulose-composites over extended periods. Future research and development efforts should focus on overcoming these challenges and exploring the full potential of cellulose-composites as corrosion inhibitors. This includes the development of new combinations of cellulose sources, reinforcements, and fillers to enhance the performance of the composites. Exploration of novel cellulose sources, such as agricultural waste or cellulose from bacteria, can provide alternative and sustainable options for corrosion inhibition applications. Furthermore, investigating long-term stability and implementing performance enhancement strategies will be crucial for practical implementation in various industries.

REFERENCES

[1] Al Jahdaly BA, Elsadek MF, Ahmed BM, Farahat MF, Taher MM, Khalil AM. Outstanding Graphene Quantum Dots from Carbon Source for Biomedical and Corrosion Inhibition Applications: *A Review. Sustainability.* 2021;13 (4).

[2] Al Kiey SA, Hasanin MS, Dacrory S. Potential Anticorrosive Performance of Green and Sustainable Inhibitor Based on Cellulose Derivatives for Carbon Steel. *Journal of Molecular Liquids.* 2021;338:116604.

[3] Alharthi AF, Gouda M, Khalaf MM, Elmushyakhi A, Abou Taleb MF, Abd El-Lateef HM. Cellulose-Acetate-Based Films Modified with Ag(2)O and ZnS as Nanocomposites for Highly Controlling Biological Behavior for Wound Healing Applications. *Materials (Basel).* 2023;16 (2):777.

[4] Al-Masoud MA, Khalaf MM, Heakal FE, Gouda M, Mohamed IMA, Shalabi K, et al. Advanced Protective Films Based on Binary ZnO-NiO@polyaniline Nanocomposite for Acidic Chloride Steel Corrosion: An Integrated Study of Theoretical and Practical Investigations. *Polymers (Basel).* 2022;14 (21):4734.

[5] Arukalam IO, Madu IO, Ijomah NT, Ewulonu CM, Onyeagoro GN. Acid Corrosion Inhibition and Adsorption Behaviour of Ethyl Hydroxyethyl Cellulose on Mild Steel Corrosion. *Journal of Materials.* 2014;2014:1–11.

[6] Azani N, Haafiz MKM, Zahari A, Poinsignon S, Brosse N, Hussin MH. Preparation and Characterizations of Oil Palm Fronds Cellulose Nanocrystal (OPF-CNC) as Reinforcing Filler in Epoxy-Zn Rich Coating for Mild Steel Corrosion Protection. *International Journal of Biological Macromolecules.* 2020;153:385–398.

[7] Azani NFSM, Hussin MH. Comparison of Cellulose Nanocrystal (CNC) Filler on Chemical, Mechanical, and Corrosion Properties of Epoxy-Zn Protective Coatings for Mild Steel in 3.5% NaCl Solution. *Cellulose.* 2021;28 (10):6523–6543.

[8] Bao Y, He J, Song K, Guo J, Zhou X, Liu S. Functionalization and Antibacterial Applications of Cellulose-Based Composite Hydrogels. *Polymers (Basel).* 2022;14 (4):769.

[9] Calegari F, Sousa I, Ferreira MGS, Berton MAC, Marino CEB, Tedim J. Influence of the Operating Conditions on the Release of Corrosion Inhibitors from Spray-Dried Carboxymethylcellulose Microspheres. *Applied Sciences.* 2022;12 (4), 1800.

[10] Chen L, Lu D, Zhang Y. Organic Compounds as Corrosion Inhibitors for Carbon Steel in HCl Solution: A Comprehensive Review. *Materials (Basel)*. 2022;15 (6):2023.

[11] Chen Y, Li J, Yang W, Gao S, Cao R. Enhanced Corrosion Protective Performance of Graphene Oxide-Based Composite Films on AZ31 Magnesium Alloys in 3.5 wt.% NaCl Solution. *Applied Surface Science*. 2019;493:1224–1235.

[12] Dalhatu SN, Modu KA, Mahmoud AA, Zango ZU, Umar AB, Usman F, et al. L-Arginine Grafted Chitosan as Corrosion Inhibitor for Mild Steel Protection. *Polymers (Basel)*. 2023;15 (2):398.

[13] De Fenzo A, Giordano M, Sansone L. A Clean Process for Obtaining High-Quality Cellulose Acetate from Cigarette Butts. *Materials (Basel)*. 2020;13 (21):4710.

[14] Vijayan P P, Tanvir A, El-Gawady YH, Al-Maadeed M. Cellulose Nanofibers to Assist the Release of Healing Agents in Epoxy Coatings. *Progress in Organic Coatings*. 2017;112:127–32.

[15] Zhu B, Xu Y, Sun J, Yang L, Guo C, Liang J, et al. Preparation and Characterization of Aminated Hydroxyethyl Cellulose-Induced Biomimetic Hydroxyapatite Coatings on the AZ31 Magnesium Alloy. *Metals*. 2017;7 (6):214.

[16] Wang Z, Hu B, Yu H, Chen GZ. Synergistic Effects of 1-Octyl-3-Methylimidazolium Hexafluorophosphate and Cellulose Nanocrystals on Improving Polyacrylate Waterborne Anti-Corrosion Coatings. *Polymers (Basel)*. 2023;15 (4):810.

[17] Devadasu S, Sonawane SH, Suranani S. Self-Healing Corrosion Inhibition Coatings with pH-Responsive Activity by Incorporation of Nano Cellulose in Two Pack Epoxy Polyamide System. *Materials Today: Proceedings*. 2021;46:5544–5549.

[18] Dibetsoe M, Olasunkanmi LO, Fayemi OE, Yesudass S, Ramaganthan B, Bahadur I, et al. Some Phthalocyanine and Naphthalocyanine Derivatives as Corrosion Inhibitors for Aluminium in Acidic Medium: Experimental, Quantum Chemical Calculations, QSAR Studies and Synergistic Effect of Iodide Ions. *Molecules*. 2015;20 (9):15701–15734.

[19] El-Lateef HMA, Albokheet WA, Gouda M. Carboxymethyl Cellulose/Metal (Fe, Cu and Ni) Nanocomposites as Non-Precious Inhibitors of C-Steel Corrosion in HCl Solutions: Synthesis, Characterization, Electrochemical and Surface Morphology Studies. *Cellulose*. 2020;27 (14):8039–8057.

[20] El-Lateef HMA, Gouda M. Novel Nanocomposites of Nickel and Copper Oxide Nanoparticles Embedded in A Melamine Framework Containing Cellulose Nanocrystals: Material Features and Corrosion Protection Applications. *Journal of Molecular Liquids*. 2021;342:116960.

[21] Shi S-C, Wang C-C, Cheng Y-C, Lin Y-F. Surface Characterization and Tribological Behavior of Graphene-Reinforced Cellulose Composites Prepared by Large-Area Spray Coating on Flexible Substrate. *Coatings*. 2020;10 (12):1176.

[22] Solomon MM, Gerengi H, Umoren SA. Carboxymethyl Cellulose/Silver Nanoparticles Composite: Synthesis, Characterization and Application as a Benign Corrosion Inhibitor for St37 Steel in 15% H (2)SO (4) Medium. *ACS Applied Materials & Interfaces*. 2017;9 (7):6376–89.

[23] Sridhara PK, Vilaseca F. Assessment of Fiber Orientation on the Mechanical Properties of PA6/Cellulose Composite. *Applied Sciences*. 2020;10 (16):5565.

[24] Tamalmani K, Husin H. Review on Corrosion Inhibitors for Oil and Gas Corrosion Issues. *Applied Sciences*. 2020;10 (10):3389.

[25] Tavakoli M, Ghasemian A, Dehghani-Firouzabadi MR, Mazela B. Cellulose and Its Nano-Derivatives as a Water-Repellent and Fire-Resistant Surface: *A Review. Materials (Basel)*. 2021;15 (1):82.

[26] Toghan A, Gouda M, Shalabi K, El-Lateef HMA. Preparation, Characterization, and Evaluation of Macrocrystalline and Nanocrystalline Cellulose as Potential Corrosion Inhibitors for SS316 Alloy during Acid Pickling Process: Experimental and Computational Methods. *Polymers (Basel)*. 2021;13 (14):2275.

[27] Umoren PS, Kavaz D, Umoren SA. Corrosion Inhibition Evaluation of Chitosan-CuO Nanocomposite for Carbon Steel in 5% HCl Solution and Effect of KI Addition. *Sustainability.* 2022;14 (13):7891.

[28] Umoren SA, AlAhmary AA, Gasem ZM, Solomon MM. Evaluation of Chitosan and Carboxymethyl Cellulose as Ecofriendly Corrosion Inhibitors for Steel. *International Journal of Biological Macromolecules.* 2018;117:1017–1028.

[29] Umoren SA, Eduok UM. Application of Carbohydrate Polymers as Corrosion Inhibitors for Metal Substrates in Different Media: A Review. *Carbohydrate Polymers.* 2016;140:314–441.

[30] Umoren SA, Solomon MM, Madhankumar A, Obot IB. Exploration of Natural Polymers for Use as Green Corrosion Inhibitors for AZ31 Magnesium Alloy in Saline Environment. *Carbohydrate Polymers.* 2020;230:115466.

[31] Wang X, Liu S, Yan J, Zhang J, Zhang Q, Yan Y. Recent Progress of Polymeric Corrosion Inhibitor: Structure and Application. *Materials (Basel).* 2023;16 (8):2954.

[32] El-Lateef HMA, Gouda M, Khalaf MM, Al-Shuaibi MAA, Mohamed IMA, Shalabi K, et al. Experimental and In-Silico Computational Modeling of Cerium Oxide Nanoparticles Functionalized by Gelatin as an Eco-Friendly Anti-Corrosion Barrier on X60 Steel Alloys in Acidic Environments. *Polymers (Basel).* 2022;14 (13):2544.

[33] Farhadian A, Assar Kashani S, Rahimi A, Oguzie EE, Javidparvar AA, Nwanonenyi SC, et al. Modified Hydroxyethyl Cellulose as a Highly Efficient Eco-Friendly Inhibitor for Suppression of Mild Steel Corrosion in a 15% HCl Solution at Elevated Temperatures. *Journal of Molecular Liquids.* 2021;338:116607.

[34] Fawzy A, Toghan A, Alqarni N, Morad M, Zaki MEA, Sanad MMS, et al. Experimental and Computational Exploration of Chitin, Pectin, and Amylopectin Polymers as Efficient Eco-Friendly Corrosion Inhibitors for Mild Steel in an Acidic Environment: Kinetic, Thermodynamic, and Mechanistic Aspects. *Polymers (Basel).* 2023;15 (4).

[35] Ferreira F, Pinheiro I, de Souza S, Mei L, Lona L. Polymer Composites Reinforced with Natural Fibers and Nanocellulose in the Automotive Industry: A Short Review. *Journal of Composites Science.* 2019;3 (2):51.

[36] Gan T, Zhang Y, Yang M, Hu H, Huang Z, Feng Z, et al. Synthesis, Characterization, and Application of a Multifunctional Cellulose Derivative as an Environmentally Friendly Corrosion and Scale Inhibitor in Simulated Cooling Water Systems. *Industrial & Engineering Chemistry Research.* 2018;57 (32):10786–1097.

[37] Gouda M, El-Lateef HMA. Novel Cellulose Derivatives Containing Metal (Cu, Fe, Ni) Oxide Nanoparticles as Eco-Friendly Corrosion Inhibitors for C-Steel in Acidic Chloride Solutions. *Molecules.* 2021;26 (22):6945.

[38] Thakur A, Kaya S, Abousalem AS, Kumar A. Experimental, DFT and MC Simulation Analysis of Vicia sativa Weed Aerial Extract as Sustainable and Eco-Benign Corrosion Inhibitor for Mild Steel in Acidic Environment. *Sustainable Chemistry and Pharmacy.* 2022;29:100785.

[39] Thakur A, Kaya S, Abousalem AS, Sharma S, Ganjoo R, Assad H, et al. Computational and Experimental Studies on the Corrosion Inhibition Performance of an Aerial Extract of Cnicus benedictus Weed on the Acidic Corrosion of Mild Steel. *Process Safety and Environmental Protection.* 2022;161:801–818.

[40] Thakur A, Kaya S, Kumar A. Recent Innovations in Nano Container-Based Self-Healing Coatings in the Construction Industry. *Current Nanoscience.* 2022;18 (2):203–216.

[41] Thakur A, Kumar A. Sustainable Inhibitors for Corrosion Mitigation in Aggressive Corrosive Media: A Comprehensive Study. *Journal of Bio- and Tribo-Corrosion.* 2021;7 (2):1–48.

[42] Thakur A, Kumar A. Recent Advances on Rapid Detection and Remediation of Environmental Pollutants Utilizing Nanomaterials-Based (Bio)Sensors. *Science of The Total Environment.* 2022;834:155219.

[43] Thakur A, Kumar A, Kaya S, Vo D-VN, Sharma A. Suppressing Inhibitory Compounds by Nanomaterials for Highly Efficient Biofuel Production: A Review. *Fuel.* 2022;312.

[44] Thakur A, Kumar A, Sharma S, Ganjoo R, Assad H. Computational and Experimental Studies on the Efficiency of Sonchus arvensis as Green Corrosion Inhibitor for Mild Steel in 0.5 M HCl Solution. *Materials Today: Proceedings.* 2022;66:609–621.

[45] Thakur A, Sharma S, Ganjoo R, Assad H, Kumar A. Anti-Corrosive Potential of the Sustainable Corrosion Inhibitors Based on Biomass Waste: *A Review on Preceding and Perspective Research. Journal of Physics: Conference Series.* 2022;2267 (1):012079.

[46] Haris NIN, Sobri S, Yusof YA, Kassim NK. Innovative Method for Longer Effective Corrosion Inhibition Time: Controlled Release Oil Palm Empty Fruit Bunch Hemicellulose Inhibitor Tablet. *Materials (Basel).* 2021;14 (19).

[47] Huang L, Chen W-Q, Wang S-S, Zhao Q, Li H-J, Wu Y-C. Starch, Cellulose and Plant Extracts as Green Inhibitors of Metal Corrosion: A Review. *Environmental Chemistry Letters.* 2022;20 (5):3235–3264.

[48] Jaafar MZ, Mohd Ridzuan FF, Mohamad Kassim MH, Abu F. The Role of Dissolution Time on the Properties of All-Cellulose Composites Obtained from Oil Palm Empty Fruit Bunch. *Polymers (Basel).* 2023;15 (3):691.

[49] Bashir S, Thakur A, Lgaz H, Chung I-M, Kumar A. Computational and Experimental Studies on Phenylephrine as Anti-Corrosion Substance of Mild Steel in Acidic Medium. *Journal of Molecular Liquids.* 2019;293:111539.

[50] Bashir S, Thakur A, Lgaz H, Chung I-M, Kumar A. Corrosion Inhibition Performance of Acarbose on Mild Steel Corrosion in Acidic Medium: An Experimental and Computational Study. *Arabian Journal for Science and Engineering.* 2020;45 (6):4773–4783.

[51] Bashir S, Thakur A, Lgaz H, Chung I-M, Kumar A. Corrosion Inhibition Efficiency of Bronopol on Aluminium in 0.5 M HCl Solution: Insights from Experimental and Quantum Chemical Studies. *Surfaces and Interfaces.* 2020;20:100542.

[52] Ji L, Xue W, Zhu L, Jiang J. An Alternative Carbon Source from Cassava Residue Saccharification Liquid for In-Situ Fabrication of Polysaccharide Macromolecule/ Bacterial Cellulose Composite Hydrogel: A Comparative Study. *Sustainability.* 2022;14 (21):14277.

[53] Ji T, Zhang R, Dong X, Sameen DE, Ahmed S, Li S, et al. Effects of Ultrasonication Time on the Properties of Polyvinyl Alcohol/Sodium Carboxymethyl Cellulose/Nano-ZnO/Multilayer Graphene Nanoplatelet Composite Films. *Nanomaterials (Basel).* 2020;10 (9):1797.

[54] Khalid MY, Imran R, Arif ZU, Akram N, Arshad H, Al Rashid A, et al. Developments in Chemical Treatments, Manufacturing Techniques and Potential Applications of Natural-Fibers-Based Biodegradable Composites. *Coatings.* 2021;11 (3):293.

[55] Khan MAA, Irfan OM, Djavanroodi F, Asad M. Development of Sustainable Inhibitors for Corrosion Control. *Sustainability.* 2022;14 (15):9502.

[56] Komartin RS, Balanuca B, Necolau MI, Cojocaru A, Stan R. Composite Materials from Renewable Resources as Sustainable Corrosion Protection Coatings. *Polymers (Basel).* 2021;13 (21):3792.

[57] Liu J, Lv C. Durability of Cellulosic-Fiber-Reinforced Geopolymers: A Review. *Molecules.* 2022;27 (3):796.

[58] Lyagin I, Maslova O, Stepanov N, Presnov D, Efremenko E. Assessment of Composite with Fibers as a Support for Antibacterial Nanomaterials: A Case Study of Bacterial Cellulose, Polylactide and Usual Textile. *Fibers.* 2022;10 (9):70.

[59] Mobin M, Rizvi M. Adsorption and Corrosion Inhibition Behavior of Hydroxyethyl Cellulose and Synergistic Surfactants Additives for Carbon Steel in 1M HCl. *Carbohydrate Polymers.* 2017;156:202–214.

[60] Nawaz M, Habib S, Khan A, Shakoor RA, Kahraman R. Cellulose Microfibers (CMFs) as a Smart Carrier for Autonomous Self-Healing in Epoxy Coatings. *New Journal of Chemistry*. 2020;44 (15):5702–5710.

[61] Oprea M, Voicu SI. Cellulose Composites with Graphene for Tissue Engineering Applications. *Materials (Basel)*. 2020;13 (23):5347.

[62] Ouarga A, Noukrati H, Iraola-Arregui I, Elaissari A, Barroug A, Ben Youcef H. Development of Anti-Corrosion Coating Based on Phosphorylated Ethyl Cellulose Microcapsules. *Progress in Organic Coatings*. 2020;148:105885.

[63] P PV, Al-Maadeed M. Self-Repairing Composites for Corrosion Protection: A Review on Recent Strategies and Evaluation Methods. *Materials (Basel)*. 2019;12 (17):2754

[64] Peme T, Olasunkanmi LO, Bahadur I, Adekunle AS, Kabanda MM, Ebenso EE. Adsorption and Corrosion Inhibition Studies of Some Selected Dyes as Corrosion Inhibitors for Mild Steel in Acidic Medium: Gravimetric, Electrochemical, Quantum Chemical Studies and Synergistic Effect with Iodide Ions. *Molecules*. 2015;20 (9):16004–16029.

[65] Poletto M, Ornaghi HL, Zattera AJ. Native Cellulose: Structure, Characterization and Thermal Properties. *Materials (Basel)*. 2014;7 (9):6105–6119.

[66] Salama A, Abouzeid RE, Owda ME, Cruz-Maya I, Guarino V. Cellulose-Silver Composites Materials: Preparation and Applications. *Biomolecules*. 2021;11 (11):1684.

[67] Sangeetha Y, Meenakshi S, Sairam Sundaram C. Corrosion Inhibition of Aminated Hydroxyl Ethyl Cellulose on Mild Steel in Acidic Condition. *Carbohydrate Polymers*. 2016;150:13–20.

[68] Youssef AM, Hasanin MS, El-Aziz MEA, Turky GM. Conducting Chitosan/ Hydroxylethyl Cellulose/Polyaniline Bionanocomposites Hydrogel Based on Graphene Oxide Doped with Ag-NPs. *Int J Biol Macromol*. 2021;167:1435–1444.

[69] Gouda M, Khalaf MM, Al-Shuaibi MAA, Mohamed IMA, Shalabi K, El-Shishtawy RM, et al. Facile Synthesis and Characterization of CeO (2)-Nanoparticle-Loaded Carboxymethyl Cellulose as Efficient Protective Films for Mild Steel: A Comparative Study of Experiential and Computational Findings. *Polymers (Basel)*. 2022;14 (15):3078.

[70] Yang J, Liu D, Song X, Zhao Y, Wang Y, Rao L, et al. Recent Progress of Cellulose-Based Hydrogel Photocatalysts and Their Applications. *Gels*. 2022;8 (5).

[71] Yang S, Qian X. Conductive PPy@cellulosic Paper Hybrid Electrodes with a Redox Active Dopant for High Capacitance and Cycling Stability. *Polymers (Basel)*. 2022;14 (13).

[72] Yi T, Zhao H, Mo Q, Pan D, Liu Y, Huang L, et al. From Cellulose to Cellulose Nanofibrils-A Comprehensive Review of the Preparation and Modification of Cellulose Nanofibrils. *Materials (Basel)*. 2020;13 (22).

[73] Zhang X, Zhang M, Li R, Feng X, Pang X, Rao J, et al. Active Corrosion Protection of Mg-Al Layered Double Hydroxide for Magnesium Alloys: A Short Review. *Coatings*. 2021;11 (11).

[74] Zhang L, Wu Y, Zeng T, Wei Y, Zhang G, Liang J, et al. Preparation and Characterization of a Sol-Gel AHEC Pore-Sealing Film Prepared on Micro Arc Oxidized AZ31 Magnesium Alloy. *Metals*. 2021;11 (5).

[75] Zhang W, Li H-J, Chen L, Sun J, Ma X, Li Y, et al. Performance and Mechanism of a Composite Scaling-Corrosion Inhibitor Used in Seawater: 10-Methylacridinium Iodide and Sodium Citrate. *Desalination*. 2020;486.

Index

For Product Safety Concerns and Information please contact our EU
representative GPSR@taylorandfrancis.com
Taylor & Francis Verlag GmbH, Kaufingerstraße 24, 80331 München, Germany

www.ingramcontent.com/pod-product-compliance
Lightning Source LLC
Chambersburg PA
CBHW060348220326
41598CB00023B/2843